Writing the Heavens

Literatur- und Naturwissenschaften

———

Publikationen des
Erlangen Center for Literature and Natural Science/
Erlanger Forschungszentrums für Literatur und
Naturwissenschaften (ELINAS)

Edited by
Aura Heydenreich, Christine Lubkoll and Klaus Mecke

Volume 10

Writing the Heavens

—

Celestial Observation in Medieval and Early Modern Literature

Edited by
Aura Heydenreich, Florian Klaeger, Klaus Mecke,
Dirk Vanderbeke and Jörn Wilms

DE GRUYTER

Gefördert durch die Deutsche Forschungsgemeinschaft (DFG) – Projektnummer 429827737 / Funded by the Deutsche Forschungsgemeinschaft (DFG, German Research Foundation) – project number 429827737.

ISBN 978-3-11-159735-5
e-ISBN (PDF) 978-3-11-161086-3
e-ISBN (EPUB) 978-3-11-161090-0
ISSN 2365-3434
DOI https://doi.org/10.1515/9783111610863

Library of Congress Control Number: 2024944468

Bibliographic information published by the Deutsche Nationalbibliothek
The Deutsche Nationalbibliothek lists this publication in the Deutsche Nationalbibliografie; detailed bibliographic data are available on the internet at http://dnb.dnb.de.

© 2024 the author(s), editing © 2024 Aura Heydenreich, Florian Klaeger, Klaus Mecke, Dirk Vanderbeke and Jörn Wilms, published by Walter de Gruyter GmbH, Berlin/Boston
The book is published open access at www.degruyter.com.

Cover image: Glaser, Hans: [Himmelserscheinung über Nürnberg vom 14. April 1561]. zu Nürmberg : Bey Hanns Glaser Brieffmaler, [1561]. Zentralbibliothek Zürich, PAS II 12/60. https://doi.org/10.7891/e-manuscripta-91896 - Public Domain Mark.
Typesetting: Integra Software Services Pvt. Ltd.

www.degruyter.com
Questions about General Product Safety Regulation:
productsafety@degruyterbrill.com

Contents

Aura Heydenreich, Florian Klaeger, Klaus Mecke, Dirk Vanderbeke,
Jörn Wilms

Introduction: Writing the Heavens

Celestial Observation in Medieval and Early Modern Literature

The present is a time of unprecedent interest in astronomy and the stars. Public observatories, founded from the late seventeenth century onwards, are available even in small villages and on school roofs. In addition, since the twentieth century, planetariums have become much frequented places to observe the stars, and mass-produced telescopes are now popular gifts, especially at Christmas, to promote a sense of wonder and awe. And yet, the heavens are disappearing, or at least the starlight is. Some fifty years after the concept of light pollution first entered scientific discourse (Riegel 1973), Aparna Venkatesan and John C. Barentine recently proposed the term 'noctalgia' to express 'sky grief' at "[o]ur diminishing ability to view the nighttime sky due to rapidly rising human-made light pollution" (2023). Formed by analogy to 'nostalgia,' the painful yearning for a lost home, the term aims to capture the pain felt at the "loss of heritage, place-based language, identity, storytelling, millennia-old sky traditions and our ability to conduct traditional practices grounded in the ecological integrity of what we call home" (ibid.). For all its hyperbole, the claim is not wrong: from the dawn of the species, the starry heavens were part and parcel of the human experience. Indeed, the classical concept of the *contemplator coeli* held that, if for nothing else, human existence was worthwhile "for the sake of viewing the heavens and the whole order of the universe" (thus the Pre-Socratic philosopher Anaxagoras, qtd. in Blumenberg 1987, 9).

Today, we are in danger of losing the perennial spectacle of the night sky. Venkatesan and Barentine rightly stress its importance for heritage, identity, and storytelling. Alongside scientists and activists, artists are rallying to demonstrate what is at stake – witness only the sublime photographic montages of Thierry Cohen in the series *Villes éteintes* ('Darkened Cities'), combining night skies in places with great atmospheric clarity with nocturnal cityscapes at the same latitude (Cohen, Kerangal and Luminet 2012). These photographs involve a layering of temporalities in space, as the sky in one place and time is superimposed on the photograph of another place at the same time. As Cohen explains, "[b]y combining two realities, I am making a third that you cannot see . . . but it exists! I am showing you the missing stars" (Brook 2014), offering a glimpse of what today's cities might look like under unpolluted heavens. In Mohsin Hamid's 2017 novel

Exit West, a character's response to these photographs is described as 'noctalgic' *avant la lettre*:

> They were achingly beautiful, these ghostly cities – New York, Rio, Shanghai, Paris – under their stains of stars, images as though from an epoch before electricity, but with the buildings of today. Whether they looked like the past, or the present, or the future, she couldn't decide. (Hamid 2017, 57)

Hamid captures the temporal vertigo caused by Cohen's photographs, and the sense of loss they instil in reminding us of a time when the heavens were very much with us. This double nature – temporal and spatial – is shared between noctalgia and nostalgia. For the latter, too, Svetlana Boym has observed that while it may seem like "a longing for a place," it is really "a yearning for a different time" and, "[i]n a broader sense, nostalgia is a rebellion against the modern idea of time, the time of history and progress": "The nostalgic desires to obliterate history and turn it into private or collective mythology, to revisit time like space, refusing to surrender to the irreversibility of time that plagues the human condition" (2001, xv). The noctalgic, too, strategically idealizes a past when the human subject was firmly located in sidereal context, observation was a commonplace activity, and the heavens were self-evidently taken to be 'about' humankind. It is this time, and its literary representations of observation, that this volume undertakes to explore.

In doing so, the chapters collected here are soundings in medieval and early modern literary astroculture. Astroculture, as defined by historian Alexander Geppert with a view to its twentieth-century manifestations, "comprises a heterogeneous array of images and artifacts, media and practices that all aim to ascribe meaning to outer space while stirring both the individual and the collective imagination" (Geppert 2018, 8). In other words, astroculture is the array of manifestations outer space takes in specific cultures: its representations, negotiations, semanticizations, ideologizations, aestheticizations, etc., at any given point in time and space. In answer to the grandiloquently Schillerian question, 'What is, and to what end do we study, European astroculture?,' Geppert highlights astroculture's commonly terrestrial concerns:

> More often than not, transforming life on earth figured even higher up on the agenda than humanizing the heavens, and the perspective from without – whether provided by envoys of humankind, satellites or an unprecedented plethora of aliens – helped to turn the world into a planet. Anything but esoteric, farfetched or obscure, it is for this reason that the historical study of astroculture and space thought goes directly to the core of modern world-making or, as some would prefer to call it, modernity. Twentieth-century imagination, spurred by technological and political revolutions alike, no longer limited the *conditio humana* to terrestrial territory, as Hannah Arendt famously observed after Sputnik. (Geppert 2020, 372)

For the century of spaceflight, the twentieth, this is immediately convincing. However, Geppert also indicates a broader sweep – indeed one that begs the question after pre- and early modern astrocultures. This is a perspective encompassed by the 'discipline,' introduced by Hans Blumenberg (with tongue firmly in cheek), of 'astronoetics' (*Astronoetik*), literally the act or process of thinking about the stars (instead of, for instance, visiting them, as in astronautics; Blumenberg 1997, 324). In essence, Blumenberg's seminal *Genesis of the Copernican World* (pub. 1975; Blumenberg 1987) is a study of astronoetics from classical Greece to the post-Apollo age. He is well aware of the refined and complex astrocultures in prehistoric and classical civilizations across the world, but the very title of Blumenberg's book indicates the central place of modernity in his argument. Something similar holds true for Wolfgang Welsch's *Homo mundanus* (2012), which aims to place modern anthropic thought in historical perspective (cp. also the contributions in Neef, Sussman and Boschung 2014, a volume with a broad historical compass but decided focus on contemporary astroculture; and contrast the more balanced volume, Rössler, Sparenberg and Weber 2016). Astroculture, in its various conceptual and critical guises, then, seems inextricably linked to modernity.

However, a focus and reference to the Copernican World may no longer offer an appropriate framework for understanding contemporary astroculture, which is increasingly moulded by the view of giant telescopes which opened up a completely new perspective on the cosmos. This contemporary point of view takes its origin from the revolutions of the theory of relativity and quantum theory. The heavens they describe are no longer populated solely by stars and orbiting planets, but by black holes and wormholes, by red giants and white dwarfs, by blazars and quasars. Today's astroculture is influenced above all by the observations of large telescopes (Dick 2020), in particular Hubble's deep view, Planck's background observation of the early universe, Event Horizon's images of black holes, James Webb's surveying of dust clouds and nebulae, and not least Kepler's discovery of planets. All these celestial observations produced images that have arguably gained an iconic status and characterise, for our present, what is 'seen' in the sky. The cosmos is not a vast empty space anymore, but a world in constant change filled with matter, energy and motion. Deep space is no longer far away, but close and relevant to us – broiling with action and teeming with life. The modern exploration of formerly dark space thus led to a new understanding of the human place in the universe. We are just waiting for the first sign of extraterrestrial biological life, which we can expect to see in our lifetime. Modern telescopes allow us to be more than '*Gedankenastronauten*' (mental astronomers) and observe the universe in a revolutionary new way. Obviously, we are on the brink of a new era, with a new understanding of 'the heavens,' and this transition has little to do with, and cannot be understood in terms of, the Copernican revolution. In a similar fashion, modern possibilities of Artificial Intelligence and quantum technologies can no longer be understood in terms of the Gutenberg universe. It is likely that humankind will look back on the present moment, the turn of the millennium, as an epochal turning point. Our modern observations allow us, by way of contrast, to recognize the specificity of the pre-modern era with its celestial observations of luminous points. We believe that the look back can help to illuminate an epistemic transformation, and perhaps help us to better understand the transition of our own moment. It is in this spirit that the present volume offers readings in pre-modern and early modern astroculture, and we hope that they illustrate the concept's heuristic and explanatory potential outside of, but with repercussions for, its usual demesne.

The following chapters explore *literary* artefacts of astroculture from the Middle Ages and early modernity, both from Europe and China. In introducing the term 'astroculture' to his discipline, Geppert expressed hope that it would "lead to the controlled import of elsewhere long-established analytical key categories such as 'language,' 'consumption,' 'representation,' 'appropriation,' 'memory,' 'materiality,' and, above all, 'meaning,' in addition to numerous others into space history, where they have played no more than a minor, dramatically undervalued role" (Geppert 2018, 9). Literary studies is certainly among the areas where all these categories have long been central, and thus, the historical study of astroculture implies a focus on literature as a key medium. The specific nature of *literary* manifestations of astroculture – what has been termed 'astropoetics' (Klaeger 2018, 27) – is part of the wider field of a poetics of knowledge and the epistemological functions of form (see, e.g., Vogl 1999, Spiller 2004, Köppe 2011, Graduiertenkolleg Literarische Form 2017, Freiburg and Lubkoll 2016, Heydenreich and Mecke 2017). As a slew of recent publications on early modern 'science' have shown, it relied extensively "on literariness to present experimental findings; the textual representation of such discoveries necessitated an extensive use of figurative language," and indeed, "the main technologies that made natural philosophy intellectually possible were so because they could be articulated in literary terms" (Chico 2018, 1; cp., e.g., Aït-Touati 2011, Preston 2015, Rössler 2020, Wragge-Morley 2020, Gorman 2021, McLeish 2022, Siebenpfeiffer 2025). The challenge, then, is to try and "see the literary and the scientific as [. . .] not-yet-differentiated disciplines—or world views" (Marchitello and Tribble 2017, xxiv), and worldmaking *poiesis* as an activity self-consciously pursued by both poets and natural philosophers (cp. Sarkar 2023). As part of this endeavour, the volume offers insight into the uses of various literary (and para- or proto-literary) forms for the representation and 'production' of the heavens in the medieval and early modern periods.

A further, more specific focus in the following is on observation, "the most pervasive and fundamental practice of all the modern sciences, both natural and human," which "has always been a form of knowledge that straddled the boundary between art and science, high and low sciences, elite and popular practices" (Daston and Lunbeck 2011, 1 and 7). In the Middle Ages and early modernity, celestial knowledge extended from the cosmological to the meteorological, with applications and implications that touched upon a wide range of discourses (theological, legal, political, medical, agricultural, to name but a few; cf. Hardie 2022). It thus partook of both the speculative and practical branches of knowledge (Park and Daston 2008, 6). Owing in part to this wide applicability, celestial observation was frequently a subject for verbal rather than numerical and geometrical recording, more readily accessible also to the non-specialist. This is a dichotomy also apparent in the Islamic world (regrettably unrepresented in the present collection), where "the obligation [. . .] to pray at specific times in a specific direction gave rise to a substantial literature," in which astronomers "proposed mathematical solutions," while "scholars of the religious law [. . .] proposed non-mathematical solutions," which, as David King attests, "is not known to have led to any strife" (King 2004, 1: ix). In the Christian early medieval discipline of *computus* (important for determining the dates of moveable feasts), observation played a comparatively small role (Park 2011, 23). Medieval academic courses in astronomy were based on texts "cast in an ostensibly literary form, far removed from the mathematical form that was essential to the ancient scientific eminence of astronomy" (North 2013, 456), such as Johannes de Sacrobosco's *De sphaera*. Even before the birth, around 1600, of observation as "an epistemic genre, especially among astronomers and physicians," with a set of quite established formal representational requirements (Daston 2011, 81), astronomical writing at the hands of natural philosophers, poets, chroniclers, travellers, geographers, educators, and others mediated knowledge of the heavens in textual form. Such texts could variously function as (mimetic) models of the universe, and simultaneously offer (pragmatic) models for specific types of behavior. They were thus deeply enmeshed in their historical, geographical, popular, religious, philosophical, and generic environments.

For the modern scholar, these records can be difficult to decode, and the question of what they address or seek to explore is frequently obscured by the respective generic traditions, tropics and imagery, and other discursive contexts. However, as tokens of pre- and early modern astroculture, they allow insight into the changing epistemic place of astronomy. How, contributors to this volume were encouraged to explore, are textual forms bound up with pre- and early modern astronomy and its institutions? What kinds of data are represented in these texts, and what are the modes in which they are communicated? How were verbal representations of celestial phenomena encoded and self-consciously placed vis-à-vis other systems of representation and knowledge? How were discourses on anthropology, aesthetics, religion, etc. entangled with astronomical observation and knowledge? How did they realize their own medial, didactic, informational, aesthetic potential? How did they reflect on the forms of knowledge they engaged (especially in terms of the epistemological purchase of 'observation' and 'imagination')? Which spatialized conceptions of human nature were recognizable before and immediately after the (alleged) 'Copernican disillusionment'? How did individual scholars, texts, and concepts travel between European and non-European cultures, both in space and in time, and which constructions of self and other arose in the process?

The volume is organized along the axes of time and space. The first three sections provide case studies from the European Middle Ages (sect. 1), the sixteenth and seventeenth centuries (2), and the long eighteenth century (3). A fourth section opens the geographical range to include early modern China. The chapters on the high and later Middle Ages, from the twelfth to the fifteenth century, dwell on the interplay between the cosmic and human realms. All texts discussed here revolve around the question after the meaning accorded, across various genres, to astronomical knowledge. In keeping with Aristotelian doctrine, the Christian Middle Ages considered the heavens immutable, but at the same time, at least partially inscrutable: according to the 'postulate of visibility' (*Sichtbarkeitspostulat*), certain things were hidden, supposedly on purpose, from the senses and human knowledge because they were irrelevant for salvation (Blumenberg 1985, 364–365). Under this dual assumption, astronomical knowledge fulfils a double role: knowledge of the order and movements of the heavens can serve as an index of an individual's intellectual and spiritual standing and, moreover, as an index of the poet's ability. Maximilian Wick's chapter, centred on twelfth-century philosophical epics, highlights the depiction of the world as perfect and man as corrupted by sin. The potential for man's improvement and understanding of the cosmic order through astronomical instruction and observation is a key concern. Johannes de Hauvilla's *Architrenius* raises the dilemma of what happens when a person cannot attain such understanding, adding an element of pessimism. Sophie Knapp's exploration of Middle High German *Sangspruch* poetry reveals how poets claimed astronomical and cosmological knowledge to exhibit their erudition. This knowledge was simultaneously portrayed as both ostentatious and arcane, serving as a way for poets to assert *meisterschaft*, or mastery, in their craft. Walker Horsfall's chapter, centred on Heinrich von Mügeln's Marian praise poem, *Der meide kranz*, challenges the assumption of that work's astronomical inaccuracy. The potential Horsfall finds for an alternative interpretation, where inaccuracy serves as an invitation for critical and innovative reflection, ties into the broader theme of how medieval texts could be creatively interpreted, even in scientific matters. Finally, Daniel Könitz's chapter examines the manuscript tradition of the treatise *Von den elf Himmelssphären* in Middle High German. This chapter emphasizes the importance of manuscript transmission and how texts evolved, the use of illustrations and the identification of manuscripts serving as windows into the evolution of medieval knowledge. Here, astronomical knowledge becomes a tool for the modern scholar in tracing textual genealogies and networks. In sum, chapters in this section highlight how astronomical and cosmological knowledge was used in the Middle Ages both to assert poetic competence and to grapple with the limitations of human understanding.

The chapters discussing the transitional period of the sixteenth and seventeenth centuries in section 2 offer readings in three very different 'genres': Jean Bodin's treatise on witchcraft; the written account, by a tradesman, of a Warsaw city parade; and the images used by early modern cartographers of the heavens to describe constellations. They are connected by an interest, widespread in the early modern period, in what links the heavens and earth. Helge Perplies' chapter explores Jean Bodin's presentation of astrology in *Démonomanie des sorciers* (1580) and in Johann Fischart's German translations of 1581 and 1586, which undertake certain revisions in the interest of Fischart's linguistic agenda. Bodin emphasizes that heavenly influences are limited to the material world and should thus be considered apart from religious matters. His use of technical language, and his critique of astrological and astronomical writings by Firmicus Maternus, Abu Ma'shar, Pierre d'Ailly and Cyprian Leowitz, are a performance of authorial self-fashioning, while Fischart's variations and additions are intended to demonstrate the potential of the German language for scientific discourse. In this way, a renunciation of astrology's influence is turned into an assertion of authorial and linguistic agency. A complementary view of the forces attributed to celestial influence is presented in Agata Starownik's innovative study of a carnival-like event in 1580 Warsaw, the parade of the planets. In Martin Gruneweg's early seventeenth-century account of this event – an 'observation' of the heavens on earth –, it showcased the seven planets associated with the zodiac signs, merging cosmology, astrology, and mythology. It was a unique blend of entertainment for both the public and the royal court, highlighting how astronomical and astrological knowledge was adapted for diverse audiences. The parade offers a glimpse into the popular astronomical consciousness of the time, filtered through the complex worldviews of its various participants. Exploring how objects of observation were rendered symbolically, and how observation was conversely guided verbally, Gábor Kutrovátz's chapter focuses on the textual descriptions of constellations (a tradition with an influential literary pedigree since at least Hesiod and Aratus) in early modern star catalogues based on Ptolemy by Tycho Brahe, Edmond Halley, and Johann Hevelius. Kutrovátz offers a quantitative analysis of how stars were described, particularly in anatomical terms related to constellation figures. This reveals differences and similarities in the usage of descriptions among catalogues, shedding light on individual author preferences, possible historical and cultural influences, and the function of constellation lore within astronomy. Describing stars as structural or anatomical elements, the chapter argues, served a cognitive purpose, aiding in memorization and recognition. The three chapter in this section offer complementary perspectives on 'observation' in their explorations of how to observe influences (and how to turn their discussion into a performance of authority), observing a literal parade of orbs on earth (which 'incarnates' the heavenly bodies and renders their supposed influences more visible), and projecting mundane objects onto the heavens to impose a verbal and conceptual order onto the observed phenomena.

The third section focuses on Enlightenment negotiations of a cosmos that is both filled with a plurality of worlds and at least potentially devoid of the metaphysical certainties of the past. In the Dutch, English, French, and German seventeenth- and eighteenth-century authors discussed here, all three chapters find a desire to confront the philosophical, anthropological, and spiritual consequences of a heavens that has dramatically opened up beyond literal observation, and that can be gauged better with the poetic imagination than with the best telescopes. A thread connecting this section to the previous is the attempt to link the celestial with the mundane, as contemporary discourses on colonialism, education, religion, and poetics become sites for the negotiation of the heavens on earth. Thus, Hania Siebenpfeiffer's chapter explores the notion of extraterrestrial life in the context of Enlightenment encounters with the non-European other. The chapter links Nicolaus Cusanus's fifteenth-century theological writings on the 'finite infinity' of the universe with multiple inhabited worlds to debates about extraterrestrial life in the long eighteenth century. In influential late seventeenth-century texts such as Bernard le Bovier de Fontenelle's *Entretiens* and Christiaan Huygens's *Kosmothéoros*, Siebenpfeiffer attests to a new awareness that an expanded universe requires a revised and comprehensive natural philosophy. Fictions of cosmic travel by Kepler, Francis Godwin, Cyrano de Bergerac, and Eberhard Christian Kindermann, Siebenpfeiffer argues, show a development towards a 'temporalized anthropology,' linking the fictional extraterrestrial alien and earthbound humankind: from aliens who are linked to the past of humanity, through those satirically reflecting present earthly abuses, to those promising humankind's future. As they pave the way towards 'science fiction' of a recognizably modern kind, these Enlightenment fictions of other worlds thus become a means of self-observation, self-reflexion, and self-critique. This analogy between astronomical knowledge and self-knowledge also informs Alexander Honold's chapter on the *Bildungsroman* genre in German-language literature around 1800. Authors like Goethe, Hölderlin, and Jean Paul are shown to use astronomic patterns and the movements of celestial bodies as metaphors for the individual's path toward self-development. The planets and their orbits around the sun serve as educational models, reflecting the balance of opposing forces and the dynamic understanding of the individual subject's development. This constitutes a modern reinterpretation of the relationship between macrocosm and microcosm, and one that highlights the changing metaphorical status of astronomy as a system of knowledge. Addressing the same issue, Reto Rössler offers an in-depth reading of a text also touched upon by Honold, Jean Paul Richter's paratextual "Speech of the Dead Christ" from the novel *Siebenkäs* (1796). Like Siebenpfeiffer, Rössler reads Kepler's *Somnium* as a foil to an Enlightenment fiction. This literary thought experiment seizes on the contradictions of an outdated harmonious world model, illustrating a decline in the idea that the world is entirely 'readable.' The short narrative portrays a cosmos that is infinite, empty, and chaotic, and like Siebenpfeiffer and Honold, Rössler detects a shift in the perception of the human condition within the universe. However, he also finds Jean Paul, in his experiments with the novel genre and several narrative and ontological frames, to foreground the role of the artist as a creator of possible worlds, going beyond the idea of mimesis. As the observed sky ceases to convey a meaningful message to its human contemplator, that contemplator is encouraged to inscribe meaning of his own making. Chapters in Section 4 collectively foreground the progressive sundering of the domains of literature and science in the long eighteenth century. Observation of the heavens, and its interpretation, increasingly acquires a metaphorical significance. The texts discussed here explore the idea of extraterrestrial life, the changing metaphors concerning the structure of the heavens (from the mechanical clockwork and the architectural *Weltgebäude* to the lush garden), and the use of astronomy as a symbolic tool in literature to reflect on human development and self-discovery.

The fourth section, finally, delves into the intersections of European and Chinese astrocultures from the seventeenth century, extending – with the survival of a particular genre – all the way into the twentieth century. Gianamar Giovannetti-Singh's chapter surveys the deployment of Chinese astral sciences in early modern European popular literature. Jesuit missionaries, Giovannetti-Singh argues, were instrumental in establishing a European trope linking East Asia to the celestial realm. Their writings on the concept of "Heaven" as an agent of historical change during the transition from the Ming to the Qing dynasties in China took hold in European popular culture. At the same time, these writings show how the Jesuits utilised the place of astronomy in mid-seventeenth-century China to gain influence with the Manchu dynasty. Discussing the reception of their accounts, Giovannetti-Singh stresses how China came to be associated with astral language in the European imaginary – a finding that is reflected, among other things, in texts discussed by Siebenpfeiffer. Andrea Bréard's contribution centres on the genre of "Complete Books of Myriad Treasures," an encyclopaedic form in seventeenth- to twentieth-century China providing readers with knowledge on heavenly patterns and their influence on daily life. In close dialogue with concerns from sections 2 and 3, the chapter traces, in a diachronic analysis, how classical astronomical knowledge was reorganized and integrated into cosmic theories, while also indicating how resistance to scientific knowledge found expression in the genre. The concept of *tianwen* is discussed as both a pattern in the sky and a written trace from heaven, emphasizing the idea that heaven's patterns can be read and interpreted like a text; but the chapter also surveys how the *Complete Books* presented their astronomical content to readers. Together, these chapters highlight how early modern Chinese 'writings of the heavens' worked as part of early modern Chinese astroculture, but also, how they were utilised in 'translating' knowledge about China to Europeans and in identifying Chinese culture with the heavens.

The chapters, especially in the third and fourth sections, illuminate a cosmos filled with a plurality of worlds, but these worlds are ultimately mere copies of the human world in a solar system with a planet orbiting around a sun. Modern observations reveal, however, that the cosmos is inhabited by totally strange celestial objects which could be named only by metaphorical neologisms such as 'white dwarfs', 'red giants', or 'black holes'. What does this entail for our self-observation, self-reflexion, and self-critique? On the one hand, the heavens really are a lush garden, illuminated by contemporary telescopes with beautiful, spectacular images, but on the other hand, they look fundamentally different from everything that was observed and imagined from the Middle Ages to the Enlightenment. And there is more to come. For instance, gravitational wave detectors opened a new window where no window openings were even suspected in the nineteenth century. We can now observe details of processes in the very early universe, of the 'big bang', and of colliding, collapsing neutron stars. At present, we have only a glimpse of phenomena and events in the heavens which will certainly be quite different from the celestial observations discussed in this volume.

In consequence, the heavens and celestial observation have regained an entirely new significance for humankind. In contrast to the frequently evoked empty, endless universe that has ceased to convey a meaningful message to its human contemplators, who are forced to inscribe their own meaning onto it, we have the opportunity to understand ourselves better in a cosmos full of hitherto unfamiliar processes and phenomena. The sky has moved closer to us and become stranger at the same time, because we have been given more to see than we could have dreamt of. If today's observations of the heavens lead to a new physical understanding that no longer is based on the old celestial mechanics of a world made up of suns and planets, but a quantum cosmos full of strange, peculiar processes, then we may be facing a new Copernican revolution.

Bibliography

Aït-Touati, Frédérique. *Fictions of the Cosmos: Science and Literature in the Seventeenth Century*. Transl. Susan Emanuel. Chicago: Chicago University Press, 2011.

Blumenberg, Hans. *The Legitimacy of the Modern Age*. 1976. Transl. Robert M. Wallace. Cambridge, MA: MIT Press, 1985.

Blumenberg, Hans. *The Genesis of the Copernican World*. Transl. Robert M. Wallace. 1975. Cambridge, MA: MIT Press, 1987.

Blumenberg, Hans. *Die Vollzähligkeit der Sterne*. Frankfurt am Main: Suhrkamp, 1997.

Boym, Svetlana. *The Future of Nostalgia*. New York: Basic Books, 2001.

Brook, Pete. "What Cities Would Look Like If Lit Only by the Stars." *WIRED*, November 13, 2014. https://www.wired.com/2014/11/thierry-cohen-darkened-cities/ (September 27, 2023).

Chico, Tita. *The Experimental Imagination: Literary Knowledge and Science in the British Enlightenment*. Stanford: Stanford University Press, 2018.

Cohen, Thierry, Maylis de Kerangal, and Jean-Pierre Luminet. *Villes Éteintes*. Paris: Marval, 2012.

Daston, Lorraine. "The Empire of Observation, 1600–1800." *Histories of Scientific Observation*. Ed. Lorraine Daston and Elizabeth Lunbeck. Chicago and London: University of Chicago Press, 2011. 81–115.

Daston, Lorraine, and Elizabeth Lunbeck. "Observation Observed (Introduction)." In *Histories of Scientific Observation*. Ed. Lorraine Daston and Elizabeth Lunbeck. Chicago and London: University of Chicago Press, 2011. 1–9.

Dick, Steven J. *Space, Time, and Aliens: Collected Works on the Cosmos and Culture*. Cham: Springer, 2020.

Freiburg, Rudolf, Christine Lubkoll, and Harald Neumeyer. "Einleitung." *Zwischen Literatur Und Naturwissenschaft*. Ed. Rudolf Freiburg et al. Berlin and Boston: De Gruyter, 2017. IX–XVI.

Geppert, Alexander C. T. "European Astrofuturism, Cosmic Provincialism: Historicizing the Space Age." *Imagining Outer Space: European Astroculture in the Twentieth Century*. Ed. Alexander C. T. Geppert. 2nd ed. Houndmills: Palgrave Macmillan, 2018. 3–28.

Geppert, Alexander C. T. "What Is, and to What End Do We Study, European Astroculture?" *Militarizing Outer Space: Astroculture, Dystopia and the Cold War*. Ed. Alexander C. T. Geppert, Daniel Brandau and Tilmann Siebeneicher. London: Palgrave Macmillan, 2020. 371–377.

Gorman, Cassandra. *The Atom in Seventeenth-Century Poetry*. Woodbridge, Suffolk: D. S. Brewer, 2021.

Graduiertenkolleg Literarische Form, eds. *Formen des Wissens: Epistemische Funktionen literarischer Verfahren*. Heidelberg: Universitätsverlag Winter, 2017.

Hamid, Mohsin. *Exit West: A Novel*. New York: Riverhead Books, 2017.

Hardie, Philip R. *Celestial Aspirations: Classical Impulses in British Poetry and Art*. Princeton and Oxford: Princeton University Press, 2022.

Heydenreich, Aura, and Klaus Mecke. *Physics and Literature: Concepts-Transfer-Aestheticization*. Berlin and Boston: De Gruyter, 2022.

King, David A. *In Synchrony with the Heavens: Studies in Astronomical Timekeeping and Instrumentation in Medieval Islamic Civilization*. 2 vols. Leiden and Boston: Brill, 2004.

Klaeger, Florian. *Reading into the Stars: Cosmopoetics in the Contemporary Novel*. Heidelberg: Universitätsverlag Winter, 2018.

Köppe, Tilmann, ed. *Literatur und Wissen: Theoretisch-methodische Zugänge*. Berlin and New York: De Gruyter, 2011.

Marchitello, Howard, and Evelyn Tribble. "Introduction." *The Palgrave Handbook of Early Modern Literature and Science*. Ed. Howard Marchitello and Evelyn Tribble. London: Palgrave Macmillan, 2017. xxiii–xliv.

McLeish, Tom. *The Poetry and Music of Science: Comparing Creativity in Science and Art*. Oxford: Oxford University Press, 2019.

Neef, Sonja A. J., Henry Sussman, and Dietrich Boschung, eds. *Astroculture: Figurations of Cosmology in Media and Arts*. Paderborn: Fink, 2014.

North, John. "Astronomy and Astrology." *The Cambridge History of Science*. Vol. 2: *Medieval Science*. Ed. David C. Lindberg and Michael H. Shank. Cambridge: Cambridge University Press, 2013. 456–484.

Park, Katharine. "Observation in the Margins, 500–1500." *Histories of Scientific Observation*. Ed. Lorraine Daston and Elizabeth Lunbeck. Chicago and London: University of Chicago Press, 2011. 15–44.

Park, Katharine, and Lorraine Daston. "Introduction: The Age of the New." *The Cambridge History of Science*. Vol. 3: *Early Modern Science*. Ed. Katharine Park and Lorraine Daston. Cambridge: Cambridge University Press, 2008. 1–17.

Pomata, Gianna. "Observation Rising: Birth of an Epistemic Genre, 1500–1650." *Histories of Scientific Observation*. Ed. Lorraine Daston and Elizabeth Lunbeck. Chicago, London: University of Chicago Press, 2011. 45–80.

Preston, Claire. *The Poetics of Scientific Investigation in Seventeenth-Century England*. Oxford and New York: Oxford University Press, 2015.

Riegel, K. W. "Light Pollution: Outdoor Lighting Is a Growing Threat to Astronomy." *Science* 179.4080 (1973): 1285–1291.

Rössler, Reto. *Weltgebäude: Poetologien kosmologischen Wissens der Aufklärung*. Göttingen: Wallstein, 2020.

Rössler, Reto, Tim Sparenberg, and Philipp Weber, eds. *Kosmos & Kontingenz: Eine Gegengeschichte*. Paderborn: Wilhelm Fink, 2016.

Sarkar, Debapriya. *Possible Knowledge: The Literary Forms of Early Modern Science*. Philadelphia: University of Pennsylvania Press, 2023.

Siebenpfeiffer, Hania. *Die literarische Eroberung des Alls: Literatur und Astronomie (1593–1771)*. Göttingen: Wallstein, 2025.

Spiller, Elizabeth. *Science, Reading, and Renaissance Literature: The Art of Making Knowledge, 1580–1670*. Cambridge: Cambridge University Press, 2004.

Venkatesan, Aparna, and John C. Barentine. "Noctalgia (Sky Grief): Our Brightening Night Skies and Loss of Environment for Astronomy and Sky Traditions." https://arxiv.org/pdf/2308.14685.pdf. (August 28, 2023).

Vogl, Joseph. "Einleitung." In *Poetologien des Wissens um 1800*. Ed. Joseph Vogl. Munich: Fink, 1999. 7–19.

Welsch, Wolfgang. *Homo mundanus: Jenseits der anthropischen Denkform der Moderne*. Weilerswist: Velbrück Wissenschaft, 2012.

Wragge-Morley, Alexander. *Aesthetic Science: Representing Nature in the Royal Society of London, 1650–1720*. Chicago and London: University of Chicago Press, 2020.

I The Middle Ages

Maximilian Wick

Mirari faciunt magis hec quam scire: Ways of (Not) Understanding the Cosmos in Johannes de Hauvilla's *Architrenius*

Abstract: The philosophical epics of the twelfth century again and again focus on the relation between man as a microcosm and the world as a macrocosm. They depict the latter as perfect and the former as in a deplorable state, corrupted by sin. Theologians and poets – or theologians as poets – such as Bernardus Silvestris and Alanus ab Insulis (Alain of Lille) approach this problem conceptually within their epics by postulating the possibility of man's perfection through his own actions (according to Bernardus' *Cosmographia*) or through a savior (according to Alanus' *Planctus naturae* and his *Anticlaudianus*). According to them, the prerequisite for such an improvement, or at least for a consolation, is deep insight into the cosmic order, which can be obtained by astronomical instruction and observation. But what is the consequence if a person is incapable of reaching such a state of understanding? This dilemma, associated with the problem of theodicy, is at the center of Johannes de Hauvilla's *Architrenius*. As this essay will show, his rather pessimistic text can be understood as a satirical response to the optimistic epics of the aforementioned poets. How this pessimism reverberates after *Architrenius* can be demonstrated by a concluding examination of the *Laborintus* by Everardus Alemannus (Eberhard the German), who elaborates on Johannes' idea using the example of a schoolmaster condemned to misery.

✶✶✶

The preoccupation with nature, the cosmos, and man's role within the latter reached an unheard of and unique intensity in twelfth-century north-western Europe, which contributed to this century being characterized as a period of protorenaissance.[1] Although the term has meanwhile been viewed critically, it is largely undisputed that an eminent change in anthropology and cosmology took place during this period – a change which has been particularly associated with the so-called School of Chartres. In the rather loosely imagined circle surrounding

---

1 Cf. Haskins' (1927) influential monograph on *The Renaissance of the Twelfth Century* and, from recent research, for example, Dinzelbacher (2017), Jaeger (2003), Swanson (1999) and Wetherbee (1992).

the cathedral school of Chartres, theologians like Alanus ab Insulis and Bernardus Silvestris developed not only a new form of cosmological speculation, they also shaped a new mode of poetic philosophy or philosophical poetry with their *integumentum* (in short: philosophical truth covered in poetry).[2] Their neo-platonically inspired works mainly revolve around the figure of personified nature, a 'Mother Nature' who, as *prodea* or *vicaria dei*,[3] brings creation and its natural order onto the stage of events.[4]

The main subject of this chapter, the *Architrenius* written by Johannes de Hauvilla in 1184, can be located among this intellectual milieu. Before discussing this text, however, I want to frame it with the help of some notes on the *Cosmographia* by Bernardus Silvestris and the *Planctus naturae* by Alanus ab Insulis, since Johannes, as I would like to show, reacts directly to these texts' cosmological conceptions. Unlike his predecessors, who claim, or at least promise, the perfection of the microcosmos, Johannes holds up a mirror to the degenerate world in his satire and thereby ridicules the two of them. This is particularly evident in the unsuccessful consolation of the protagonist by Nature, who, in the sense of *contemplatio caeli*, presents the astronomical wonders of the macrocosm to the protagonist, which the latter, however, cannot comprehend and ultimately cannot draw any consolation from. I will conclude with a brief outlook on an even more pessimistic astronomical observation in a later text, the *Laborintus* by Everardus Alemannus, that also offers a satirical response to the concepts of its predecessors.

1 The *Cosmographia* and Bernardus' optimism

Written between 1143 and 1148, the *Cosmographia* (*De mundi universitate*) is the first of the Chartrian philosophical epics and thus a model for the later texts.[5] Bernardus aligns himself formally with Boethius' *Consolatio Philosophiae* and Martianus Capella's *De Nuptiis Philologiae et Mercurii*, and content-wise with Plato's *Timaeus*. The latter, the Platonic dialogue concerning the cosmogony, was the only Platonic dialogue available in the twelfth century in a partial translation and

---

2 Cf. Bezner (2005) and, furthermore, for example Brinkmann (1980, 169–214), Stock (1972, 11–62), and Wetherbee (1972, 36–48).

3 On the concept of *vicaria dei* in the twelfth century with regard to the *Cosmographia*, the *Planctus naturae*, the *Anticlaudianus*, and the *Architrenius*, cf. White (2000, 69–109).

4 Cf. Modersohn (1997, 23–44), Huber (1992), Dronke (1980), Economou (1972).

5 For an introduction to this text cf. Kauntze (2014), Wetherbee (in Bernardus Silvestris 1990,1–62; Wetherbee 1972, 158–186), and Dronke (1978, 1–91).

with a commentary by the Neoplatonist Calcidius.[6] Since Plato was the philosophical authority in matters of creation, it was necessary to harmonize him with the other great authority, the Mosaic report – and that was Bernardus' main task in the *Cosmographia:* to reconcile Plato with Moses, or Moses with Plato.

The text deals with the creation of the cosmos in the form of a prosimetrum of alternating parts in verse and prose: with the creation of the macrocosm in a first book and with that of man as a microcosm in a second book.[7] The poem begins with Mother Nature's lament before Noys, the first and highest emanation of God. Nature complains to her about the shapeless and therefore ugly form of Silva, the primordial matter, who herself demands her *informatio,* her forming. Noys agrees and thereby causes Silva to form the perfect macrocosm. Consequently, as the crowning glory of her creation, Noys wants to shape man from the remains of Silva, but since she has already used the better parts for the macrocosm, only the remnants ("non elementa, sed elementorum reliquiae"; Bernardus Silvestris 1978, II.XIII.4) remain for the creation of man. To make matters worse, Noys does not herself lend a hand for this. She delegates this task to Nature instead, who, after a long journey through various regions of the heavens to seek help from Urania (the celestial principle), in turn entrusts Physis (the material principle) with it, another and even lower emanation.

As Ratkowitsch (1995) has shown, this chain of delegation serves to shift the responsibility for man, who is imperfect in contrast to the perfect macrocosm, 'downwards' and thus to provide an answer to theodicy. Not only is God not mentioned during the creation of man; by inserting further instances between him and the authority that creates the imperfect human being, Bernardus creates rhetorical distance between the creator and creation itself. Once Physis, who, as a lower emanation, is visibly overstrained with her delicate work on the corrupt remains of Silva, has finished her task, the entire problem is simply left unsolved.[8] Instead, the text tips into an enthusiastic praise of man as the crown of creation. In the course of this praise, all parts of his body are now individually interpreted and praised – concluding with the male genitalia fighting Lachesis by

---

6 On Calcidius as translator of the *Timaeus,* cf. Ratkowitsch (1996).

7 On the relation of microcosm and macrocosm according to Bernardus, cf. Finckh (1999, 116–158).

8 The Fall of Man is briefly alluded to only once, which is why it can hardly be considered a reason for the not yet achieved perfection of the microcosm (cf., with an opposite assessment, Kauntze 2009, 22): "Hos, reor, incoluit riguos pictosque recessus / Hospes – sed brevior hospite – primus homo" (Bernardus Silvestris 1978, I.III,335–336) ["In this well watered and richly colored retreat, I believe, the first man dwelt as a guest – but too brief a time for a guest" (Bernardus Silvestris 1990, 83).

knotting the thread of life anew.[9] With this somewhat peculiar and, as the use of the obscene term *mentula* suggests, possibly also satirically intended allegory, man's own responsibility for his main problem as an incomplete being is emphasized toward the end of the *Cosmographia*: that is, his mortality, which he must counter himself with perpetuated procreation.

This laudatory description of the crown of creation is initiated by emphasizing the analogy between man as a microcosm and the perfect macrocosm. A few explanations about the functioning of the brain follow, and then the praise of the five senses, the first and highest of which is both traditionally and specifically in this text the sense of sight:

> Sol, oculus mundi, quantum communibus astris
> Preminet, et celum vendicat usque suum,
> Non aliter sensus alios obscurat honore
> Visus, et in solo lumine totus homo est.
> Querenti Empedecles quid viveret, inquid: 'Ut astra
> Inspiciam; celum subtrahe – nullus ero.'
> Ceca manus detractat opus, pes ebrius errat,
> Quando opus in tenebris et sine luce movent.
> (Bernardus Silvestris 1978, II.XIV,11–18)

> [Just as the sun, the world's eye, excels its companion stars and claims as its own all below the firmament, even so the sight overshadows the other senses in glory; the whole man is expressed in this sole light. To one who asked why he was alive, Empedocles replied, 'That I may behold the stars; take away the firmament, I will be nothing.' The unseeing hand spoils its work, the foot strays drunkenly, when they perform their tasks in darkness, without light. (Bernardus Silvestris 1990, 123–124)]

Bernardus might have taken the quote from Calcidius' commentary on Plato's *Timaeus*, the central reference text of the *Cosmographia*, although Calcidius ascribes the quote to Anaxagoras in this case (Pfeiffer 2001, 262). Yet the attribution is not all that important; what is more striking is that here, at a prominent point in the text, at the beginning of the description, observation of the heavens is asserted as a central human task in the cosmos. This fits perfectly with the unfolding of the universe in the first book of the *Cosmographia*, in which the created stars are immediately interpreted astrologically and thus tell the history of mankind. Amazingly, this version of history concerns mainly ancient authorities like Plato and Thales and characters from the Trojan War. It contains almost no elements of salvation history and even Christ is mentioned only once and briefly: "Exemplar speciemque dei virguncula

---

9 "Militat adversus Lachesin sollersque renodat / Mentula Parcarum fila resecta manu" (Bernardus Silvestris 1978, II.XIV,165–166).

Christum / Parturit, et verum secula numen habent" (Bernardus Silvestris 1978, I.III.53–54) ["A tender virgin gives birth to Christ, at once the idea and the embodiment of God, and earthly existence realizes true divinity" (Bernardus Silvestris 1990, 76)]. Not a single word about the crucifixion, the resurrection, or the apocalypse. How can this, as Stock puts it, "poetic dexterity, even audacity" (1972, 133), be explained?

I would like to propose to consider the answer to this question by way of the conception of man that Bernardus outlines in his *Cosmographia* and in the astonishing sudden change from the description of the problematic creation of man into his laudatory praise.[10] Simply put: the human delineated here, Bernardus' *homo silvestris*, as I would like to call him, is not necessarily a being in need of redemption. Hence, only the Incarnation and not the death on the cross is mentioned in regard to the birth of Christ (which can easily be harmonized with Plato's cosmogony). Instead of being redeemed, the *homo silvestris* must compensate for the flaw in creation himself. Being infected with Silva's "ingenitum contagium" (Bernardus Silvestris 1978, II.XIII.1), her 'innate plague', a flaw that could not be eradicated by Physis, man must take his perfection into his own hands. However, he is equipped excellently for this task, as the praise shows: with his highest sense analogized with the sun in the macrocosm, man can observe the sky and thus make astrological deductions about his own role in the cosmos and how to behave in it, as well as recognize the perfection of the macrocosm and find comfort in these perceptions. Assuming such a *contemplatio caeli*[11] as the main human task besides procreation explains why the poem ends with a praise of the (male) genitals: with only a briefly mentioned Messiah and no mention at all of the Apocalypse, Bernardus' rather optimistic historical model does not run linearly toward an end. Instead, it regards the reform of man as an ongoing task, to the fulfillment of which he, as a mortal creature, must necessarily procreate.

2 *Contemplatio caeli* as consolation in Alanus' *Planctus naturae*

Around 1160/1165, Alanus raises the cosmological question of the gulf between what should be and what is again in his first epic, the *Planctus naturae*, which is distinctly inspired by Boethius' *Consolatio philosophiae*, as well as the *Cosmogra-*

---

10 In light of this interpretation, cf. my dissertation (Wick 2021, 113–118) and on the *recreatio* as an ongoing task Dronke (2008, 144), Finckh (1999, 152–155), and Stock (1972, 144–147).
11 For a detailed discussion of the concept, cf. Pfeiffer on the role of *contemplation caeli* in the *Cosmographia* (2001, 259–282).

phia.[12] The text, also a prosimetrum, focuses primarily on Nature, who appears to the complaining poet as *vicaria dei* in a vision and is introduced in an extensive and astonishing *descriptio*.[13] As the poet puts it, humanity is in a deplorable state due to sexual as well as linguistic deficiencies.[14] The culprit, as he learns from the cosmic mother, is Venus, who was entrusted with the supervision of mankind in her service, but preferred to commit adultery, leading to the treacherous vices.[15] As a consequence, humans are the only creatures in the cosmos who rebel against nature,[16] whose order in this instance is an even more distinctly moral one than in Bernardus' cosmology. Since Nature cannot solve the problem of morally depraved and therefore 'denatured'[17] humanity by herself (Bezner 2020, 104), she calls Genius for help, who at the end of the poem speaks the anathema over the vices in a quite messianic gesture.[18] However, this solution in no way absolves mankind of its responsibility for the cosmos which it must fulfill – with the help of God – by way of a *secunda nativitas* (Köhler 1991, 60–65).

When Nature approaches the poet at the beginning of his vision, he sinks ecstatically to the ground, "stricken by stupor" ("stupore vulneratus," Alanus 2013, 6.1) and is erected and rebuked by her for his being stunned:

'Heu,' inquit, 'quae ignorantiae caecitas, quae alienatio mentis, quae debilitas sensuum, quae infirmatio rationis, tuo intellectui nubem opposuit, animum exulare coegit, sensus hebetavit potentiam, mentem compulit aegrotare, ut non solum tuae nutricis familiari cogni-

---

12 For an introduction to this text, cf. Wetherbee (in Alanus 2013, XVI–XXVIII; Wetherbee 1972, 188–211).

13 On Nature's relation to God in the process of creation cf. Urban (2019), on Nature's *descriptio*, Kellner (2020), Kirakosian (2020), Stolz (2019).

14 On the meaning of grammar and its relation to sexuality in Alanus' works cf. Kellner (2020) and Ziolkowski (1985).

15 As Kellner (2017, 136), has pointed out, Alanus here pursues a similar rhetorical strategy for the solution of the theodicy as Bernardus. While God himself does not appear in the *Planctus naturae*, Nature takes his place as the central and therefore responsible agent of creation, but delegates this responsibility to Venus, with whom Alanus can present a convenient culprit for the miserable status quo.

16 This idea is already encountered in the Middle High German *Annolied* about 80 years before the *Planctus*. In the third section of this poem, man and Lucifer are contrasted with the rest of creation, which adheres to their natural order given by God (*Das Annolied*, 3).

17 Nature accuses man of "Naturae naturalia denaturare pertemptans" [seeking "to put the natural gifts of Nature to an unnatural use"] (Alanus 2013, 8.3). All references to the *Planctus naturae* including the translation are to the bilingual edition by Wetherbee (Alanus 2013).

18 On Genius, cf. most recently Urban (2021, 41, 44–50). Alanus' second allegorical epic, the *Anticlaudianus*, contains more distinct messianic overtones, concerning the creation of a perfect man (*homo novus*) who faces the vices in a Psychomachia (cf. Urban 2021, 163–169).

tione tua intelligentia defraudetur, verum etiam, tanquam monstruosae imaginis novitate percussa, in meae apparitionis ortu tua discretio patiatur occasum? [. . .]'.

['Alas,' she said, 'what blind ignorance, what disorientation of mind, what defect of perception, what weakness of reason has imposed this cloud on your intellect, driven your mind into exile, dulled the power of your senses, caused your thought to become so diseased that not only has your understanding been robbed of its intimate acquaintance with your nurse, but your rational powers succumb to darkness at my very appearance, as though shocked by some strange monstrous form? (. . .)'] (Alanus 2013, 6.3)

Analogous to Bernardus' conception, man, according to Alanus, also is excellently equipped for the task of his *recreatio*. After all, Nature has endowed man not only with a worthy and, above all, perceptive body, but she also generously gave "valuable powers" ("virtuales potentiae") to his spirit: "the power of active intelligence" ("ingenialis virtutis potentia"), "the seal of reason" ("rationis signaculum") and "the power of memory" ("memorialis potentia") (Alanus 2013, 6.5). In consequence, she makes it quite clear that being so richly endowed and the microcosmic image of the cosmic realm, there was no reason for him to complain, even if he was a mixture of conflicting forces, spirituality striving for heaven as much as for corrupting sensuality:

Nec in hac re hominis natura meae dispensationis potest ordinem accusare. De rationis enim consilio tale contradictionis duellum inter hos pugiles ordinavi ut, si in hac disputatione ad redargutionem sensualitatem ratio poterit inclinare, antecedens victoria praemio consequente non careat. Praemia enim victoriis comparata caeteris muneribus pulcrius elucescunt.

[And human nature cannot blame my ordering and management of all this. For it is on the advice of reason that I ordained such a war of contradiction between these two antagonists, so that if Reason is able to force sensuality to yield to refutation in this dispute, the antecedent victory will not lack its consequent reward. For rewards obtained by victories shine more brightly than other gifts.] (Alanus 2013, 6.8)

After explaining the possibility of and the reason for the necessary assumption of responsibility by man, Nature proceeds to praise the macrocosm and invites her interlocutor to behold its perfection. Following the concept of *contemplatio caeli*, she begins to praise the heavens as the residence of God and the angels as his ministers, who govern the cosmos and especially man residing there like a noble city.[19] Then Nature explains the analogy between the perfect macrocosm and the

---

19 Nature distinctly asks her interlocutor to turn his gaze to the sky: "Attende qualiter in hoc mundo velut in nobili civitate quaedam reipublicae maiestas moderamine rato sancitur. In caelo enim, velut in arce civitatis humanae, imperialiter residet Imperator aeternus" ["Observe how,

ideal state of the microcosm, starting with the analogy between the heavens and the human head, from which wisdom governs the body, while the heart below is the seat of magnanimity, which takes the function of the angels. Finally, the limbs, the localization of desire, in the macrocosm correspond to human beings as obedient subjects of God.

Even if this explanation obviously does not solve the fundamental problem (as in the case of Boethius' *factum brutum*, his impending execution, which remains imminent), Nature's instructive consolation is successful, and the protagonist can at last properly greet his cosmic mother:

> Dum per haec verba michi Natura naturae suae faciem develaret, suaque ammonitione quasi clave praeambula cognitionis suae mihi ianuam reseraret, a meae mentis confinio stuporis evaporavit nubecula. Et per hanc ammonitionem, velut quodam potionis remedio, omnes fantasiae reliquias quasi nauseans stomachus mentis evomuit. A meae mentis igitur peregrinatione ad me reversus ex integro, ad Naturae devolutus vestigia, salutationis vice pedes osculorum multiplici impressione signavi.

> [While Nature with these words was unveiling for me the face of her nature, and through her instruction, as if with a preambulatory key, was opening the door of my recognition of her, the cloud of dullness evaporated from within my mind. [. . .] Thus wholly restored to myself after my mind's journeying, I fell at Nature's feet and, by way of greeting, stamped them with the impression of many kisses.] (Alanus 2013, 6.19)

3 Failing on *contemplatio caeli* in Johannes' *Architrenius*

In contrast to Bernardus' completely allegorical *Cosmographia* and just like the *Planctus naturae* by Alanus, whose "most discerning reader and most devastating critic" (Godman 1995, 64) Johannes was, his *Architrenius*, an epic poem of about 4,300 hexameters, has a human protagonist: the eponymous hero, whose name one could translate as 'Arch-Lamenter,' 'Arch-Weeper,' or 'Arch-Mourner.'[20] One day,

---

in this universe as in a noble city, a kind of majestic civil order is ensured by well considered governance. For in the heavens, as in the citadel of the human city, the eternal Emperor dwells in imperial state"] (Alanus 2013, 6.9). Even if Nature subordinates her doctrine (and thus *contemplatio caeli*) to theology expressis verbis, the order of precedence, as Urban has shown (2021, 36–37), is anything but clear; as a consequence, *contemplatio caeli* is also not necessarily a clearly inferior alternative to theology.

20 For an introduction to this text cf. Wetherbee (in Johannes de Hauvilla 2019, VII–XXVI; Wetherbee 1972, 242–255).

when he reaches manhood, Architrenius reflects on his past behavior and concludes that "he has never devoted a single day to virtue" ["nec se virtutibus unum / impendisse diem," Johannes de Hauvilla 2019, 224–225].[21] According to him, the blame for this lies with Nature, who had abandoned him like the rest of mankind and did not take sufficient care of them, even though she was downright omnipotent:[22] she creates whatever pleases her,[23] is able to perform miracles, ordered and decorated the macrocosm and even created numerous 'monsters' (*monstra*),[24] which Architrenius lists in a catalogue of creatures with special properties based on Pliny (Johannes de Hauvilla 2019, 478–480). At the end of this catalogue, he lists his own person, being not sufficiently equipped by Nature to ward off sins and therefore labelled a 'monster.' Architrenius sees Nature as the one responsible for these calamities in the world. She consequently appears to him as somewhat malevolent, maybe even in the Manichaeist sense simply as an evil counter-principle to the good Creator, or at least as a malignant stepmother, a *tristis noverca*, as Pliny pessimistically describes her in his *Natural History* (Carlucci 2021, 205–212). At this

---

21 References to the *Architrenius* including the translation are to the bilingual edition by Wetherbee (2019).

22 In light of such a conception of Nature and her human accuser, the *Architrenius* clearly stands out from the mainstream of philosophical epics of the time and their classical predecessor: "Compared to the human characters in Boethius' *Consolatio* Alan's *De Planctu*, Architrenius has a significantly more confrontational attitude. His final invectives against Nature do not find any correspondence in the works of Boethius, Bernardus Silvestris or Alan of Lille. Nowhere in the latter writings do we find man accusing Nature of being hateful to mankind" (Carlucci and Marino 2019, 64).

23 "Illud enim supraque potest. Nullaque magistras / non habet arte manus, nec summa potencia certo fine coartatur [. . .]. / Natura est quodcumque vides; incudibus illa / fabricat omniparis, quidvis operaria nutu / construit, eventusque novi miracula spargit" ["Nature can do all of this and more. There is no art that her hand has not mastered, and her supreme power knows no limit. [. . .] Whatever you behold is Nature; all-creating, she works at her forge, produces anything at will, and spreads abroad a miraculous array of novelties [. . .]."] (Johannes de Hauville 2019, 234–243). Architrenius' description shows Nature even more powerful than she describes herself in Alanus' *Planctus naturae*, where she subordinates herself to God concerning creation (whom Architrenius does not mention at all) – even if this submission seems like a lip service (Kellner 2017, 127). While one might assume that this is just Architrenius' possibly exaggerated point of view, Nature's own description at the end of the text confirms this position: as she puts it, her goodness (*bonitas*; as the *causa finalis* of creation) is not only known to God, but also related to His goodness. Additionally, the two attributes are rhetorically connected by paronomasia (*cognata* / *cognita*), which further reduces their distinction. Nature's goodness is, as it were, God's goodness (Johannes de Hauville 2019, 220).

24 This is another point by which Johannes clearly contradicts Alanus' conception of Nature, who according to the *Planctus naturae* did not create any monsters, thus is not responsible for them (White 2000, 101–104).

point, Johannes "[inverts] the direction of complaint in *De planctu*" (White 2000, 101), by having his protagonist set out to find Nature in the earthly realm, on the optimistic assumption that she will surely comfort him with her motherly compassion; he may even be able to "repair the broken bonds of love" ["rupti forsan amoris / restituam nodos"] (Johannes de Hauville 2019, 323–324).

On his journey, which comprises the bulk of the story, Architrenius is confronted with the flaws of his degenerate age and regularly bursts into tears. The road takes him to the palace of Venus, where he marvels at a particularly beautiful maiden, described in over 300 verses. He encounters Cupid as a fashion fool, also described in detail, on whom the superficiality of external beauty is demonstrated. From there on, he travels to the nearby house of Gluttony, then to Paris, where he laments the students' miserable lives as future *philosophi*. His journey continues to the palace of Ambition near the hill of Presumption, then to a battlefield where he meets Sir Gawain fighting alongside King Arthur against the forces of Avarice in a Psychomachia, and finally to the island of Tylos, where he is welcomed and instructed by a whole coterie of philosophers.

At last, Architrenius gets to meet Nature on this mythical island of Tylos, an earthly paradise. Yet instead of presenting his complaint, he throws himself crying at her feet, giving her the opportunity to open the conversation:

> venit, affandique negatur
> Copia, de mundo Genesi texente loquelam.
> "Omnigene partus homini famulantur, eique
> Et domus et nutrix ancillaque, machina mundi,
> Omne bonum fecunda parit, maiorque minori
> Obsequitur mundus. tibi discors unio rerum
> Eternum statura cohit, fractoque tumult
> Pax elementa ligat. gaude tibi sidera volvi
> Defigique polos, mundique rotatilis aule
> Artificem gratare Deum, dominumque ministro
> Erexisse domum, cuius molicio summum
> Actorem redolet [. . .]."

> [A rush of words emerges, but is stayed, as Genesis herself fashions a discourse about the universe. "Creatures of every kind attend on man. For him this universal frame is home and nurse and handmaid, brings forth every good in its fertility; the greater world obeys the will of the lesser. It is for you that that [sic] the union of its discordant parts coheres in eternal stability, that peace has put down the conflict of the elements and united them. Be glad that the stars revolve for you; that the poles stand firm; that they are pleasing to God, the maker and ruler of this whirling court, the universe; that the Lord has constructed for his minister a home, whose very composition tells of its supreme author [. . .]."] (Johannes de Hauville 2019, 322–333)

Although Nature has no way of knowing with what concerns Architrenius had come to her at this stage, it seems as if she is creating a decent starting point for herself in the discussion with this praise of creation for man's sake. While Architrenius had only dealt very superficially with Nature's work on the macrocosm in his description of her power, preferring instead to list earthly 'monsters,' Nature focuses on certain celestial circles and some complex things that occur because of the changing position of the horizon, e.g., the increase of daytime in Cancer (Johannes de Hauville 2019, 417).

After a while Architrenius interrupts his cosmic mother, praises her for what she has explained so far and tells her that despite all his admiration, he cannot understand her:

> "Mirari faciunt magis haec quam scire, rudisque
> ingenii non est" ait Architrenius "astris
> intrusisse stilum vel, quae divina sigillant
> scrinia, deciso dubii cognoscere velo.
> Trans hominem sunt verba deae. Miracula caecus
> audio, nam lampas animi subtilia pingui
> celatur radio. Nec mens sublimia visu
> vix humili cerni; [sic; instead of cernit, M.W.], nec distantissima luce
> fumidula monstrat [. . .]."

> ["These things create wonder rather than knowledge," says Architrenius. "It is not for the rude intellect to impose a pattern on the stars, or rend the veil of doubt and learn what divine caskets keep sealed. The words of a goddess are beyond human understanding. I hear of such miracles as if blind, for the lamp of my mind obscures subtle matters by its dull light. My mind can scarcely discern sublimity with its lowly gaze; its smoky little lamp cannot reveal things so remote [. . .]."] (Johannes de Hauville 2019, 1–9)

Nature's reaction to Architrenius' admission that he lacks comprehension is startling: instead of explaining the content once more in a simpler manner, she just continues with her explanations and now addresses even more complicated matters like the *circulus draconis*, the two lunar nodes, the points at which the orbit of the moon intersects the ecliptic (Johannes de Hauville 2019, 110–119).

Nature's explanations tend to be based on the Arab astronomer Alfraganus, representing the state of art at that time (Wetherbee in Johannes de Hauville 2019, 521–522). This is not entirely unimportant, since some recipients could end up doing the same as Architrenius, who is even more overwhelmed with this continuation than before and finally interrupts Nature a second time:

> "Quam procul eloquii fluvius decurret et aures
> influet exundans," ait Architrenius, "utre
> iam duplici pleno? Satis est hausisse referto

vase ; nec auriculae pelagi capit alveus undam."
Haec fatus rumpitque moras, pedibusque loquentis
irruit, et genuum demissos complicat artus,
et cubitos sternens, iunctis iacet infimus ulnis.

["How much longer," says Architrenius, "will this river of eloquence run on, filling my ears to overflowing though the sack has already been filled twice over? Enough has been swallowed when the vessel is full; the little channel of my ear cannot contain an ocean's flood." Having said this he delays no longer, and rushes toward the speaker's feet, folds the joints of his lowered knees, and spreading his elbows, lies prostrate with his arms together.] (Johannes de Hauville 2019, 149–155)

Not only does Architrenius admit his inability to follow the cosmic mother in her explanations a second time; here, he is also given the opportunity to present his concerns to her. Finally, he complains that humanity is morally depraved and calls for Nature's assistance. Amazingly however, Mother Nature readily agrees and takes Architrenius to her breast, providing him at least with a basic moral upbringing by breastfeeding him,[25] and then marries him to Moderantia, an apparently ideal bride.[26] The text ends with the wedding feast at which even Fortuna, who is all of a sudden and unconventionally portrayed as the just goddess of luck, gives her blessing to the couple:

Respicit et blandis epulas percurrit ocellis
et vultus adhibet animi cum melle favorem:
Sors inopum vindex, regum Tuchis ulta tumores,
Rhamnus opum terror, Nemesis suspecta tirannis,
Casus agens mitras, tribuens Fortuna curules.
[. . .] absit
meta deum clausura dapes; connubia Virtus
sanctiat et dempto convivia fine perhennet!

[There is one looking on who surveys the feast with kindly eyes, and matches the favor in her expression with the sweetness of her thoughts: Chance, that protects the poor and avenges the overbearing pride of kings, the Rhamnusian scourge of wealth, the Nemesis feared by tyrants, the Accident that creates Prelates, the Fortune that allots high office. [. . .]

---

25 On the traditional metaphor of the nursing-teaching personification, cf. Dronke (1980, 28–29).
26 In my dissertation I tried to show that the marriage with Moderantia does not at all mean a successful purification of the protagonist: on the one hand, she is not really flawless, especially with regard to her marital gifts; on the other hand, she is closer to the rejected beautiful girl from the Venusian palace than it appears at first sight (cf. Wick 2021, 145–150), and with an opposite assessment Roling (2003, 213–214) and Wetherbee (1972, 254–255).

Let that term be far off which will bring to an end this feast of the gods; let Virtue hallow their marriage and perpetuate their feasting without end!][27] (Johannes de Hauville 2019, 452–462)

How to understand this 'happy ending'? As Roling has shown, Nature's astronomical instruction once again aims at the *contemplatio caeli* (2003, 211): Architrenius should improve himself by understanding the secrets of the perfect macrocosm. However, this attempt fails for a distinct reason: the Arch-Mourner, as he himself admits, cannot understand the secrets of the macrocosm and consequently draws neither consolation nor improvement from it. As Godman puts it, Architrenius is "a puer of lacrimose stupidity," an "anti-hero of 'Unbildung.' What Architrenius learns is nothing" (1995, 65),[28] hence, he is in no way capable of comforting himself by contemplating the perfection of the macrocosm. His "reaction is undeniably telling and seems to suggest the insufficiency of the argument from design for solving the initial dilemma of the protagonist" (Carlucci 2021, 208–209).

Accordingly, the *Architrenius* provides a vivid example of how the *contemplatio caeli* and in turn Bernardus Silvestris' optimistic conception of a self-perfecting *homo silvestris* does not work for everyone; moreover, it offers a satire on Alanus' messianism. The problem of the imperfect human, for which Alanus in the *Planctus naturae* blamed Venus and made man responsible, shifts back to Nature, but she hides it by at least providing individual help to the Arch-Mourner.[29] However, this

---

27 On a tapestry in the palace of Ambition however, there was previously the traditional warning of Fortuna as a fickle goddess of fortune (cf. Ratkowitsch 1991, 271–294, and furthermore Wandhoff 2003, 216–222).

28 Unlike Alanus' "vir perfectus, reformed by the gifts of learning in the 'Anticlaudianus'," Johannes' protagonist is a "perfect idiot, incapable of profiting from them" (Godman 1995, 67). Piehler (1971, 93) comes to a comparable conclusion: "Her [i.e., Nature's] cosmological explications of the order and benevolence of God's universe seem admirable to the hero, but he complains that they are too far above his head to assuage his doubts and anxieties, as if implying that the cosmic order contemplation which had played such a large part in the spiritual enlightenment of Boethius and Alan was too exalted for him."

29 As Payen (1984, 393) also emphasizes, this does not offer a supraindividual solution: "La leçon majeure du texte est qu'un bonheur individuel est possible, qui soit stable et serein, et qui s'atteigne par une réforme intérieure fondée sur le savoir et sur la mesure: humanisme utopique, plus individualiste que social, qui préfigure celui des érudits de la Renaissance. Il lui manque bien évidemment – et cela réduit considérablement sa portée – une dimension collective qui élargirait à l'humanité toute entière l'espérancede promotion contenue dans son message" ["The major lesson of the text is that individual well-being is possible, which is stable and serene, and which is reached by an interior reform based on knowledge and on moderation: utopian humanism, more individualistic than social, which prefigures that of the scholars of the Renaissance. It obviously lacks – and this considerably reduces its scope – a collective dimension which would extend to the whole of humanity the hope of promotion contained in its message" (my transl.)].

does not really work either: the completely exaggerated end of the poem, in which Nature infantilizes Architrenius in a grotesque way and, to a certain extent, silences him by way of the wedding, rather underlines Johannes' severe criticism of the conception of his predecessors. Using the example of a person who is completely incapable of taking care of himself in a cosmological way, Johannes plays out Bernardus' conception and satirically unmasks his project as elitist and Alanus' to be an unworldly one. Looking up at the sky may astonish people like Architrenius, but it does not alleviate their hardships, for which more tangible solutions are required.[30]

4 Everhardus' *Laborintus*: 'Do not reach for the stars!'

The *Laborintus* from the first half of the thirteenth century, written in about 500 elegiac couplets, consists of a poetic instruction in the sense of the *artes poeticae*, which is framed by the story of a most pitiful schoolmaster.[31] Like Architrenius, he too is confronted with Nature personified, and again the story concerns a complaint. However, this time it is not the protagonist who laments, but Nature herself, as in Alanus' epics, who complains about the schoolmaster's fate even before he is born:

> Exhorret Natura parens dum matris in alvo
> Elimat miseri parvula membra viri.
> Si sub membrana praesentit membra magistri,
> Interrumpit opus officiosa suum;
> Inspirat dicit: "Operis lex pauset in isto!
> Exopto mea sit desidiosa manus.
> Si me non alia regeret lex quam mea, vellem
> Inceptum limae deseruisse meae.
> Sed Natura jubet naturans ne manus illic
> Cesset ubi fuerit materiale bonum;
> Et quia lege regor regis, quia legor ab Alto,
> Consummabit opus linea nostra suum [. . .]."
> (Everardus 2020, ll. 11–22)

---

30 These solutions are not necessarily genuinely Christian and may be an expression of Johannes' attempt at a reconciliation "of the pagan and Christian *Weltanschauung* [. . .] Nature's final admonition to Architrenius retains a distinctively pagan *Stimmung*: the hero must abide to the *religio nativa* of procreative-oriented marriage, rather than to contemplate the mysterious action of Divine Providence through his *ratio*." (Carlucci 2021, 210).

31 For an introduction to this text, cf. Vollmann (2020, 9–23) and Haye (2013).

[Mother Nature trembles while in a mother's womb she perfects tiny frame of wretched man. If under the membrane she divines the frame of a teacher she ingeniously interrupts her work. With a sigh she says, "O that the law of work might permit delay in this! Would that my hand could be idle! If no other law than mine were directing me I might wish to abandon the task of my file. But creative Nature commands the hand not to cease where there is matter; and because I am ordered by the King's decree, because I am appointed by the Most High, my task shall reach its end [. . .]."] (Everardus 1930, 5–6)[32]

This complaint does not thematize a general flaw of humanity, but concerns an innate flaw that leads to this particular human becoming a schoolmaster, the most pitiful creature in the cosmos according to the *Laborintus*. While Johannes concluded his text with an individual solution, Everardus raises an individual problem. Moreover, amazingly, this problem is not kept from God, but rather more or less blatantly reproached to him by Nature, who limits her own competencies to the execution of what is laid out in the *materia*. At the same time, she rejects any responsibility for the unborn's further fate, so that a later intervention, as in the *Architrenius*, no longer seems possible. Nature goes as far as demonstrating her powerlessness with a horoscope that confirms his fate:

[. . .] Me tua Parca vocat: tibi non vult parcere; filum
Jam nevit; nostras arguit illa moras.
Nasceris ergo, miser; misero tibi signa figurant
Sidereusque vigor officiale malum.
Scribitur in stellis paupertas, copia rerum,
Vitae commoditas, acre laboris onus;
Scribitur in stellis famae discrimen, honoris
Culmen, livoris flamma, favoris amor;
Scribitur in stellis virtutis laus, vitiorum
Dedecus, aetatis longa brevisque mora.
Omnem perlegi seriem caeli, nec in illa
Inveni sidus quod tibi mite meat:
Ecce Dyonaeum tibi flammas non vomit astrum,
Nec tibi scintillat Mercuriale decus;
Saturni sed curva tuos falx fascinat annos,
Et tibi fax Martis insidiosa rubet:
Est caeli virtus tibi tota propheta laboris,
In quo ditari non tua cura potest
(Everardus 2020, ll. 23–40)

---

32 References to the *Laborintus* are to the edition by Vollmann (2020), the translations given are from those by Carlson (1930). It is remarkable that Nature explicitly blames God in her lament over the schoolmaster's inevitable fate, while the other philosophical epics place the blame either on her, on a third entity such as Venus, or on man (Wick 2021, 155).

[(. . .) Your Fate calls me: she is unwilling to spare you; she has already spun your thread; she censures my delay. Therefore, you shall be born, wretched one; for you, wretch, the signs and activity of the constellations are shaping the misfortune attendant upon your profession. Written in the stars are poverty, abundance, life of ease, the irritating burden of toil. Written in the stars are a hazard of fame, a pinnacle of glory, a flame of envy, love of applause. Written in the stars are the renown of virtue, the shame of vice, and the long and brief sojourn of life. I have scanned the entire course of the sky and have failed to find wandering there a constellation kind to you: behold, the Dyonaean star emits no blaze for you; and for you, Mercury's splendor does not glitter; but Saturn's curved scythe casts a spell upon your years; and for the reddens the treacherous torch of Mars. The whole character of the sky foretells for you hardship by which for all your care you cannot profit. (Everardus 1930, 6–7)][33]

After Nature has predicted the future of the unborn child, doomed to be a schoolmaster, she hands the books he will need for his life to his mother: among them chiefly the small school grammar by Donatus. However, the catalogue of the books that he does not receive is described much more broadly. These are books on the seven *artes liberales* and the higher faculties, medicine, law and theology.[34]

With her selection of books, Nature once again consolidates the schoolmaster's predetermined path in life. He will not make a career in any of the more lucrative subjects, but will be giving basic grammar lessons forever. Thus, contrary to the case of Architrenius, the origin of the teacher's misfortune does not lie in a lack of talent, but in predestination. Accordingly, the gaze that Nature lifts up into the sky does not serve as consolation or edification, but only as confirmation of his fate. In an even more drastic manner than the example of Archmourner, the example of the schoolmaster shows that reaching for the stars is only an opportunity for those who are destined to do so. For others, the stars are just signs of their own limitations – they come to inspire wonder rather than knowledge and bring despair rather than remedy.

---

[33] The stellar script, highlighted by anaphoric repetition (*Scribitur in stellis* . . .), clearly refers to the *Cosmographia*, which gives Noys as its author (Vollmann 2017, 60–62). Accordingly, viewed across texts, she would be to blame for the schoolmaster's mischief as the highest emanation of God. Moreover, the stellar script in the *Cosmographia* is central to the comforting *contemplatio caeli* (cf. Pfeiffer 2001, 268–271), while its astrological reading by Nature in the *Laborintus* leads only to the devastating realization of an ongoing *miseria*.

[34] It is precisely this devastating horoscope that is confirmed shortly afterwards once again by Fortuna (Everardus 2020, ll. 81–118), who just like at the end of the *Architrenius*, does not appear as a fickle goddess and instead describes herself as a powerful entity that is merely portrayed as powerless and under whose influence only the corrupt flourish: "[. . .] Ecce caeca probor, quia caecos tollo, videntes / Deprimo, degeneres nutria, sperno bonos" (Everardus 2020, ll. 117–119) ["Behold I am represented as blind because I exalt the blind; I suppress those with clear vision; I foster degenerates; I spurn those who are good" (Everardus 1930, 10)].

Bibliography

[Alanus ab Insulis. *Planctus naturae*] Alan of Lille. "The Plaint of Nature." *Literary Works*. Ed. and transl. Winthrop Wetherbee. Cambridge, MA, and London: Harvard University Press, 2013. 21–217.

Das Annolied. Ed. and transl. Eberhard Nellmann. Stuttgart: Reclam, 2010 [1975].

Bernardus Silvestris. *Cosmographia*. Ed. Peter Dronke. Leiden: Brill, 1978.

Bernardus Silvestris. *The* Cosmographia *of Bernardus Silvestris*. Transl. Winthrop Wetherbee. New York: Columbia University Press, 1990 [1973].

[Everardus Alemannus] Eberhard der Deutsche. *Laborintus*. Ed. and transl. Justin Vollmann. Basel: Schwabe Verlag, 2020.

[Everardus Alemannus]. *The Laborintus of Eberhard*. Transl. Evelyn Carlson. Ithaca: Cornell University Press, 1930.

Johannes de Hauvilla. *Architrenius*. Transl. Winthrop Wetherbee. Cambridge, MA, and London: Harvard University Press, 2019.

Bezner, Frank. "Hybridisierung der Natur – Literatur als Hybridisierung. Natur, Kosmos und Sexualität im 'Planctus Naturae' des Alanus ab Insulis." *Natur – Geschlecht – Politik. Denkmuster und Repräsentationsformen vom Alten Testament bis in die Neuzeit*. Eds. Beate Kellner and Andreas Höfele. Paderborn: Wilhelm Fink, 2020. 81–108.

Bezner, Frank. Vela veritatis: *Hermeneutik, Wissen und Sprache in der* intellectual history *des 12. Jahrhunderts*. Leiden and Boston: Brill, 2005.

Brinkmann, Hennig. "Verhüllung ('*Integumentum*') als literarische Darstellungsform im Mittelalter." *Der Begriff der* repraesentatio *im Mittelalter. Stellvertretung – Symbol – Zeichen – Bild*. Ed. Albert Zimmermann. Berlin and New York: Walter de Gruyter, 1971. 314–339.

Carlucci, Lorenzo. "A *mise en scene* of 'pagan pessimism' in the Twelfth Century: Book 7 of Pliny's *Naturalis historia* in Johannes of Hauvilla's *Architrenius*." *Rivista di cultura classica e medioevale* 63 (2021): 187–212.

Carlucci, Lorenzo, and Laura Marino. "Echoes of Manichaeism in the *Architrenius*." *Dianoia* 28 (2019): 49–80.

Dinzelbacher, Peter. *Structures and Origins of the Twelfth-Century 'Renaissance.'* Stuttgart: Anton Hiersemann, 2017.

Dronke, Peter. *The Spell of Calcidius. Platonic Concepts and Images in the Medieval West*. Florence: Sismel, 2008.

Dronke, Peter. "Bernard Silvestris, Natura, and Personification." *Journal of the Warburg and Courtauld Institutes* 43 (1980): 16–31.

Economou, George D. "The Two Venuses and Courtly Love." *In Pursuit of Perfection. Courtly Love in Medieval Literature*. Ed. Joan M. Ferrante and George D. Economou. New York and London: Kennikat Press, 1975. 17–50.

Finckh, Ruth. *Minor Mundus Homo. Studien zur Mikrokosmos- Idee in der mittelalterlichen Literatur*. Göttingen: Vandenhoeck & Ruprecht, 1999.

Godman, Peter. "*Opus consummatum, omnium artium . . . imago*. From Bernard of Chartres to John of Hauvilla." *Zeitschrift für deutsches Altertum und deutsche Literatur* 124.1 (1995): 26–71.

Haskins, Charles Homer. *The Renaissance of the Twelfth Century*. Cambridge and London: Harvard University Press, 1927.

Haye, Thomas. "Der *Laborintus* Eberhards des Deutschen. Zur Überlieferung und Rezeption eines spätmittelalterlichen Klassikers." *Revue d'histoire des textes* 8 (2013): 339–369.

Huber, Christoph. "Die personifizierte Natur. Gestalt und Bedeutung im Umkreis des Alanus ab Insulis und seiner Rezeption." *Bildhafte Rede in Mittelalter und früher Neuzeit. Probleme ihrer Legitimation und ihrer Funktion*. Eds. Wolfgang Harms and Klaus Speckenbach. Tübingen: Max Niemeyer Verlag, 1992. 151–172.

Jaeger, C. Stephen. "Pessimism in the Twelfth-Century 'Renaissance'." *Speculum* 78 (2003): 1151–1183.

Kauntze, Mark. *Authority and Imitation. A Study of the* Cosmographia *of Bernard Silvestris*. Leiden and Boston: Brill, 2014.

Kauntze, Mark. "The Creation Grove in the *Cosmographia* of Bernard Silvestris." *Medium aevum* 78 (2009): 16–34.

Kellner, Beate. "Wider die Natur? Normierung der Sexualität und Transgressionen der Geschlechterordnung bei Alanus ab Insulis und Konrad von Würzburg." *Natur Geschlecht Politik*. Eds. Andreas Höfele and Beate Kellner. Paderborn: Wilhelm Fink Verlag, 2020. 109–148.

Kellner, Beate. "Allegorien der Natur bei Alanus ab Insulis – mit einem Ausblick auf die volkssprachliche Rezeption." *Schriftsinn und Epochalität. Zur historischen Prägnanz allegorischer und symbolischer Sinnstiftung*. Eds. Bernhard Huss and David Nelting. Heidelberg: Winter, 2017. 113–143.

Kirakosian, Racha. "Intertextuelle Textilien. Imaginäre Kleider und Temporalität bei Alanus ab Insulis und Gertrud von Helfta." *Beiträge zur Geschichte der deutschen Sprache und Literatur* 142.2 (2020): 236–266.

Köhler, Johannes. "Natur und Mensch in der Schrift 'De Planctu Naturae' des Alanus ab Insulis." *Mensch und Natur im Mittelalter*. Eds. Albert Zimmermann and Andreas Speer. Vol 1. Berlin and New York: Walter de Gruyter, 1991. 57–66.

Modersohn, Mechthild. *Natura als Göttin im Mittelalter. Ikonographische Studien zu Darstellungen der personifizierten Natur*. Berlin: Akademie Verlag, 1997.

Payen, Jean-Charles. "L'utopie chez les Chartrains." *Le Moyen Âge* 90 (1984): 383–400.

Pfeiffer, Jens. *Contemplatio Caeli. Untersuchungen zum Motiv der Himmelsbetrachtung in lateinischen Texten der Antike und des Mittelalters*. Hildesheim: Weidmann, 2001.

Piehler, Paul. *The Visionary Landscape. A Study in Medieval Allegory*. London: Edward Arnold, 1971.

Ratkowitsch, Christine. "Die Timaios-Übersetzung des Chalcidius. Ein Plato christianus." *Philologus* 140 (1996): 139–162.

Ratkowitsch, Christine. *Die Cosmographia des Bernardus Silvestris. Eine Theodizee*. Cologne, Weimar and Vienna: Böhlau Verlag, 1995.

Ratkowitsch, Christine. *Descriptio Picturae. Die literarische Funktion der Beschreibung von Kunstwerken in der lateinischen Großdichtung des 12. Jahrhunderts*. Vienna: Verlag der Österreichischen Akademie der Wissenschaften, 1991.

Roling, Bernd. "Das *Moderancia*-Konzept des Johannes de Hauvilla. Zur Grundlegung einer neuen Ethik laikaler Lebensbewältigung im 12. Jahrhundert." *Frühmittelalterliche Studien* 37 (2003): 167–258.

Stock, Brian. *Myth and science in the twelfth century. A Study of Bernard Silvester*. Princeton: Princeton University Press, 1972.

Stolz, Michael. "Bewegtes Beiwerk. Ästhetische Funktionen der Kleiderthematik bei Alanus ab Insulis und Giovanni Boccaccio." *Ästhetische Reflexionsfiguren in der Vormoderne*. Eds. Annette Gerok-Reiter, Anja Wolkenhauer, Jörg Robert and Stefanie Gropper. Heidelberg: Winter, 2019. 357–392.

Swanson, R. N. *The Twelfth-Century Renaissance*. Manchester: Manchester University Press, 1999.

Urban, Alexandra. *Poetik der Meisterschaft in 'Der meide kranz'. Heinrich von Mügeln auf den Schultern des Alanus ab Insulis*. Berlin and Boston: Walter de Gruyter, 2021.

Urban, Alexandra. "Zwischen kosmologischer Anthropologie und Tugendethik – Transformationen der Natur in den poetischen Entwürfen des Alanus ab Insulis und Heinrichs von Mügeln." *Limina – Natur – Politik. Verhandlungen von Grenz- und Schwellenphänomenen in der Vormoderne.* Eds. Annika von Lüpke, Tabea Strohschneider and Oliver Bach. Berlin and Boston: Walter de Gruyter, 2019. 93–114.

Vollmann, Justin. "Kosmopoetologie. Koordinaten der Selbstverortung mittelalterlichen Dichtens." *Literaturwissenschaftliches Jahrbuch* 58 (2017): 49–67.

Wandhoff, Haiko. *Ekphrasis. Kunstbeschreibung und virtuelle Räume in der Literatur des Mittelalters.* Berlin and New York: Walter de Gruyter, 2003.

Wetherbee, Winthrop. "Philosophy, Cosmology, and the Twelfth-Century Renaissance." A *History of Twelfth-Century Western Philosophy.* Ed. Peter Dronke. Cambridge and New York: Cambridge University Press 1992 [1988]. 21–53.

Wetherbee, Winthrop. *Platonism and Poetry in the Twelfth Century. The literary influence of the School of Chartres.* Princeton: Princeton University Press, 1972.

White, Hugh. *Nature, Sex, and Goodness in a Medieval Literary Tradition.* New York: Oxford University Press, 2000.

Wick, Maximilian. *Kosmogenetisch erzählen: Poetische Mikrokosmen in philosophischer und höfischer Epik des Hochmittelalters.* Berlin: Peter Lang, 2021.

Ziolkowski, Jan. *Alan of Lille's Grammar of Sex. The Meaning of Grammar to a Twelfth-Century Intellectual.* Cambridge: The Medieval Academy of America, 1985.

Sophie Knapp

Between Nigromancy and Erudite *meisterschaft*: Astronomical-Cosmological Knowledge in Middle High German *Sangspruchdichtung*

Abstract: In the process of genre development, the Middle High German *Sang-spruch* poets increasingly tended to participate in learned discourses in order to exhibit their poetic competence and to distinguish themselves. References to astronomical and cosmological knowledge represent a particularly interesting area of such appropriation. On the basis of two contrasting textual examples from the second half of the thirteenth century, this article traces how such knowledge in Middle High German *Sangspruchdichtung* is on the one hand unproblematically and ostentatiously claimed as proof of greatest erudition, while on the other hand it is stylized as so arcane that the attempt to acquire it is condemned as a sacrilegious advance into the mysteries of God – though at the same time the authors nevertheless implicitly suggest expertise in this field. Thus, we can observe a changing functionalization of the astronomical-cosmological discourse in Middle High German *Sangspruchdichtung*, which is closely connected to the central intention of the poets to prove their own *meisterschaft*.

Knowledge and erudition play a key role in the vernacular literature of the Middle Ages; they serve as proof of poetic qualification and thus claim validity. This can be observed especially in the so-called *Sangspruchdichtung* [*Sangspruch* poetry], a strongly didactic lyrical genre of travelling professional poets. In general, their texts are characterized by a certain fixation on consensus and repetition of widely accepted tenors as well as practical life advice (cf. Stackmann 1958, 69–121). Nevertheless, the authors – likely due to their precarious social status – repeatedly emphasized their teachings' relevance and highlighted their specific intellectual and moral aptitude for the self-imposed task of imparting knowledge (cf. Stackmann 1958, 69–121). To this end, they decidedly and often polemically distinguished themselves from other travelers and minstrels (cf., e.g., Stackmann 1958, 119, 121; Wachinger 1973). Moreover, they increasingly sought to include learned knowledge in their poetry to prove their own expertise, their "meisterschaft" [mastery] – as they term it (cf. Grubmüller 2009, 710–711). As the genre developed, the bodies of knowledge they referred to became increasingly advanced, although it is not always possible to assess how profound their learnedness really was (cf., e.g., Gade 2005; Grubmüller 2009; Knapp 2021).

The field of astronomical-cosmological knowledge, which was referenced in Sangspruch poetry since the mid-thirteenth century, is particularly dazzling in this respect. On the one hand, the authors presented their knowledge as proof of highest erudition: among the *septem artes* astronomy served as the ultimate touchstone of true "meisterschaft"; on the other hand, the Sangspruch poets stylized astronomical-cosmological knowledge as arcane and subsequently condemned attempts to probe it too deeply as illegitimate and sinful advances into the divine mysteries (cf. Gade 2005, 181; Ragotzky 1971, 71–72; Siebert 1939, S. 222–224[1]). Discussing astronomical-cosmological knowledge could therefore lead to moral devaluation (which, in the case of the Sangspruch poets, also meant artistic devaluation), and it could even spark suspicions of forbidden black magic.[2]

I want to focus on this area of conflict using two textual examples illustrating these opposite poles. This will also touch upon the question of what the term *astronomical-cosmological knowledge* denotes in the context of vernacular literature by laymen for laymen – and on what contemporary recipients may have gathered from such a *writing of the heavens*.

The first example consists of two stanzas by the Kanzler, a Sangspruch poet of the late thirteenth century, whose telling name (deriving from "cancellarius", i.e., scribe in a medieval chancery) possibly already indicated an attempt to imply *litterati* status with the accompanying claim to education (cf. Kornrumpf 1983, 986).[3] Unlike most texts with astronomical-cosmological contents in Sangspruch poetry, which predominantly reproduce a recurring repertoire of certain phrases and generally understandable terms (cf. Gade 2005, 180), in the case of the Kanzler we also find unusually specific astronomical-cosmological questions as well as – and this is especially unusual – technical terms. However, it is exactly here that both manuscripts transmitting the text display serious defects.[4] The more recent version can hardly be transformed into a comprehensible text anymore.

---

1 On astronomy and its ambivalent position between *septem artes* and a problematic proximity to magical practices/nigromancy, cf. Haage and Wegner (2007, ch. B. IV); Fürbeth (2020).
2 This can be observed particularly in the *Wartburgkrieg*, as discussed below, cf. esp. Kellner and Strohschneider 1998, 157–165; Kellner and Strohschneider (2007).
3 For the following, see also Knapp (2021).
4 The older manuscript containing these stanzas is the *Codex Manesse* (UB Heidelberg, Cpg 848), dating from 1300–1340, where they are part of the Kanzler corpus tradition. The younger transmission is Basel Cod. O IV 28, dating from 1430 (Schanze 1987, 346; Marner 2008, 9; Bartsch [1886, 275], however, dates the manuscript to the second half of the fifteenth century). Here the stanzas are completed to a *Bar* of three stanzas by way of a further original Kanzler stanza. Apart from this, the younger manuscript transmits a diverse mix of *Meisterlieder* of spiritual and secular

With some editorial interventions, the older manuscript's first stanza reads as follows in the left column[5] (juxtaposed on the right is the more recent variant[6] in order to illustrate the deviations, especially regarding technical terms and their presumed textual corruption):

Codex Manesse	Basel Cod. O IV 28
Hat ieman sin so snellen,	Hatt iemen sin so schnelle,
der tûte ein ellich zenter mir,	der tût mir die vier eccenter mas
da nach die para*l*ellen,	und die bar elle,
zwen orienten – dest min gir –,	zuen uracentran, das ist min begir,
5 des vúnfte*n* wesen*s* schin,	das fûnft er wider schin,
	wie sich die dy gemone
wie sich dar inne mane	birgz klein und ist doch gross
nu klein erzeiget unde nu gros,	und dar by wandel sône.
doch wandelunge ane.	der oberhimel saget mir,
da nah den himel lu*f*tgenos,	der under ein cristellin,
10 unde danne den kristall*i*n.	wie zod iacob
	zwôlff der sunnen waltend,
der tút ôch, wie zediacus	wie polus und oracius
zwelf*v*alt die sunnen halt*e*	das firmament enthaltent,
unde wie polus enpireus	wie disen hohen sache
des sunnentaches walte[ ]	bezaichent den ursprung,
15 unde wie dù erste sache	und sich ouch umme wachet
in schepf[ ]ungen dur die welt	der himel ob dem nûten ist,
wùrk unde wunder mache	as mir du gschriffet vergicht.
unde wie sich wege der erste ring,	
der beslûzet ellù ding.	
3 parabellen; 5 der vúnfte wesent;	
9 lust genos; 10 kristallen; 11 tût;	Kanzler – stanza 1 ([1]Kanzl/2/10[7])
12 welf walt [. . .] halt; 14 waltet;	
16 schepfe iungen	

---

content by different authors (cf. Bartsch 1886, 275–301; Schanze 1987, 346–347), some of which also touch on astronomical-cosmological topics (see below).

5 As I want to present a version of the text that was actually available in and received during the Middle Ages, but yet give a comprehensible text I follow the latest edition by Stephanie Seidl in *Lyrik des deutschen Mittelalters, online*, which tries to reconstruct the stanzas from the manuscript's sometimes obscure wording with as little intervention as possible. Siebert, on the other hand, has undertaken a sophisticated reconstruction of the text with the aid of technical literature of the time in order to advance to the presumed 'original' (cf. Siebert 1938, 1–14); the validity and reliability of such a comprehensive philological reconstruction, however, remains disputed.

6 It follows the manuscript Basel Cod. O IV 28's wording without editorial intervention, apart from the insertion of basic punctuation and careful orthographical normalization.

7 The numbers here and below refer to the common designation of all stanzas in the RSM (*Repertorium der Sangsprüche und Meisterlieder*).

[He, who has a truly sharp mind, / may explain the center of all, / then the parallels, / the two † orients † – that is my desire –, / the fifth essences' glow, / and how in it the moon presents / soon small and soon large / and yet does not change. / then the heaven † luftgenos † [the term is reconstructed; the manuscript writes even less clearly, *lust genos*, S. K.], / and then the crystalline. / he shall also explain, how zodiacus / holds the sun twelvefold, / and how the polus enpireus / rules over the sun roof, / and how the first cause / wielded creation for the sake of the world / and created amazing things, / and how the first circle moves, / that encloses everything.][8]

The stanza sets in with a sort of challenge: "He, who has a truly sharp mind, may interpret . . ." (Kanzler, st. 1, v. 1–2). It thus dares listeners to prove their intelligence via their ability to explain what is presented to them. In this way, the narrator stylizes his subject as an intellectual challenge. By creating a kind of examination or riddle setting (cf. Tomasek 1994, 220), he himself acts as someone with superior knowledge, thereby creating a knowledge gap.[9] The astronomical-cosmological questions and technical terms raised thereafter are generally based on an Aristotelian-Ptolemaic model of the spheres as first conveyed to a broader audience by Honorius Augustodunensis[10] and, more nuanced, by John de Sacrobosco in the mid-thirteenth century.[11]

---

8 My transl. of the *Codex Manesse* version. Translating the stanzas is difficult, due, on the one hand, to the verses' debatable syntax (as medieval manuscripts lack punctuation), and on the other hand – in this particular case – due to some obscured technical terms. The above translation thus does not attempt to create a poetic text but tries to reproduce the Middle High German text as literally as possible for the sake of intelligibility. The meaning of some Middle High German terms is unclear here, they are marked with cruces and will be discussed in the interpretation below.

9 Cf. Bulang (2005, 44–45), who described this strategy in the case of the Sangspruch poet Boppe. The strategy can be observed frequently in the genre, as it serves the implicit presentation of the poets' intellectual abilities and *meisterschaft*.

10 Especially Honorius Augustodunensis' *Imago mundi*, dating from the first half of the twelfth century and compiling the knowledge of ancient school authors as well as ecclesiastical authorities in a concise and simplified manner for educational purposes (Freytag 1983, 122–123). The astronomical-cosmological knowledge conveyed here was received and translated into Middle High German in the anonymously transmitted *Lucidarius* around 1190–1195, which was widely received at that time and thus made the body of knowledge handed down by Honorius also accessible in the vernacular (Freytag 1983, 122, 129–130; Steer 1985; *Der deutsche Lucidarius* 1994).

11 Referencing ancient and Arabic sources, Sacrobosco's treatise *De Sphaera mundi*, which dates from about 1230, conveys astronomical-cosmographical knowledge in a clearly comprehensible and systematic manner, thus making specific cosmological expertise more comprehensively available to recipients than Honorius (Brévard 1983, 731–733; Thorndyke 1949 [*The Sphere of Sacrobosco*]). This treatise was also translated into the vernacular, though not until the mid-fourteenth century. The first German translation is Konrad of Megenberg's *Die deutsche Sphaera* (Brévard 1983, 732–735).

The Kanzler's enumerations begin with the question of the "ellich zenter" (Kanzler, st. 1, v. 2), the cosmic center, where, according to this model, the earth as the heaviest element rested (cf. *Sphaera mundi*, 84–85). Next, he names the term "parabellen" (Kanzler, st. 1, v. 3), apparently referring to the *paralleli*, that is the summer and winter solstice lines running in parallel to the equator, as well as the northern and southern polar circles (cf. *Sphaera mundi*, 93; Siebert 1938, 4). The following third 'technical term,' the two "orienten" (Kanzler, st. 1, v. 4), is less identifiable. It could refer to different times of sunrise, for example at the summer and winter solstices (cf. Krieger 1931, 82–83); however, in the classical astronomical-cosmological discourse these do not usually appear in topical duality. The parallel transmission reads "uracentren" here (Basel, Kanzler, st. 1, v. 4). Philological text reconstruction proposed conjecturing from both readings the term "horizons" representing a dyad likely to be expected in the discourse, namely the geocentric *orizon rectus* and the *orizon obliquus* depending on the observer's viewpoint (Siebert 1938, 4–5).[12]

Following this prelude of technical terms, the narrator turns beyond earth. Employing enigmatic formulations, he mentions the "vûnfte wesens schin" (Kanzler, st. 1, v. 5), that is the etheric realm's glow, the *quinta essentia*, which begins above the elemental spheres (cf. *Sphaera mundi*, 79), and he refers to the moon, which appears large and small alternately, nevertheless without ever changing (Kanzler, st. 1, v. 6–8).[13] Once more requesting interpretation, the narrator then proceeds to the outer celestial spheres. The initially introduced concept of "himel luftgenos" (Kanzler, st. 1, v. 9), remains unclear, as it has no equivalent in the cosmological discourse of the time,[14] in contrast to the following "kristallin" (Kanzler, st. 1, v. 10), the crystal heaven (cf. Gade 2005, 211).[15]

---

12 *Sphaera mundi*, 91: "Est autem duplex orizon, rectus scilicet et obliquus sive declivis" (which the treatise elaborates on in more detail below.)

13 The Middle High German translation "vûnfte wesen" for the *quinta essentia* also appears later in Konrad von Megenberg's *Die Deutsche Sphaera* (cf. Siebert 1938, 5), who elaborated it in more detail: 'it is called the fifth because it follows the four elements [in the center of the cosmos, S. K.] and has a nature different from them' (*Deutsche Sphaera*, 10, 10–14, my transl.). Siebert has considered the possibility of the statement about immutability relating to the *quinta essentia*, thus indicating that the ether's supralunar sphere was conceived as immutable in the contemporary astronomical-cosmological discourse (cf. Siebert 1938, 5).

14 The younger transmission (Basel Cod. O IV 28) here refers to the "oberhimel" apparently referencing the firmament, which is possibly also referred to by "luftgenoz" in the *Codex Manesse* transmission. The Basel manuscript thus shares noticeable proximity with the wording in Honorius Augustodunensis' *Imago mundi* (I, 87): "superius caelum dicitur firmamentum" (cf. Siebert 1938, 6).

15 The term *coelum cristallinum* does not appear in Sacrobosco's treatise, who only mentions the 'outermost sphere' (cf. *Sphaera mundi*, 79). Only Konrad's von Megenberg translation explan-

The stanza then jumps back again one sphere to the "zediacus" [zodiac],[16] which 'holds the sun twelvefold' (Kanzler, st. 1, v. 11–12), an enigmatic description of the fixed star sky, before moving on to the last sphere, the *empyreum* ("polus enpireus", Kanzler, st. 1, v. 13).[17] This reference illustrates that the stanza is based on a Christianized Ptolemaic model of the world that considers the realm of God to be located in a further heaven, above the crystal heaven, whereas according to ancient conception the latter is the outermost sphere (cf. Gade 2005, 211–214). A cause for irritation, however, can be found in the statement that this "polus enpireus" (Kanzler, st. 1, v. 13) rules over the 'sun roof' ("sunnentach", Kanzler, st. 1, v. 14), which presumably means the fixed star sky, as the latter is earlier said to 'hold the sun twelvefold' (zodiacus) (Kanzler, st. 1, v. 12). In contrast to the preceding verses this suggests that the *firmamentum* is followed directly by the *empyreum* rather than by the crystal heaven. Further contradictions are generated at the end of the stanza with the question addressing the first cause, the *prima causa/primum mobile*, and how the outer ring moves ("wie sich wege der erste ring"), that encloses 'all things' ("der beslûzet ellû ding", Kanzler, st. 1, v. 19). This, in return, suggests that the outermost sphere is in motion. However, the *empyreum*, which was mentioned before as outermost sphere, is considered fixed (cf. Siebert 1938, 8–9). According to this model, the penultimate sphere, here equated with the crystal heaven, was in motion – according to the ancient conception, however, this was considered to be the outermost sphere (named *primus motus;*[18]

---

atorily added the term in the fourteenth century (*Deutsche Sphaera*, 7), however the sermons of Berthold von Regensburg, among others, demonstrate that it was indeed already common to the vernacular in the thirteenth century (cf. Berthold von Regensburg *Predigten*, 179–180; 235); cf. also note 18.

16 Cf. *Sphaera Mundi*, 87–89; *Lucidarius*, I, 79.

17 Besides 'coelum' 'polus' can also mean 'heaven,' for example in Alanus' *Anticlaudianus*; Siebert has raised the question, whether this could be indicating a dependence on this specific source here (Siebert 1938, 8).

18 Sacrobosco's *Sphaera mundi* also follows this ancient conception. Konrad von Megenberg's translation, on the other hand, added the *empyreum* above the *primum mobile* in the fourteenth century, equating the latter with the crystal heaven: "Und ob dem setzen die kristen und die juden ainen himel, der haizzet der feurein himel [. . .]. Und der hat kainen lauf, sunder got rût mit seinen lieben darinne. Aber unser Johannes sagt von dem selben himel niht, noch kain ander haidenisch sternseher" (*Deutsche Sphaera*, 7) ["And above this the Christians and the Jews place another heaven, which is called the fiery heaven [. . .]. And it does not move, but God rests in it with his beloved ones. But our John [de Sacrobosco, S. K.] says nothing of this heaven, nor do any of the pagan star observers," my transl.]. The fact that this concept of the tripartite model of heaven already occasionally appeared in the vernacular in the thirteenth century can be observed, for example, in the sermons of Berthold von Regensburg (cf. Berthold von Regensburg, *Predigten*, vol. 1, 179–180).

cf. Siebert 1938, 8–9; concerning the *primum mobile*, also cf. Gade 2005, 250–251). The author thus confuses different variants of the cosmic model. On the one hand, this reflects the contemporary competition and intermingling of the ancient model and its Christianized adaptation. On the other hand, it indicates that the author either did not distinguish between crystal heaven, *primum mobile* and *empyreum*, or did not clearly grasp the last spheres' sequence (cf. Siebert 1938, 9). In any case, it seems to show that he was not as well informed about his subject as he purported, which also casts doubt on the reliability of his over-all knowledge and thus raises the question whether the use of these erroneous technical terms might rather indicate pretentious self-fashioning (cf. Runow 2015, 94–95). Be this as it may, the transmitted text presents grave metrical and grammatical errors (cf. Runow 2015, 94), which cannot possibly have originated with the author himself. In this respect, the stanzas' corruption can presumably be ascribed to a large extent to the scribes and editors. The significantly worse condition of the text in the more recent manuscript (especially with regard to technical terms) also points to an increasing corruption process through transmission.[19] This indicates that contemporary recipients already faced substantial difficulties in understanding it (cf. Runow 2015, 94). The fact that it was nevertheless continuously transmitted, however, also indicates the attraction and fascination of the subject.

The first stanza, which is thus obviously in need of further explanation, is followed by a second one. But this one now, rather than offering resolutions and explanations, moves on to poetological self-conception and self-legitimation.

---

[19] The Basel manuscript provides a further parallel case of astronomical-cosmological terminology, a stanza by Heinrich von Mügeln (*Kleinere Dichtungen*, La 7 [¹HeiMü/500,7]). Next to correctly transmitted terms, this stanza, which is clearly younger, presents similar defects as the Kanzler stanza, when for instance the term "zodiacus" is corrupted to "sodegicus," v. 4 (in the Kanzler stanza in the same manuscript "zod iacob," st. 1, v. 11) and the "orizon" to "arisa," v. 5 (in the Kanzler stanza "uracentren," st. 1, v. 4). Furthermore, Mügeln's stanza mentions "colarie" instead of "colure," v. 4; "ante artigus" instead of "antarticus," v. 9; "sturm nebel das zaichet blindes Coiades" instead of "sturm, nebel das zeichen Pliades und Hyades," v. 16 (cf. Heinrich von Mügeln, *Kleinere Dichtungen*). These mistakes point to problems of transmission concerning this kind of specialized terminology and indicate that these terms could not be assumed to be generally known at the time, regardless of whether the errors can be traced back to this manuscript's scribe or whether he was already working with a flawed copy.

Codex Manesse
Swie swerer last sich neiget
ze der erden zenter, wa das stat,
unde wie sich umbeweiget
der himel, ob dem niht enist,
5 des mir dù schrift vergiht:

klar hitzig sunnenblig,
des manen kelte, des regens sprat,
der bernde wint erkiket.
planeten kraft, ir lôfes vrist[ ]
10 min kunst vermisset niht –

wie lùhtet himel, sternen kraft
der erde hilfe bringet,
daz es ir sûze hohgeschaft
durh blût in vrûhte dringet.
15 wa elemente sich rûrent,
wa swer, wa liht, wa heis, wa kalt,
wies us nature fûrent
lebendig geschepfde manigvalt
gar wunderlich gestalt.
9 vristen; 14 vûhte; 17 fûret

Basel Cod. O IV 28
[V. 1f. are missing]
[V. 3–5 of the Codex Manesse *version*
are here at the end of stanza 1 instead
of Codex Manesse *V. 17–19]*

Clar, heiss ist sunnen bliczgin.
des mones schin, des regen sprat,
der berndu winde quikge,
planeten lauff, und auch ir list
min kunst vermisset nicht;
wie himel lûfftig sternekrafft
der erde hilffe bringet,
O hohu sûssu maisterschifft
durch blut die frûchte dringet.
wo elementen sich rûrent,
wo heiss, wol kalt, wo licht, wo schwer
in die nature fûret
lebent geschôpfft gar manigfalt
so wunneclich gestalt etc.

Kanzler – stanza 2 (^{1}Kanzl/2/11)

[In which way heavy mass leans towards / the center of the earth, / where that is located, / and how the sky turns over, / above which is nothing, / that is what the scriptures tell me. / the clear hot rays of the sun, / the moon's cold, the rain's sprinkling, / the fruitful wind's refreshing, / the power of the planets, the duration of their orbit / my art (i.e. poetic skill) does not fail to grasp; / how heaven shines because of the power of the stars, / which helps earth so that / it pushes its lovely, noble creation / via blossom into fruit. / where the elements mingle – / where heavy, where light, where hot, where cold –, / how they create from nature / manifold living creatures / of quite astonishing shape.] (my transl. of *Codex Manesse* version)

In the first *Stollen* of the stanza (v. 1–5) the narrator refers to scripture (the "schrift" tells me, Kanzler, st. 2, v. 5) regarding his knowledge on how all things strive toward the center of the cosmos and how the heavens move, thus addressing the origin of his knowledge and endowing it with validity by decisively asserting his participation in written, *litterate*, erudition.

This arcane knowledge of the cosmic composition is followed in the second *Stollen* (v. 6–10) by an – in comparison to the preceding *Stollen* and stanza – rather simple enumeration of astronomical-astrological influences which are directly perceptible on earth: the heat of the sun, the cold of the moon, wind and rain, the planets' orbital periods and powers as well as the force of the stars influencing growth and decay on earth and the miraculous origination of all things down there, from a mixture of the four elements and their qualities.[20] The narrator thus reproduces in this second part of the stanza typical astronomical and cosmological 'knowledge' as it was to be expected in Sangspruch poetry – and accordingly does now decisively not refer to books for the specific knowledge given in this second enumeration, but to his own "kunst," his art, his central qualification as a Sangspruch poet, that makes him capable of grasping all these things. The stanza thus creates a two-fold poetological legitimation of erudition on the one hand and poetic ability on the other ("schrift" and "kunst"). Possessing both of these is thus presented as necessary prerequisite to fully address astronomical-cosmological subjects. Implicitly, the narrator thus claims for himself the "sin so snelle" (Kanzler, st. 1, v. 1), the truly sharp mind demanded at the beginning, an extraordinary intellectual disposition that can be tested and proven in the ability to speak about learned expertise. The actual extent of this knowledge, however, is cleverly concealed by the challenging, keyword-like series of questions, which implies more than it explains (cf. Runow 2015, 92). The use of exotic technical terms and deliberately obscure phrasings creates a calculated knowledge gap vis-à-vis the recipients, and – anticipating that they will not be able to explain these terms – outmaneuvers potential competitors. In contrast to the genre's stated claim to instruct, this series of stanzas obviously does not aim to impart astronomical-cosmological knowledge to laymen, but rather to exhibit the author's learnedness and qualifications. Astronomical-cosmological knowledge is thus functionalized to implicitly claim "meisterschaft" via participation in this learned discourse and its validity.

Significantly, in the context of the genre – and this takes us to the second example – the Kanzler does not qualify astronomical-cosmological knowledge religiously, in contrast to the religious frame of almost all other Sangspruch stanzas touching upon astronomical knowledge, for these predominantly embed it in the praise of God (and sometimes of the *septem artes*, cf. Gade 2005, esp. 180–182). They present the creation of the cosmos and its secret laws as an astonishing miracle revealing God's incomprehensibility and insist that man should not venture

---

20 Similar bodies of knowledge can be encountered in the vernacular in the *Lucidarius* (I. 79–90, 93, 99, 102) and in the sermons of Berthold von Regensburg, who, among other things, described the influence of the planets on the formation of wind and rain and the powerful influence of the stars on the living beings of the earth (Berthold von Regensburg, *Predigten*, 50–51) (cf. Siebert 1938, 11–12). The Sangspruch poets received this more general knowledge about God's creation, elements, and planets and incorporated it into their poetry, preferably in the context of praising the Creator (see also below).

inappropriately deep into God's secrets.[21] This is particularly striking in the so-called *Wartburgkrieg* ['war at the Wartburg'], a complex conglomeration of stanzas, which in many respects can be regarded a metatext of Sangspruch poetry (cf. Tervooren 2001, 37; differently Hallmann 2015, 117–130). Fictitious minstrel contests form the central theme of this varying text, which was subsequently expanded over the decades, featuring partly fictional, partly historical poets (cf. Wachinger 1973, 5–89; 1999). Among others, the eminent Middle High German poet Wolfram von Eschenbach and his fictional character, the sorcerer Klingsor from his *Parzival* novel, are confronted here (cf. Ragotzky 1971, 45–64; Bulang 2015). Especially in the oldest part of the text, the so-called *Rätselspiel* ['riddle game'/'mystery game'], the roles of the antagonists are clearly contrasted:[22] Wolfram appears as a pious layman and positive role model for Sangspruch poets, while Klingsor is introduced as a haughty learned "meisterpfaffe," a master-priest or master-cleric who puts Wolfram's vaunted prudence to the test with allegorical-theological riddles (Ragotzky 1971, esp. 45–64; Wachinger 1973, 87–89). This constellation presents the two men as exponents of different concepts of knowledge, which are thus narratively negotiated in their value: On the one hand, the knowledge of laymen, who remain humble in the face of God's omniscience despite all that is revealed to them; on the other hand the knowledge of the scholars who cunningly touch the mysteries of God (cf. Ragotzky 1971, esp. 62–63, Kellner and Strohschneider 1998, 2007). Wolfram excels through his capacity to solve all riddles, whereupon Klingsor accuses him of being in league with supernatural powers as well as – and this is where it becomes especially interesting for my subject – of possessing astronomical knowledge (*Rätselspiel*, st. L 8, L 11[23]). Astro-

---

21 Cf., e.g., the Sangspruch poets Marner ([1]Marn/6/16), Meißner ([1]Mei/1/2), Rumelant ([1]Rum/1/5), Walther of Breisach ([1]WaltBr/1/1–2), of whom Rumelant ([1]Rum/3/3) is particularly clear in his condemnation of the discussion of astronomical-cosmological knowledge by "gar gelerte[] leiebere[] pfaffen" (Rumelant, *Sangsprüche*, st. III,3,v. 1), thus apparently attacking Sangspruch poets who "act like extremely learned clergymen, speculate about the nature of heaven, hell, and the cosmos, and thereby presume knowledge that God possesses, but not man" (Kern [Rumelant, *Sangsprüche*] 2014, 354 [my transl.]). Kern also points out the biblical tradition of this criticism and its reception not only in Sangspruch poetry but elsewhere in Middle High German literature too (cf. Kern [Rumelant, *Sangsprüche*] 2014, 345–346; Kern also offers numerous references to primary texts here).

22 The poem's oldest assumed 'core' (on which I lay my focus here) likely dates to the second quarter of the thirteenth century (Wachinger 1999, 749). However, the *Rätselspiel* was apparently also expanded over time, including additional stanzas, thus slightly blurring the characters' polarity (see below).

23 With its strongly varying transmission, the text, especially in the *Rätselspiel*'s case, makes editioning an almost unsolvable task, particularly since the transmission is also corrupted at times due to the text being partially deliberately obscured (e.g., concerning its specific riddle-like allegories; cf. Bulang and Runow 2016). Hallmann (2015) synoptically edited the manuscripts trans-

nomical knowledge thus here appears shady and discredits the expert. Accordingly, Klingsor tries to prove Wolfram guilty: He summons the devil Nasion to examine Wolfram by means of astronomical questions in order to force him to reveal his supposedly dark sources of knowledge:

(Nasion speaks to Wolfram)
Nu sage – hastu meisterschaft –
wie das firmamentum mit so hoher craft
gein den siben planeten muge kriegen
Oder wie der polus Articus
ste und der hohe meisterstern Antarticus.
nu sage mir – zwar, du kanst mich niht betriegen:
Saturnus, wan der osten stat, waz diutent uns die wunder?
(*Rätselspiel*, L 14, v. 1–7)

[Now tell me, if you are endued with mastery, / how the firmament wages war with such great power / against the seven planets, / Or how the polus Arcticus stands / and the high master star Antarticus, / tell me – you truly cannot deceive me – / when Saturn stands in the east, what do these miracles indicate/signify?, my transl.][24]

Much like the Kanzler, the devil challenges his opponent to prove *meisterschaft* by answering astronomical and astrological questions, namely regarding the opposite movement of planets and fixed stars, why the star of the north and south pole stand still, and which power Saturn unfolds in the east.[25] At the same time,

---

mitting the text (namely the *Codex Manesse* [Heidelberg, UB, cpg 848]; the *Büdinger Fragmente* [Wolfenbüttel, HAB, Cod. 326 Novissimi 8]; the *Jenaer Liederhandschrift* [Jena, ULB, Ms. El. F. 101]; the *Lohengrinhandschrift A* [Heidelberg, UB, cpg 364] and the *Kolmarer Liederhandschrift* [Munich, BSB, cgm 4997]. The citations used here mostly follow his *Lohengrin* manuscript edition, as it offers the most comprehensible and coherent text. The Lohengrin manuscript is henceforth referenced with the abbreviation L, the *Codex Manesse* with C.

24 Although it presents astronomical-cosmological knowledge as inscrutable, this stanza interestingly also makes use of some technical terms. Even more interestingly, the oldest manuscript transmitting the stanza, the *Codex Manesse*, i.e., the very manuscript that also contains the Kanzler stanzas discussed above, here presents similar defects regarding those terms. Instead of "polus Articus" (*Rätselspiel*, L 14, v. 4) the *Codex Manesse* reads "polus Artanticus", so here we can observe that the manuscript uses "Antarticus" (which the *Lohengrin* manuscript mentions only in the following verse) instead of "Articus" and just slightly misspells it, whereas the 'master star' (the "Antarticus" according to the Lohengrin manuscript, *Rätselspiel*, L. 14, v. 5) is corrupted to "Antribilus" in the *Codex Manesse*. This indicates once again that the specific astronomical content was not completely understood and that these terms therefore cannot be assumed to have been common knowledge at the time (cf. note 19), while on the other hand they might also illustrate the author's intention to make the devil's speech sound exclusive and learned.

25 Also cf. note 27; Hallmann (2015, 166–167) furthermore points out that this knowledge can also be found in other (earlier) literary vernacular texts of the thirteenth century. The opposite move-

Nasion tries to seduce Wolfram with this knowledge by promising profound enlightenment in these questions, if he can answer only one of them ("Kanst du mir einez der gesagen, / mine müe, die wil ich gar gein dir verdagen, / sint ich dirz allez han genant besunder", L 14, v. 8–10 ["If you can tell me but one of these, / I will not speak of my efforts / until I have explained them to you one by one," my transl.]).

In this 'devils' exam,' Wolfram vehemently refuses to know – or desire to know – anything about these things – "ich enruoch, wiez osten, western stat" (*Rätselspiel*, C 52, v. 4)[26] ["I don't care, how things are (literally 'what stands/what is') in the east or west (. . .)," my transl.] – and states that God alone rules over them (in strong contrast – by the way – to the historical author Wolfram, who proved to be rather astronomically versed in his epic poetry, especially in *Parzival*[27]). Wolfram banishes the devil by striking the cross in front of him, and Nasion grudgingly reports to Klingsor that Wolfram's *meisterschaft* is irrefutable (*Rätselspiel*, L 18). The text thus sets a clear boundary between revealed and esoteric, corrupt knowledge. 'Righteous' knowledge becomes a category of moral-religious exemplarity, it is proven precisely not in boundlessly striving for erudition, but in the renunciation of knowledge, where learned knowledge would delve too deeply into the divine secrets (cf. Ragotzky 1971, 53–55; Kellner and

---

ment of the firmament and the planets, for example, is discussed in *Lucidarius* (cap. I,21–23) and in Berthold's von Regensburg *Predigten*, S. 392 (on the overall debate about the motus duplex, cf. Gade 2005, 247–252). The southern pole star being fixed is mentioned for example in Albrecht von Scharfenberg's *Jüngerer Titurel* (v. 4806,1–3) and refers to the idea (which is also evidenced elsewhere in vernacular literature) that the stars do not move, because the celestial axis is attached to them (Bauer 1937, 45).

26 For this passage, I follow the transmission in C by way of exception, since the text in L is not quite as pointed here: "Mir ist niht kunt ir underscheid. / daz du mich drumbe fragest vil, daz ist mir leit! / für war ich waiz niht rehte waz ir meinet. / Ich weiz was osten, westen stat, / wo iegelich stern nach sinem zikel sunder gat. / der sie beschuf, der hat ir ganc vereinet" (L 16, v. 1–6). With different punctuation than Hallmann's edition (e.g., a colon after v. 3, or comma after v. 5), L however, can be read as aiming at the same statement as C (with a colon after v. 3, one could paraphrase: 'I don't know what you mean by saying that I know these things,' or with a comma after v. 5: 'I know about it only so much, as that the creator has arranged these things the wondrous way they are').

27 A sophisticated play with allusions is revealed here in particular, that aims at the recipients' familiarity with Wolfram's writings, seeing that exactly the astronomical knowledge Nasion quizzes Wolfram with is mentioned (among others) in his works: The 'war' of the firmament against the planets (*Parzival*, v. 782,14–18; *Willehalm*, v. 216,9–11 [Nellmann (2003, 62) points out here, that Wolfram is the first in Middle High German literature to use the metaphor "kriegen," i.e. 'waging war,' in this context]), as well as "Antarticus" as the name of a star (*Parzival*, v. 715,16; *Willehalm* v. 216,6) and mentioning the special effect Saturn has in certain astronomical positions (*Parzival*, v. 489,24–29; 492,27–493,1) (cf. Hallmann 2015, 155).

Strohschneider 2007, 354f.).[28] In this respect, astronomic-cosmological knowledge here expressly turns into the touchstone of true erudition, albeit in reverse. It is branded as devilish and erudition is fundamentally discredited by the negatively portrayed character Klingsor. The latter disavows himself as versed in black magic with Nasion's incantation and finally even brags about his esoteric expertise (cf. Krohn 1993, 104): "Nigramanciam weiz ich gar. / der astronomie nim ich an den sternen war" (*Rätselspiel*, L 19, v. 1–2) ["I know everything about nigromancy (i.e., black magic), / astronomical knowledge I derive from the stars (. . .)," my transl.].[29]

I end my observations at this point and summarize them briefly: The two examples show, firstly, that the texts obviously do not try to really explain and convey astronomical-cosmological knowledge. Rather, the keyword-like lists of fragments function as a cipher for a specific form of learnedness. Secondly, a change in the functionalization of the astronomical-cosmological discourse in Middle High German Sangspruch poetry can be observed, which is closely connected to the intention to prove mastery, and this, thirdly, speaks of a changing evaluation of such knowledge, which reflects the contemporary theological discussion about the legitimacy of *astronomia* – and of *curiositas* in this field (cf. Gade 2005, 176–179): While astronomical-cosmological examination questions in the *Wartburgkrieg* appear as evidence for black magic activities, the Kanzler instrumentalizes them some decades later as proof of superior erudition, apparently without a problem. Tendencies of such a de-problematization can also be observed in the *Wartburgkrieg* itself, when more recently added stanzas continue the contest between Klingsor and Wol-

---

28 Hallmann, in contrast, sees a fundamental reservation against the usefulness of learned education in this, rather than the problematic crossing of a boundary in terms of knowledge (Hallmann 2015, 166–168). Yet on the one hand, Wolframs' interpretation of the preceding riddle (*Quaterrätel*) makes it very clear that he is preserving a boundary in terms of arcane knowledge here (*Rätselspiel*, L 10, v. 9f.), and thus rather clearly introduces the topic of a 'knowledge boundary' (this seems all the more significant seeing that he decidedly refers back to this situation in the Nasion scene [Rätselspiel, L 16, v. 10; even more clearly in manuscript C 52, v. 8–10]). On the other hand, Hallmann's parallel example, which he uses as evidence for his interpretation, is itself ambivalent. It is a critique uttered by Armer Hartmann (*Rede vom Glauben*), which exemplarily sets off the exploration of astronomical-scholarly knowledge as striving for useless erudition against the mortality of man and the true wisdom of Christ, without however foregoing an enumeration of this seemingly useless knowledge, by which the author, in performative contradiction to his statement, also displays wisdom in these matters and implicitly presents himself to his audience as erudite. On the other hand, this astronomical-cosmological enumeration could also be an intended theological reference to the Book of Job (chapter 38–39).

29 Nevertheless, the text and its author respectively also implicitly exhibit knowledge of the astronomical-cosmological discourse here and thus to some extent play with the fascination the subject has on the audience by creating such a dazzling figure as Klingsor.

fram after the 'devil's exam' and thereby increasingly turn it into a knowledge contest, in which both opponents try to surpass one another with their expertise in various fields, especially astronomy (also in *Zabulons Buch*, a younger part of the *Wartburgkrieg*; cf. Ragotzky 1971 esp. 70–75; Krohn 1993, 105; Kellner and Strohschneider 2007, 342–343). A development towards a purely positive valuation of astronomical-cosmological knowledge, however, cannot be concluded from this (cf. Gade 2005, 180–182). This becomes obvious, for example, in the *Hort von der Astronomie*, an even younger part of the *Wartburgkrieg* complex, in which an ambivalent attitude towards astronomical knowledge emerges once more. A closer look at the more recent transmission of the Kanzler stanzas (Basel Cod. O IV 28) reveals a similar development: It extends the song discussed above by an additional stanza, preceding the text with a praise of the creator, which celebrates God's power over the cosmos ([1]Kanzl/2/1). Consequently, the established 'challenge' as to who is clever enough to explain the cosmos, is rendered rhetorical, as all knowledge is entrusted to God alone. Like Wolfram in the *Wartburgkrieg*, the narrator is thus posthumously transformed into a humble admirer of God. However, even this remains ambivalent, seeing that the poets' ostensible humility regarding knowledge performatively contradicts their astronomical-cosmological testimonies in the end, which thus – despite all pious protestations – primarily serve their self-auratization. The long-lasting productivity of astronomical-cosmological themes in the Middle High German poetry up to the *Meistergesang* (cf. Gade 2005) speaks of the success of this practice and shows one thing above all: the immense potential for fascination of this 'writing of the heavens.'

Bibliography

Albrecht von Scharfenberg. *Jüngerer Titurel*. Eds. Werner Wolf and Kurt Nyholm. 4 Vols. Berlin: Akademie Verlag, 1955–1995.

Bauer, Georg-Karl. *Sternkunde und Sterndeutung der Deutschen im 9.-14. Jahrhundert unter Ausschluß der reinen Fachwissenschaften*. Berlin: Ebering, 1937.

Bartsch, Karl. *Beiträge zur Quellenkunde der Altdeutschen Literatur*. Strasbourg: Trübner, 1886.

Berthold von Regensburg. *Vollständige Ausgabe seiner Predigten*. Ed. Franz Pfeiffer [1862]. 2 vols., here vol. 1. Reprint. Berlin: De Gruyter, 1964.

Brévard, Francis B., and Menso Folkerts. "Johannes de Sacrobosco." *Die deutsche Literatur des Mittelalters. Verfasserlexikon*. 2nd, fully rev. ed. Eds. Kurt Ruh et al. Vol. 4. Berlin and New York: De Gruyter, 1983. l73–736.

Bulang, Tobias. "*wie ich die gotes tougen der werlte gar betiute*. Geltungspotentiale änigmatischen Sprechens in der Sangspruchdichtung." *Geltung der Literatur. Formen ihrer Autorisierung und Legitimierung im Mittelalter*. Eds. Beate Kellner et al. Berlin: Erich Schmidt, 2005. 43–62.

Bulang, Tobias. "Intertextualität und Interfiguralität des 'Wartburgkriegs'." *Sangspruchdichtung um 1300. Akten der Tagung in Basel vom 7. bis 9. November 2013*. Eds. Gert Hübner and Dorothea Klein. Hildesheim: Weidmann, 2015. 127–145.

Bulang, Tobias, and Holger Runow. "Allegorie und Verrätselung in der mittelhochdeutschen Sangspruchdichtung." *Verrätselung und Sinnzeugung in Spätmittelalter und Früher Neuzeit*. Ed. Beatrice Trînca. Würzburg: Königshausen & Neumann, 2016. 27–46.

Der deutsche Lucidarius. Bd. 1: Kritischer Text nach den Handschriften. Eds. Dagmar Gottschall and Georg Steer. Tübingen: Niemeyer, 1994.

Fürbeth, Frank. "Propheten, Beschwörer, Nigromanten, Märchenzauberer, Illusionisten, Automatenbauer. Phänotypen des Zauberers in der deutschen Literatur des Mittelalters unter diskursgeschichtlichem Aspekt." *Der Begriff der Magie in Mittelalter und früher Neuzeit*. Eds. Jutta Eming and Volkhard Wels. Wiesbaden: Harrassowitz, 2020. 47–80.

Gade, Dietlind. *Wissen, Glaube, Dichtung: Kosmologie und Astronomie in der meisterlichen Lieddichtung des vierzehnten und fünfzehnten Jahrhunderts*. Tübingen: Niemeyer, 2005.

Grubmüller, Klaus. "Autorität und *meisterschaft*. Zur Fundierung geistlicher Rede in der deutschen Sangspruchdichtung des 13. Jahrhunderts." *Literarische und religiöse Kommunikation in Mittelalter und Früher Neuzeit. DFG-Symposium 2006*. Ed. Peter Strohschneider. Berlin and New York: De Gruyter, 2009. 689–711.

Haage, Bernhard D., and Wolfgang Wegner. *Deutsche Fachliteratur der Artes im Mittelalter*. Berlin: Erich Schmidt Verlag, 2007.

Hallmann, Jan. *Studien zum mittelhochdeutschen Wartburgkrieg. Literaturgeschichtliche Stellung – Überlieferung – Rezeptionsgeschichte. Mit einer Edition der Wartburgkrieg-Texte*. Berlin and Boston: De Gruyter, 2015.

[Heinrich von Mügeln] *Die kleineren Dichtungen Heinrichs von Mügeln*. Ed. Karl Stackmann, with contributions by Michael Stolz. Zweite Abteilung. Berlin: Akademie-Verlag, 2003.

[John de Sacrobosco] *The Sphere of Sacrobosco and its Commentators*. Ed. Lynn Thorndike. Chicago: University of Chicago Press, 1949.

Der Kanzler. Eds. Manuel Braun and Stephanie Seidl. *Lyrik des deutschen Mittelalters, online*. Gen. eds. Manuel Braun et al. http://www.ldm-digital.de (May, 28 2022).

Kellner, Beate, and Peter Strohschneider. "Die Geltung des Sangs. Überlegungen zum Wartburgkrieg C." *Wolfram-Studien XV: Neue Wege der Mittelalter-Philologie. Landshuter Kolloquium 1996*. Eds. Joachim Heinzle et al. Berlin: Erich Schmidt, 1998. 143–167.

Kellner, Beate, and Peter Strohschneider. "Poetik des Krieges. Eine Skizze zum Wartburgkriegkomplex." *Das fremde Schöne. Dimensionen des Ästhetischen in der Literatur des Mittelalters*. Eds. Manuel Braun and Christopher Young. Berlin and Boston: De Gruyter, 2007. 335–356.

Knapp, Sophie. *Intertextualität in der Sangspruchdichtung. Der Kanzler im Kontext*. Berlin and Boston: De Gruyter, 2021.

Konrad von Megenberg. *Die Deutsche Sphaera*. Ed. Francis B. Brévart. Tübingen: Niemeyer, 1980.

Kornrumpf, Gisela. "Der Kanzler." *Die deutsche Literatur des Mittelalters. Verfasserlexikon*. 2nd, fully rev. ed. Eds. Kurt Ruh et al. Vol. 4. Berlin and New York: De Gruyter, 1983. 986–992.

Krieger, Harald. *Der Kanzler. Ein mittelhochdeutscher Spruch- und Liederdichter um 1300*. Bonn: Bonner Univ.-Druckerei, 1931.

Krohn, Rüdiger. *"ein phaffe der wol zouber las.* Gesichter und Wandlungen des Zauberers Klingsor." *Gegenspieler*. Eds. Thomas Cramer and Werner Dahlheim. Munich and Vienna: Hanser, 1993. 88–113.

Der Marner. *Lieder und Sangsprüche aus dem 13. Jahrhundert und ihr Weiterleben im Meistersang*. Ed. Eva Willms. Berlin and New York: De Gruyter, 2008.

Nellmann, Eberhard. "Der *Lucidarius* als Quelle Wolframs." *Zeitschrift für deutsche Philologie* 122 (2003): 48–72.

Ragotzky, Hedda. *Studien zur Wolfram-Rezeption. Die Entstehung und Verwandlung der Wolframrolle in der deutschen Literatur des 13. Jahrhunderts*. Stuttgart: Kohlhammer, 1971.

Repertorium der Sangsprüche und Meisterlieder des 12. bis 18. Jahrhunderts. Eds. Horst Brunner and Burghart Wachinger. 16 vols. Tübingen: Niemeyer, 1986–2009.

Runow, Holger. "*Hât ieman sin sô snellen . . .* Rezeptionsbedingungen des Sangspruchs um 1300 zwischen Mündlichkeit und Schriftlichkeit." *Sangspruchdichtung um 1300. Akten der Tagung in Basel vom 7. bis 9. November 2013*. Eds. Gert Hübner and Dorothea Klein. Hildesheim: Weidmann, 2015. 89–108.

Rumelant. *Die Sangspruchdichtung Rumelants von Sachsen. Edition – Übersetzung – Kommentar*. Ed. Peter Kern. Berlin and Boston: De Gruyter, 2014.

Schanze, Frieder. "Meisterlieder." *Die deutsche Literatur des Mittelalters. Verfasserlexikon*. 2nd, fully rev. ed. Eds. Kurt Ruh et al. Vol. 6. Berlin and New York: De Gruyter, 1987. 342–356.

Siebert, Johannes. "Die Astronomie in den Gedichten des Kanzlers und Frauenlobs." *Zeitschrift für deutsches Altertum* 75 (1938): 1–23.

Siebert, Johannes. "Himmels- und Erdkunde der Meistersänger." *Zeitschrift für deutsches Altertum* 76 (1939): 222–253.

Stackmann, Karl. *Der Spruchdichter Heinrich von Mügeln. Vorstudien zur Erkenntnis seiner Individualität*. Heidelberg: Winter, 1958.

Steer, Georg. "Lucidarius". *Die deutsche Literatur des Mittelalters. Verfasserlexikon*. 2nd, fully rev. ed. Eds. Kurt Ruh et al. Vol. 6. Berlin and New York: De Gruyter, 1985. 939–947.

Tervooren, Helmut. *Sangspruchdichtung*. Stuttgart and Weimar: Metzler, 2001.

Tomasek, Tomas. *Das deutsche Rätsel im Mittelalter*. Tübingen: Niemeyer, 1994.

Wachinger, Burghart. „Der ‚Wartburgkrieg'." *Die deutsche Literatur des Mittelalters. Verfasserlexikon*. 2nd, fully rev. ed. Eds. Kurt Ruh et al. Vol. 10. Berlin and New York: De Gruyter, 1999. 740–766.

Wachinger, Burghart. *Sängerkrieg. Untersuchungen zur Spruchdichtung des 13. Jahrhunderts*. Munich: Beck, 1973.

Wolfram von Eschenbach. *Parzival*. Based on the ed. by Karl Lachmann, rev. and comm. Eberhard Nellmann, transl. Dieter Kühn. 2 vols. Frankfurt am Main: Deutscher Klassiker Verlag, 1994.

Wolfram von Eschenbach. *Willehalm* . *Nach der Handschrift 857 der Stiftsbibliothek St. Gallen*. Frankfurt am Main: Deutscher Klassiker Verlag, 1991.

Walker Horsfall

Astronomical (In)accuracy in Heinrich von Mügeln's *Der meide kranz*

Abstract: Heinrich von Mügeln's fourteenth-century Marian praise poem *Der meide kranz* appears to contain an astronomical inaccuracy: the order of the signs of the zodiac are listed in two places in jumbled order, while a third list re-establishes the correct sequence. Because such a fundamental error seems uncharacteristic of an author as learned as Mügeln, this article explores alternative explanations for this apparent confusion. An analysis with other contemporary Middle High German texts suggests that, while there was great importance placed on scientific accuracy, there may also be a potential literary function of inaccuracy as an invitation for critical and innovative reflection. Building on this idea, this article offers an experimental reading of the jumbled order as indicative of the poet's invocation of a unique cosmological model, whereby the order of the zodiac is determined not merely by the position of the Sun, but by a theoretical Sun–Venus complex. This system, while able to provide an explanation for the jumbled order, and in line with Mügeln's hierarchical and scholastic worldview, is otherwise unattested in other medieval astronomical works, and indeed quite complicated to identify and conceptualize, prompting questions about the extent of interpretability of problematic medieval texts, and about the methodology required to determine the upper limit of poetic creativity.

The present study represents an experimental reading of an apparent inaccuracy made in the astronomy sections of the fourteenth-century poem *Der meide kranz* ["The Garland of the Virgin"] by the Middle High German poet Heinrich von Mügeln (fl. mid-fourteenth c.), in order to ascertain whether it is in fact inaccurate, and to highlight the implications of its veracity for the interpretability of objective errors in literary texts. Because *Der meide kranz* is quite a long and involved poem, a brief summary of the narrative is useful. The first part of the text features each of the personified liberal arts, along with the disciplines of Philosophia, Alchimia, Theologia, Phisica, and Metaphisica, presenting themselves to the Emperor in turn and arguing their case as to why they deserve to wear the garland of the Blessed Virgin Mary. After hearing their claims, the Emperor nominates Theologia and sends her to be crowned in the domain of Natura, personified nature. At the coronation, Natura calls upon the Virtues to help crown Theology, but a dispute quickly forms with them about whether the Virtues derive their being from Nature or from God. In the second part of the text, all the Virtues (Justice, Humility, Generosity, Love, etc.) present their case, whereafter Theologia determines that they derive their being from God. Natura then attempts one last time to prove that she existed before the Virtues, which she does first and foremost by a delineation of astronomical material, including information on the order of the spheres and the twelve signs of the zodiac, and their astrological effects on people born under those signs. In spite of this, the poem ends with the "meister dises buches" ["the Master of this Book"] delivering his final say in a brief statement, namely that Natura was created by virtue, which in turn was created by God.

The apparent astronomical inaccuracy under examination here involves the order of the signs of the zodiac. The order of the zodiac is presented at three points in *Der meide kranz*. First, in the speech of personified Astronomia, she lists some of the salient discoveries of her art, including the influence of the stars on the fate of humanity, and the motion and order of the celestial spheres, before turning to the signs of the zodiac:

> ouch ler ich miner künste kint,
> wie das der zeichen zwelfe sint,
> und wie die sunn darinne get.
> Aquarius der erste stet,
> Ster, Fische, Ochse, Gemini,
> nach dem krouch der Krebeß ie,
> der Leuwe, Meit und Scorpio,
> die Wage mit dem Schütze so,
> der Steinbok muß der letzte sin:
> das ist Saturni hüselin.

> [I also teach the children of my art how there are twelve signs, and how the Sun moves in them. Aquarius is the first one, then Aries, Pisces, Taurus, Gemini, after which Cancer always comes creeping; Leo, Virgo, and Scorpio, then Libra with Sagittarius; the last one must be Capricorn, which is the house of Saturn.] (Mügeln 1997, 131–133, ll. 501–510)[1]

Natura gives the same order of the zodiac signs in her final defense, though beginning with Aries rather than Aquarius. This section opens with a rubric that is quite germane to this investigation, here italicized to distinguish it from the rest of the text:

> *Hie nent die zeichen alle gar*
> *Natur nach ordenlicher schar.*
> Der Ster das erste zeichen heißt,
> nach dem die Fische, darnach reist

---

1 All further citations from *Der meide kranz*, including Middle High German text and English translation, are from this edition, given in the following as (*MK*, page, line).

der Ochse, darnach Gemini,
nach dem krouch der Krebeß ie,
der Leuwe, Meit und Scorpio,
die Wag, darnach der Schütze jo,
der Bok, darnach der Waßerman . . .

[*Here Natura names all the signs as an orderly group*. Aries is the name of the first sign; after that Pisces; after that comes Taurus, then Gemini, after which Cancer always comes crawling, Leo, Virgo and Scorpio, Libra, after that Sagittarius, Capricorn, and then Aquarius . . .] (*MK* 339–340, ll. 2355–2361)

Despite the rubric's claim, both Astronomia and Natura seem to present the zodiac signs out of order, placing Pisces after Aries and Libra after Scorpio, in contradiction to the order in which the sun passes through them and their conventional arrangement in astrological theory. Finally, Natura immediately follows this section by discussing astrological prognostics, noting the influence of each sign on the physical and behavioural traits and characteristics of those born under it, as well as how each sign affects the world and human fate in general.[2] What is intriguing is that in this final section the zodiac signs are discussed in their conventional order[3] – the only time in the three delineations of the zodiac in the text where this is so.

So what is to be made of the two other lists featuring an apparently inaccurate order? To consider them as mere errors would seem improper, since the order of the signs in Natura's prognostics clearly demonstrates that Mügeln was not ignorant of the conventional arrangement. One solution could be to view it as a poetic convenience, as the poet does retain some of the rhymes and line construction between both lists. This solution is not particularly flattering to Mügeln as an accomplished poet and author; not only could this be considered a bit lazy, but also a violation of the basic commandment in Matthew of Vendome's rhetorical manual the *Ars versificatoria* against the improper orderings of words.[4] Annette Volfing suggests a nearly opposite interpretation of the zodiac error, namely that we could view the jumbled order "as a manifestation of Mügeln's self-confidence as an expert in astronomy." According to Volfing, while a less ambitious author might simply provide the correct order of the zodiac as a sign of their own learnedness, Mügeln's mixed order shows that he "regards the knowledge of the correct order

---

2 For instance, Natura suggests that when the sun is in Taurus, anything planted will remain undamaged for a long time. Similarly, when the moon is in Cancer, Natura recommends that one should abandon all lawsuits. Cf. *MK* (340–341, ll. 2395–2398 and ll. 2433–2434).
3 That is, Aries, Taurus, Gemini, Cancer, Leo, Virgo, Libra, Scorpio, Sagittarius, Capricorn, Aquarius, and finally Pisces.
4 "Debent etiam evitari improprie verborum positiones" [The improper positions of words should also be avoided]. (Vendôme 1988, 196).

as so rudimentary that he, the *meister*, can afford to play around with it." (*MK* 358). However, it is clear that some of Mügeln's readers were frustrated about the inaccuracy of the lists, and perhaps lacked an appreciation for such a subtle display of *meisterschaft*. In one manuscript, currently housed in Göttingen, the scribe even adds his own ending to Natura's jumbled order to correct the mistake: "den visch den zele ich auch darzu / mit dem ist ir zwelfe nu" ["I also count Pisces thereafter, and with that they are twelve"].[5]

In some respects, this scribe is not an outlier in their instinct to correct such an obvious error, as vernacular poetry was certainly not entirely exempt from criticisms of scientific inaccuracy. For instance, Dante Alighieri (1265–1321) was subjected to a vicious attack by his contemporary Cecco d'Ascoli (1257–1327) for the former's description of Fortune in his *Commedia* (*Inferno* VII), specifically on why Dante had felt the need to add Fortune as a separate celestial intelligence to those which move the celestial spheres when, in Cecco's eyes, any academic astrologer can explain the same causes and changes of earthly material bodies with reference to predictable celestial movements without modification of the traditional cosmology (see Erculei 2018).

In the Middle High German context, a similar concern for scientific accuracy can be seen in the apparent disagreement between the *Sangspruchdichter* (itinerant strophic poets) known as Der Marner (d. *ante* 1287) and Der Meissner (fl. thirteenth c.), about the proper behavior of various animals and the moral allegories derived therefrom. In one strophe, for example, Der Marner offers up a description of the ostrich: "der strus mit sinen ögen rot / drie tage an sinú eiger siht, des werdent us gebrûtet die" ["The ostrich looks at its eggs for three days with its red eyes, and because of this they are incubated"] (Willms ed. 2008, 253).[6] In turn, Der Meissner attacks an unnamed fellow poet – very likely Der Marner – who claimed that the ostrich incubates its eggs with its eyes. Instead, Der Meissner calls the ostrich a forgetful animal, whose eggs are hatched not by the ostrich itself, but by the heat of the Sun (Willms ed. 2008, 255). In reality, both Der Marner and Der Meissner are each correctly following parallel bestiary traditions of the ostrich, with Der Meissner following the older version of the ostrich's incubation, as expressed in the Bible (Job 39:13–15) and the older *Physiologus* tradition, while Der Marner is working from a newer version that had been spreading since the end of the twelfth century (Willms ed. 2008, 253).[7] Still, Der Meissner takes the perceived error to be so egregious as to warrant a personal attack: "Swer sanc, daz

---

5 Göttingen, Universitätsbibliothek, cod. Ms. Philos. 21. Cited from *MK* (358).
6 Translations into English from this edition are my own.
7 Also cf. Grubmüller (1978, 160–177) and Goldstaub (1905, 165).

der struz se dri tage an sin eier / der sanc unrecht, her si ein Swabe oder ein Beier" (Willms ed. 2008, 255) ["Whoever sang that the ostrich gazes for three days at its eggs, he sang incorrectly, and is a Swabian or a Bavarian"]. That such a seemingly trivial matter as how the ostrich incubates its eggs could spill over into politics and prejudice suggests how seriously scientific inaccuracy and misinformation in literature could be taken, at least by those with hats in the ring like the *Sangspruchdichter*, who made their livelihood through didactic poetry.

On Volfing's suggestion that Mügeln jumbles the Zodiac order as a demonstration of superior wisdom, we can also consider the role of purposeful and meaningful errors in Middle High German literature – a question that has been approached by John Greenfield (2018) in his commentary on a passage in the *Willehalm* of Wolfram von Eschenbach (c.1170–1220). In the Old French *chanson de geste* known as the *Aliscans*, a major source for the *Willehalm*, the author makes an apparent blunder in continuity: on his way to the French court, the Christian leader Guillelme is described as wearing a shabby coat overlaid with stoat-fur, while earlier in the text, his outfit was highly ornately and exotically dressed, having taken fine armour from the slain Persian king Ariofle (Holtus, ed. 1985, ll. 2566–2567 and l. 1529, respectively; cf. Greenfield 2018, 223–224). Wolfram calls out this blunder directly in the *Willehalm*:

> Kristjâns einen alten timît
> im hât ze Munlêûn an geleget:
> dâ mit er sine tumpheit reget,
> swer sprichet sô nâch wâne.
> er nam dem Persâne,
> Arofei, der vor im lac tôt,
> daz vriundîn vriunde nie gebôt
> sô spaeher zimierde vlîz,
> wan die der künec Feirafiz
> von Sekundillen durh minne enpfie:
> diu kost vür alle koste gie. (Eschenbach 1994, 125, ll. 20–30)

[*Kristjâns* had laid an old cloth on him at Laon, but whoever speaks out of such uncertainty reveals his ignorance. He had [instead] taken from the Persian, Arofel, who had lain dead before him, such skillfully made and precious adornment as no lover had ever bestowed on her beloved, except for the one that King Feirefiz received from Secundille. Its value exceeded that of all other valuable things.] (my transl.)

Kristjâns is likely referring to Chrétien de Troyes, the author of the Arthurian romance *Perceval* on which Wolfram would base his most influential work *Parzival*. The problem is that Chrétien de Troyes is, as far as we can tell, not the author of *Aliscans* – a fact of which Wolfram would have been aware. The error of continuity in the Old French *Aliscans* is then replaced with a "fake" error of attribution by the narrator of the *Willehalm*.[8] For Greenfield, Wolfram's exaggeration of the error in the *Aliscans* through his own exaggerated error "shifts the focus from the text internal to the text external level and thereby opens up a space for reflection on illogical or false narration, on narratological problems, on literary fiction and on nonsensical 'fable'" (2018, 225). This conclusion builds off Sarah Kay's work on the productive potential of errors and inconsistencies in Anglo-Norman, Old French and Occitan literary cycles, which she argues fostered the growth of courtly literature by mirroring "the multiple overlapping and conflict values and rules of court life, the overlapping and conflicting possibilities of self-identification," which "inevitably result in contradictory subject positions and incomparable demands on the object" (2001, 306–307).

We could perhaps consider Mügeln's jumbled order of the zodiac in the same way, as exaggerated errors which invite questions of fiction and narration, or as the mistakes of Astronomia and Natura as imperfect candidates undeserving of the Virgin's garland or of a loftier position in the divine hierarchy. Indeed, such an obvious inaccuracy draws attention to itself, and invites audience reflection on deeper truth and meaning. This would not be out of step with medieval conceptions of error: Eileen C. Sweeney, in a survey of philosophies of knowledge acquisition and error in the twelfth and thirteenth centuries, notes "the co-existence of multiple modes of knowing and methods of coming to truth and avoiding error" (2018, 37). Authors like Hugh of St. Victor, Albertus Magnus, and Bonaventure conceive of error not as something foundationally 'wrong,' but rather as a type of 'partial knowing' located at an early or incomplete place in the dialectic of reading and understanding, and from which more complete or perfect knowledge must arise. Such claims echo from Aristotle's observation in Book I.1 of the *Physics* that universals become known to us before individuals, as when children make the error of first referring to all men as 'father' and all women as 'mother' (Aristotle 1991, 2–3 [184a22–b14]; cf. Sweeney 2018, 30). Errors can be a symptom of positive intellectual development, and a stepping stone on the path to complete comprehension.

An examination of other confusing sections and apparent errors in *Der meide kranz* may help us determine a theoretical framework for reading errors in Mügeln, and reveal if the erroneous Zodiac order in *Der meide kranz* may in fact be purposeful as a way of provoking audience reflection on some deeper meaning or truth. To start, we should take note of Mügeln's poetic style and reputation. Some recent academic criticism on Mügeln has identified him more as a defender of established foundations rather than as an innovator or updater. Christoph Huber argues that

---

8 It should also be noted that Wolfram is no stranger to inventing fake sources for his material. In *Parzival*, for instance, he attributes the history of the Grail to a Provençal poet called Kyot, whose existence is attested in no other sources, and whose existence was almost certainly constructed by Wolfram to lend authority to his own invention. Cf. Eschenbach (2003, 457).

Mügeln "does not develop an act of renewal, but rather a static system" (1988, 305, my transl.); his text does not innovate in its descriptions of natural philosophy, but rather merely clarifies what has already been established. Susanne Köbele (2003, 252) further notes the distinction between Mügeln and his predecessor Frauenlob: whereas Frauenlob's poetics revels in mystic ambiguity, Mügeln follows a more common late medieval tendency towards differentiation, whereby knowledge may be reached through deeper and deeper levels of subordinated and distinguished terminology. Because of this obsession with hierarchized knowledge, Mügeln maintains a strict division between what is the domain of theology, and what is the domain of reason and of those arts which proceed via rationality. As Michael Stolz puts it:

> Mügeln operiert dabei auf dem Fundament scholastischer Philosophie und Theologie und strebt [. . .] nach einer Synthese von Glauben und Vernunft. Die rationalitätskritischen Tendenzen, wie sie in der zeitgenössischen Mystik begegnen, sind Mügeln fremd. Und doch zeigt er in seiner poetischen Praxis durch kalkulierte Maßnahmen die Grenzen des Rationalen auf. (Stolz 2008, 206–207)

> [Mügeln both operates on the foundation of scholastic philosophy and theology and strives [. . .] for a synthesis of faith and reason. He does not present those tendencies found in contemporary mysticism, critical of rationality. Simultaneously in his poetic practice he shows the limits of the rational through calculated measures.] (my transl.)

This drive for differentiation and emphasis on hierarchy are useful for explaining other types of apparent scientific inaccuracies in *Der meide kranz*. For instance, personified Geometria's speech gives the following account for her creation of a circle:

> kurz unde lank nach minem sin
> die lingen von dem zentrum hin
> ich leite zu dem ummesweif
> uf aller speren zirkelreif
> von punt zu punt in rechter saß
> die ling uf alle winkelmaß.

> [Long and short as it pleases me, I draw lines from the centre to the outer periphery, through the orbits of all the spheres; [I draw] the lines through all angles, from point to point in proper order.] (*MK* 115–116, ll. 399–404)

As Volfing observes (in *MK* 118–119), the description of long and short lines conflicts with the conventional medieval definition of a circle as a series of points equidistant from the centre. Yet Volfing suggests the description does make sense when considered in reference to cosmological diagrams, specifically the one associated with the *Commentarii in somnium Scipionis* of Macrobius. This diagram, highly popular in Mügeln's time, features the concentric circles of the heavenly spheres, with radiating lines from the centre dividing the cosmos into the twelve equal sections of the zodiac. Geometria's points would therefore be the intersections of these lines and the circles, and the long and short lines referring to how many spheres each line passes through. In this case, an apparent deviation from convention can be explained through reference to a popular cosmological model of the day, betraying a preference for authority over superficial clarity.

Another type of error can be found in Metaphisica's speech, where she outlines the powers which move the heavenly spheres:

> ouch ist das von der lere min,
> wie das der engel achte sin,
> die alle speren wegn in tat.

> [My teaching also states that there are eight angels who actually move all the spheres.] (*MK* 165, ll. 657–659)

Metaphisica refers to the celestial movers of Aristotelian metaphysics as *engel*, angels, which move the spheres of the seven planets and the sphere of fixed stars. The problem is her word choice: how are these *engel* different from the *engel* she invokes a few lines later, namely the ones which serve the Virgin?

> des mag ich in der kronen stan
> der maget und der mutter klar,
> der genzlich dint der engel schar.

> [For that reason I may stand in the crown of the maid and glorious mother, whom all the hosts of angels serve.] (*MK* 165, 666–668)

In Mügeln's strict hierarchization of knowledge, this confusion cannot stand, as Metaphisica now threatens to encroach on the realm of theology. Indeed, when the Emperor is giving his final verdict on who should wear the garland of the Virgin, Mügeln has his Emperor rebut Metaphisica's presentation of the *engel*:

> die eilften lobt ich immer me:
> nu dunkt mich, wie ir tichten ste
> swerlich gein dem gelouben min:
> sie lert mich, wie acht engel sin:
> die letzte engel ane zal setzet: der ich geleuben sal.

> [The eleventh one I always praised: yet it now seems to me that her teaching stands squarely in opposition to my faith. She teaches that there are eight angels; the last one puts forward angels without number: Her I shall believe.] (*MK* 186, 853–858)

The problem here is that Metaphisica does not mention only eight *engel* in her speech but instead mentions both the eight celestial movers and the hosts which attend the Virgin. It would seem that the Emperor objects that Metaphisica should only mention the eight *engel* of the celestial movers, and not conflate them with the heavenly hosts of *engel*, which properly belong to Theologia. Each art has its own domain, and their differentiation is paramount.[9]

This is not necessarily to say that all superficial oddities in *Der meide kranz* must be the product of scholastic differentiation. A prime example of an exception to this rule may be Mügeln's identification of Aries as *ster* rather than the much more common *widder*. Indeed, variations of *ster* are used in other Middle High German manuscripts to refer to Taurus, generally interchangeable with variations of *ochse*, in line with Mügeln.[10] Volfing notes that while some manuscripts can use *stier* to translate the Latin *aries* ("ram"), they are not used in an astrological context (*MK* 139).[11] Because Mügeln's use of *ochse* in clear reference to Taurus leaves little doubt that the *ster* is meant to refer to Aries, this does not seem to be as clear a case of error or inaccuracy as the jumbled order of the Zodiac; using *ster* for Aries appears more an issue of personal or dialectal nomenclature than a corruption of a fundamental astrological principle.

But based on the predilections for authority and the differentiation of knowledge, and the precedent set by the (non-)errors of the cosmological circle and the angelic movers, it is plausible that if there is an explanation for Mügeln's apparently erroneous order of the zodiac signs, then it is one which may necessitate reference to existing cosmological models, and one which will have to be drawn from the domain of astronomy itself, rather than from theology or any other distinct and differentiated art or science. A deeper analysis of Astronomia's speech may then yield clues as to Mügeln's astronomical source material. Immediately before she lists the erroneous order of the zodiac, Astronomia provides the orbit times of the planets.

---

9 Cf. Stolz (2004, 574): "An der Metaphysik tadelt der kaiserliche Richter, dass sie zu wenig genau zwischen den Bewegern der acht Himmelssphären und der zahllosen Engelschar der himmlischen Hierarchien unterscheide; dies impliziert, dass die Metaphysik ungerechtfertigterweise von ihrem Gebiet der Seinsordnungen (hier der Sphärenbeweger) zu jenem der Theologie (hier der Engelshierarchien) übertrete (vv. 853–858)" ["The imperial judge rebukes Metaphysics for not differentiating precisely enough between the movers of the eight heavenly spheres and the innumerable host of angels of the heavenly hierarchies; this implies that Metaphysics unjustifiably trespasses from her domain of the orders of being (here the movers of the spheres) to that of theology (here the angelic hierarchies) (vv. 853–858)" (my transl.)].
10 See the chart of manuscripts in *MK* 142.
11 Volfing cites the manuscripts included in Schnell and Grubmüller (1988–2001, A690).

die achte spere sunder spar
leuft sechs und drißik tusent jar,
Saturnus drißik, zwelf Jupiter,
Mars zwei, gelöube mir der mer,
ein jar die sunne loufen muß,
acht stunden minner hat Venus,
Mercurius dri virtel jar,
der man vir wochen sunder spar.

[It is certain that the eighth heaven takes 36,000 years to rotate, Saturn thirty, Jupiter twelve, Mars two, believe me in this matter; the Sun must move around for one year, and Venus takes eight hours less than that; Mercury three quarters of a year and the moon definitely takes four weeks.] (*MK* 131, ll. 491–498)

A cross-comparison of these orbit times, however, seems to yield more problems than solutions. Mügeln's orbit times for the superior planets (Mars, Jupiter, and Saturn), as well as for the Sun and Moon, are consistent with the times given by most other ancient and medieval encyclopedias and astronomical manuals, and indeed with modern observations. The inferior planets Mercury and Venus are much more problematic, as I am unable to locate any ancient or medieval astronomical manuals which share Mügeln's exact orbit times for these two planets. With few exceptions, ancient and medieval accounts for the orbit times of Mercury and Venus generally fall into two different groups: those which list both their orbits as the same as the Sun (i.e., one year), and those which list them in cascading order (i.e., Venus as a few weeks less than the Sun, and Mercury as a few weeks less than Venus).[12] Mügeln's account seems to fall into this later category, but his orbit time for Mercury is radically shorter than any of the other accounts. Similarly, the eight-hour difference between Venus and the Sun is remarkably specific, and also without precedent. It seems we will need to sort out the peculiarities of these two plan-

---

12 See for instance the following orbit times set out by some of the more influential and widely read authors (data taken from *MK* 141, and Lorimer 1925, 129):

	Mercury	Venus
Cicero	1 year (*fere*)	1 year
Pliny	339 days	348 days
Martianus Capella	1 year (*paene*)	1 year (*circa*)
Macrobius	1 year (*plus minusve*)	1 year (*plus minusve*)
Vincent of Beauvais	339 days	348 days
Bartholomaeus Anglicus	338 days	348 days
Sacrobosco	365 days and 6 hours	365 days and 6 hours
Mügeln	3/4 year	1 year less 8 hours

ets, as presented in *Der meide kranz* and in the broader medieval astronomical context, in order to proceed.

The general confusion in Mügeln and the other sources over the orbit times of Mercury and Venus, and their shared deviation from the modern calculation of 88 days for Mercury and 225 days for Venus, is understandable. During those times of year when Mercury and Venus are visible, from the position of the observer they remain very close to the Sun, which makes following their individual paths difficult. This is further compounded by the existence of competing astronomical models in the ancient and medieval periods. In the Ptolemaic geocentric universe – the primary astronomical model of the day – Mercury, Venus, and the Sun all revolve around the Earth on their own paths, in a model which required liberal use of epicycles in order to account for the location of the inferior planets when they inevitably passed behind the Sun. An alternative model for the universe, however, was transmitted in the Middle Ages through the *De nuptiis Philologiae et Mercurii* (*The Marriage of Philology and Mercury*), a highly popular liberal arts manual by the late antique author Martianus Capella (fl. early fifth c.). This model still employed epicycles in order to account for retrograde motion of the planets, but contested that Mercury and Venus in fact orbited the Sun, which in turn orbited the Earth, forming a larger Sun complex.

> Nam Venus Mercuriusque licet ortus occasusque quotidianos ostendant, tamen eorum circuli terras omnino non ambiunt, sed circa Solem laxiore ambitu circulantur; denique circuiorum suorum centron in Sole constituent [. . .]. (Capella 1836, 668)

> [Now Venus and Mercury, although they have daily risings and settings, do not travel about the earth at all; rather they encircle the Sun in wider revolutions. The center of their orbits is set in the Sun.] (Capella 1977, 333)

Several of the depictions of the other liberal arts in *Der meide kranz* can be traced back to the *De nuptiis*, and so it is reasonable to assume that Mügeln was familiar with the Martianus model as well (cf. *MK* 60 and Stolz 2002, 184–185).

So which astronomical model does Mügeln adopt? At first glance, this question seems perfectly answerable, as Natura lays out the order of the celestial spheres in her final defense. She first lists the procession of the four elemental spheres, beginning with earth at the centre, followed by water, air, and fire, before moving on to the planets:

> das für beslüßt des manden kreiß
> und ouch die elementen gar.
> darnach Mercurius fürwar
> den manden ummereifen muß.
> die vorgenannten slüßt Venus.

die sunne muß zu mittelst gan,
dri under ir, dri oben lan.
Mars darnach sunder lügenmer:
den ümmeslüßt her Jupiter.
Saturnus muß der höchste sin:
der ist der vorgenannten schrin [. . .]

[The orbit of the Moon encircles the fire and all the elements. Then Mercury must in truth encircle the moon. The aforementioned are then enclosed by Venus; the Sun must go in the middle, allowing three to be below it and three above it. Mars next, and that is no lie; Lord Jupiter encircles it. Saturn must be the highest one: he is a case around the others . . .] (*MK* 338, ll. 2308–2318)

This reads a lot like the Ptolemaic model, with its placement of Mercury and Venus between the Moon and the Sun. The problem is that Martianus Capella also has no problem placing Mercury and Venus between the Moon and the Sun, and indeed simply notes the existence of the debate:

post cujus orbem alii Mercurium Veneremque, alii ipsius circulum Solis esse concertant; deinde Martis, Jovis ac Saturni [. . .] (Capella 1836, 669)

[Immediately above (the moon's) orbit some authorities place the orbits of Mercury and Venus; others argue that the sun's orbit comes next. Then come the orbits of Mars, Jupiter, and Saturn.] (Capella 1977, 333)

Martianus' model, with Mercury and Venus orbiting the Sun, would mean that both planets would alternatively be above and below the Sun with respect to Earth, such that either order is acceptable. Placing Mercury and Venus beneath the Sun is then merely a rhetorical convention and is not entirely useful in ascertaining which astronomical model is being followed.

Our only clues then for determining Mügeln's preferred astronomical model are the orbit times for Mercury and Venus given by Astronomia. Yet it is also hard to put too much stock in these numbers as indicative of one system or another. For instance, just because most ancient and medieval authors maintain orbit times for Mercury and Venus which are identical to the Sun, it does not mean that all these authors were adopters of the Martianus model; the orbit times of the inferior planets are given as equal to the Sun not because they travel along with it in the unified Sun complex, but because they always appear so close to the Sun as to make their orbit times functionally identical, with more precise calculations of their orbit times being too difficult and tedious. In light of this, it unfortunately becomes impossible to say with certainty which astronomical mode is at play in *Der meide kranz*.

What is interesting, however, is that the Martianus model is in fact capable of explaining the strange order of the zodiac signs given by both Astronomia and Natura in a very particular way. This explanation hinges on the idea of the Sun complex, whereby it is not only the Sun which moves through the zodiac signs, but the unified whole of the Sun plus Mercury and Venus which moves through them. In this way, it is not the Sun itself which reaches Aries before Pisces, and Scorpio before Libra, but rather one of the two other members of the Sun complex, in this case Venus, which reaches these constellations first. This is possible thanks to Venus's status as the morning star, and its position relative to the Sun at its point of greatest elongation – the point at which a celestial body (in this case, Venus) is at its furthest angular separation from another (in this case, the Sun). Using the numbers provided by Martianus in *De nuptiis* (1836, 685; transl. Capella 1977, 342), Venus has a greatest elongation of 46 degrees, which means that Venus can be at most 46 degrees away from the Sun. Venus has two points of greatest elongation, one eastern point when it is the evening star, and one western point when it is the morning star. Venus's synodic period, which is the time it takes Venus to return to the same position relative to the Sun and Earth, i.e., the time between two consecutive western or eastern points of greatest elongation, is 584 days, or about 19.2 months.

The strange order of the zodiac can be replicated by considering Venus as part of the Sun complex when it is at its western point of greatest elongation. If Venus reaches its western point of greatest elongation near to the vernal equinox, then Venus will rise in the morning in Aries, while the Sun, still below the horizon, is in Pisces.[13] Therefore, the observer on the ground will see the Sun complex (Venus plus Sun) in the constellation of Aries before they see it in Pisces. The next time Venus reaches its western point of greatest elongation in just over 19 months (i.e., a full calendar year plus seven months), it will rise in Scorpio while the Sun is still in Libra, close to the autumnal equinox. The fact that *Der meide kranz* picks zodiac signs near to the equinoxes may heighten the apparent unity of the Sun complex, since the ecliptic along which the Sun travels is at its most vertical (relatively speaking) at the equinoxes, which causes the planets to seem to rise

---

13 Martianus Capella notes that the greatest elongation of Mercury and Venus from the sun is one and a half signs ("signo uno et parte") (Capella 1836, 668, transl. Capella 1977, 333). This refers to the totality of the Sun complex, as he also notes that Mercury "will never be able to depart from the sun by more than 22 degrees of elongation; never will it be able to be two signs away" (Capella 1977, 683) ["Sed idem Stilbon, licet Solem ex diversis circulis comitetur, ab eo tamen nunquam ultra viginti tres partes poterit aberrare, nec duobus signis absistere" (Capella 1836, 341)]. The difference in degrees of elongation here is a case of variation in the reference manuscripts.

faster, thereby making the totality of the Sun complex appear more quickly and thus appear to act as a more unified whole.[14]

There are several problems with this reading, however. The synodic period of Venus extends over a single calendar year, which means that even if Venus was in Aries and the Sun in Pisces at the vernal equinox, the Sun complex would still appear to reach Libra before Scorpio in that same year, as Venus would no longer be at its point of greatest elongation. As a result, this interpretation places a great amount of importance on the position of Venus, perhaps even over and above the position of the Sun, in calculating the order of the zodiac signs. It is a powerful critique of this interpretation that there is no direct precedent for this type of elevation of Venus as the determiner of zodiac order in any ancient or medieval text I can locate. A possible mitigation of this critique is to point out the importance of Venus *qua* morning star in Middle High German poetry, especially its importance as a time designator in *Tagelieder* [morning songs] and as a praise analogy in *Minnesang* [courtly love songs]. Even some pieces of didactic *Sangspruchdichtung* seem to suggest a primacy of Venus, as in a strophe by Der Meissner:

> Ich klage, daz sich die elementen hant vurkeret.
> sit der planeten louf unstete wart, sint meit mich ie geluckes rat.
> sunne unde mane, dar zu Venus, sit geeret,
> Jupiter, Mars, Mercurius, Saturnus, ob ir min genade hat.
> (Notle and Schupp 2011, 166–167)

> [I lament that the elements have changed. Since the orbit of the planets became unstable, since then the wheel of fortune has avoided me. Sun and Moon, plus Venus, be praised, Jupiter, Mars, Mercurius, Saturnus, if you show me your favor.] (my transl.)

While the privileged position of Venus in relation to the Sun in this strophe is perhaps noteworthy in this context, it is certainly not sufficient as evidence for the existence of a unique and entirely separate "poet's model" of the universe, one in which Venus takes primacy as a (sole?) member of the Sun complex, especially for the purpose of zodiac calculation.

But perhaps the unusualness of this Venus-based interpretation helps to explain Mügeln's unique but bizarre orbit times. Even if we cannot say that orbit

---

14 Cf. the description given by the French poet philosopher Alan of Lille (c. 1128–1202/03) in his *Anticlaudianus*: "Qua ratione meant obliquo signa meatu; / cur signum proprior directus exit in ortus / opposito, furans nascendi tempora, tempus / perdit in occasu quod plus expendit in ortu" ["For what reason the signs of the Zodiac wander in a slanting path; why a sign emerges in its rising more quickly than its opposing sign, stealing from its rising time; or why is loses time in setting because it spends more in its rising"] (Alan 2013, 332–333).

times indicate allegiance to a particular astronomical model, since this would imply a far greater acceptance of the Martianus model even among ancient authors, we can perhaps say something about the function of the novel and untraceable orbit times within *Der meide kranz*. The delineation of the strange orbit times of Venus and Mercury in Astronomia's speech immediately before the first mention of the alternative order of the zodiac signs may function as a signal to the audience for the reasons behind the zodiac order. That is to say, by suggesting a strong difference between the orbit times of the Sun and Mercury – one which contrasts with the vast majority of other contemporary calculations – while simultaneously retaining the similarity between the orbit times of the Sun and Venus, Mügeln may be priming his audience to consider a unified Sun-Venus complex when seeking to understand the rest of Astronomia's speech. The close association of the Sun and Venus in the orbit times may encourage his audience towards reading the *sunn* of l. 503 as a synecdoche of the Sun-Venus complex, thereby hinting at the process by which he reached his unorthodox zodiac order.

Moreover, a major advantage of the Sun–Venus interpretation is that it explains why the signs of the zodiac are restored to their conventional order in the prognostics section of Natura's final defense. The difference between the two competing orders in *Der meide kranz* can be understood as a difference between astronomy and astrology, as delineated for instance by Isidore of Seville (c. 560–636):

> Inter Astronomiam autem et Astrologiam aliquid differt. Nam Astronomia caeli conversionem, ortus, obitus motusque siderum continet, vel qua ex causa ita vocentur. Astrologia vero partim naturalis, partim superstitiosa est. Naturalis, dum exequitur solis et lunae cursus, vel stellarum certas temporum stationes. Superstitiosa vero est illa quam mathematici sequuntur, qui in stellis auguriantur, quique etiam duodecim caeli signa per singula animae vel corporis membra disponunt, siderumque cursu nativitates hominum et mores praedicare conantur. (Isidore 1911, III.xxvii)

> [There is some difference between astronomy and astrology. Astronomy concerns itself with the turning of the heavens, the rising, setting, and motion of the stars, and where the constellations get their names. But astrology is partly natural, and partly superstitious. It is natural as long as it investigates the courses of the sun and the moon, or the specific positions of the stars according to the seasons; but it is a superstitious belief that the astrologers follow when they practice augury by the stars, or when they associate the twelve signs of the zodiac with specific parts of the soul or body, or when they attempt to predict the nativities and characters of people by the motion of the stars.] (Isidore 2006, 99)

Natura's final prognostics section is an astrological section, describing how the prevailing zodiac sign influences birth characteristics and bestows fortune or misfortune. This section is therefore based on the Western astrological model, which is fundamentally tropical rather than sidereal, emphasizing seasons and times of year over precise positions of celestial bodies. In this tropical system, the vernal equinox will always mark the beginning of the period of Aries, even if the Sun still remains in Pisces. Conversely, the two other mentions of the signs of the zodiac in *Der meide kranz*, where the order is apparently jumbled, can be understood as astronomical sections, emphasizing more precise positions of the celestial bodies. The exact position of the Sun becomes much more important, and, if understood as the Sun-Venus complex, produces the jumbled zodiac order when its movements through the signs are plotted out. It is therefore possible to produce two different orders of the signs of the zodiac if the orders are allocated to their respective sciences – an act which would seem to fit right in with Mügeln's insistence on differentiation of hierarchies of knowledge.

Like *Der meide kranz* itself, my conclusion demands a verdict, and it is not one easily given. The Sun-Venus interpretation does work to explain the seemingly jumbled order of the signs of the zodiac, and does so through the emulation of authority and differentiation of fields of knowledge with which Mügeln is critically associated. However, the Sun-Venus model does not have any direct precedent in any other ancient or medieval astronomical sources as a method of zodiac calculation. A model so unattested, and so full of mathematical and geometrical complexities, does not seem believable in the context of vernacular poetry, particularly for a poet like Mügeln who, although undeniably talented and erudite, is nevertheless not associated with grand scientific innovation. My conclusion, hence, is one largely of bewilderment: what does it mean, then, that the Sun-Venus complex model *works*; what does it mean when an error, or an obscurity, in a text, can be explained scientifically, with reference to existing contemporary natural philosophical models and observable phenomena, but only in a way which not only seems beyond the capacity of the poet, but beyond the capacity of plausible accessibility? Certainly if the Sun–Venus model was employed by Mügeln, it was beyond the capacity of the scribe of the Göttingen manuscript, drastically complicating the issue of intended audience or audiences. We must ask how many epicycles must be added to a poem to rescue it from its own inaccuracies, and how many can be added until logic and proportion are threatened. A functional methodology, or a set of criteria, for determining the upper limit of a poet's creativity with their source material in this way, remains a *desideratum*.

Bibliography

Alan of Lille. *Literary Works*. Ed. and trans. Winthrop Wetherbee. Cambridge, MA: Harvard University Press, 2013.

Aristotle, *Physics*. Ed. Jonathan Barnes. Princeton, NJ: Princeton University Press, 1991.

Capella, Martianus. *De nuptiis philologiae, et Mercurii, et de septem artibus liberalibus libri novem, etc.* Ed. Ulricus Fridericus Kopp. Frankfurt: Varrentrapp, 1836.

Capella, Martianus. *Martianus Capella and the Seven Liberal Arts, Volume II: The Marriage of Philology and Mercury*. Ed. and trans. William Harris Stahl with E. L. Burge. New York: Columbia University Press, 1977.

Erculei, Ercole. "Frogs' Fairy Tales and Dante's Errors: Cecco d'Ascoli on the Florentine Poet and the Issue of the Relationship between Poetry and Truth." *Irrtum – Error – Erreur*. Eds. Andreas Speer and Maxime Mauriège. Berlin and Boston: De Gruyter, 2018. 669–680.

Eschenbach, Wolfram von. *Parzival: mittelhochdeutscher Text nach der sechsten Ausgabe von Karl Lachmann, mit Einführung zum Text der Lachmannschen Ausgabe und in Probleme der Parzival-Interpretation*. Trans. Peter Knecht, introd. Bernd Schirok. Berlin and New York: De Gruyter, 2003.

Eschenbach, Wolfram von. *Willehalm: Nach der Handschrift 857 der Stiftsbibliothek St. Gallen*. Ed. Joachim Heinzle. Tübingen: Niemeyer, 1994

Goldstaub, Max. "Physiologus-Fabeleien über das Brüten des Vogels Strauß." *Festschrift Adolf Tobler zum siebzigsten Geburtstage*. Ed. Berliner Gesellschaft für das Studium der neueren Sprachen. Braunschweig: George Westermann, 1905. 153–190.

Greenfield, John. "Would I Lie to You? A Mistake and a Fake Mistake in Wolfram's *Willehalm*." *Cultura, Espaço, & Memória* 9 (2018): 221–226. Web. https://ojs.letras.up.pt/index.php/CITCEM/article/view/6226/5857 (25 September 2023).

Grubmüller, Klaus. "Überlegungen zum Wahrheitsanspruch des Physiologus im Mittelalter." *Frühmal. Studien 12*. Ed. Karl Hauck. Berlin and New York: De Gruyter, 1978. 160–177.

Holtus, Günter, ed. *La versione franco-italiana della «Bataille d'Aliscans»: Codex Marcianus fr. VIII [=252]*. Tübingen: Niemeyer, 1985.

Huber, Christoph. *Die Aufnahme und Verarbeitung des Alanus ab Insulis in mittelhochdeutschen Dichtungen*. Munich: Artemis, 1988.

Isidore of Seville. *Isidori Hispalensis Episcopi Etymologiarum sive Originum Libri XX*. Oxford: Clarendon Press, 1911.

Isidore of Seville. *The Etymologies of Isidore of Seville*. Trans. Stephen A. Barney et al. Cambridge: Cambridge University Press, 2006.

Kay, Sarah. *Courtly Contradictions: The Emergence of the Literary Object in the Twelfth Century*. Stanford: Stanford University Press, 2001.

Köbele, Susanne. *Frauenlobs Lieder: Parameter einer literarhistorischen Standortbestimmung*. Tübingen and Basel: A. Francke, 2003.

Lorimer, W. L. *Some Notes on the Text of Pseudo-Aristotle "De Mundo"*. Oxford: Oxford University Press, 1925.

Notle, Theodor, and Volker Schupp, eds. *Mittelhochdeutsche Sangspruchdichtung des 13. Jahrhunderts. Mittelhochdeutsch / Neuhochdeutsch*. Stuttgart: Reclam, 2011.

Mügeln, Heinrich von. *Heinrich von Mügeln 'Der meide kranz': A Commentary*. Ed. Annette Volfing. Tübingen: Niemeyer, 1997.

Schnell, Bernhard, and Klaus Grubmüller, eds. *Vocabularius Ex quo: Überlieferungsgeschichte Ausgabe*. Tübingen: Niemeyer, 1988–2001.

Stolz, Michael. "Die Artes-Dichtungen Heinrichs von Mügeln. Bezüge zwischen 'Der meide kranz' und dem Spruchwerk. Mit Texteditionen." *Studien zu Frauenlob und Heinrich von Mügeln. Festschrift für Karl Stackmann zum 80. Geburtstag*. Eds. Jens Haustein and Ralf-Henning Steinmetz. Fribourg: Universitätsverlag, 2002. 175–209

Stolz, Michael. *Artes-liberales-Zyklen: Formationen des Wissens im Mittelalter*. Tübingen and Basel: A. Francke, 2004.

Stolz, Michael. *"Vernunst*. Funktionen des Rationalen im Werk Heinrichs von Mügeln." *Wolfram-Studien XX. Reflexion und Inszenierung von Rationalität in der mittelalterlichen Literatur. Blaubeurer Kolloquium 2006*. Ed. Klaus Ridder. Berlin: Erich Schmidt, 2008. 205–228.

Sweeney, Eileen C. "When Is It Wrong? Models of Argument and Interpretation from the 12th to the 13th Century." *Irrtum – Error – Erreur*. Eds. Andreas Speer and Maxime Mauriège. Berlin and Boston: De Gruyter, 2018. 19–37.

Vendôme, Matthew of. "Ars Versificatoria – Liber Quartus." *Mathei Vindocinensis Opera*, vol. 3. Ed. Franco Munari. Rome: Edizioni di Storia e Letteratura, 1988.

Willms, Eva, ed. *Lieder und Sangsprüche aus dem 13. Jahrhundert und ihr Weiterleben im Meistersang*. Berlin and New York: De Gruyter, 2008.

Daniel Könitz

The Astronomical Treatise *Von den 11 Himmelssphären* and Its Relation to the *Iatromathematisches Hausbuch*

Abstract: The article examines the manuscript tradition of the treatise *Von den elf Himmelssphären*, written in Middle High German. The treatise is preserved in a total of 29 manuscripts and existed in two different versions in the Middle Ages. The older version is more extensive and is dated to the second half of the fourteenth century on the basis of the oldest textual witness. The second, broader surviving version was most likely written at the beginning of the fifteenth century and in connection with the conception of the *Iatromathematisches Hausbuch*. The latter, younger version of the treatise *Von den elf Himmelssphären* was shortened in comparison to the older one by some text sections and, according to the conception of the *Iathromathematisches Hausbuch*, regularly equipped with an illustration. The illustrations often show scholars looking at or pointing to the stars. The presented study confirms, on the one hand, the thesis formulated by researchers that the author or better compiler of the *Iathromathematisches Hausbuch* also used already existing texts and edited them for his work, which was especially widespread in the fifteenth century. On the other hand, it shows that both versions of the treatise were copied at the same time but in clearly separate contexts of transmission. It has turned out that the abridged version of the treatise can be used as an indicator helpful in the identification of the *Iatromathematisches Hausbuch*.

A medieval parchment folio[1] was offered for auction by the London auction house Bernard Quaritch Ltd. in 2019. This single folio was presented in the accompanying sales catalog as number 7 under the rubric "Manuscript Fragments, Leaves and Cuttings" (Bernard Quaritch 2019). Its contents were classified under the keyword "Astronomy." The fragmentary specimen transmits a text concerning the Eleven Heavenly Spheres and is skillfully penned in a painstakingly written Textualis script. The sales catalog's description, which includes an image of the fragment's recto side (Figure 1), reads:

---

[1] The fragment is presently held by a private collection in Luxembourg and was sold in 2019. Cf. https://handschriftencensus.de/26150. – Thanks to Max Schmitz (Luxembourg) for valuable information regarding the fragment as well as for his permission to publish its digital reproduction.

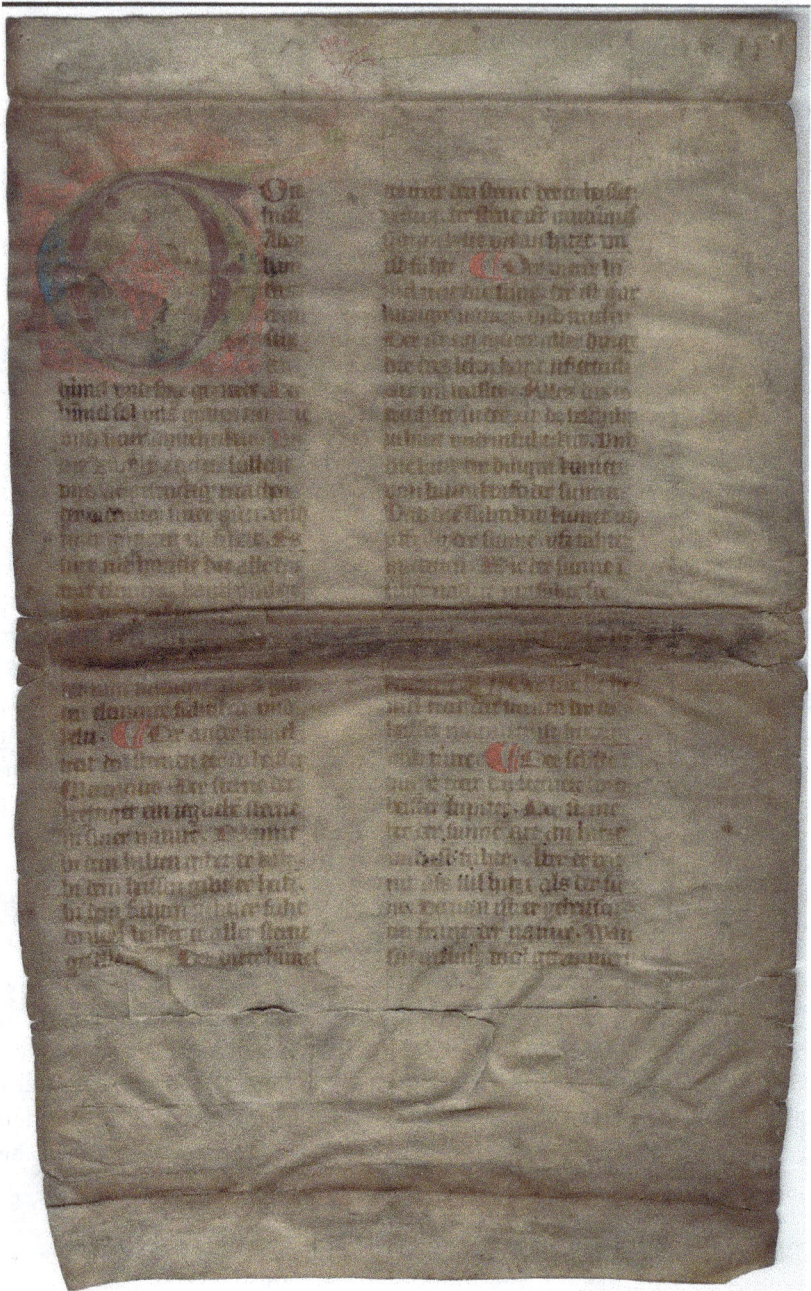

Figure 1: Private Collection Max Schmitz, Luxembourg, no shelfmark, fol. 1r.

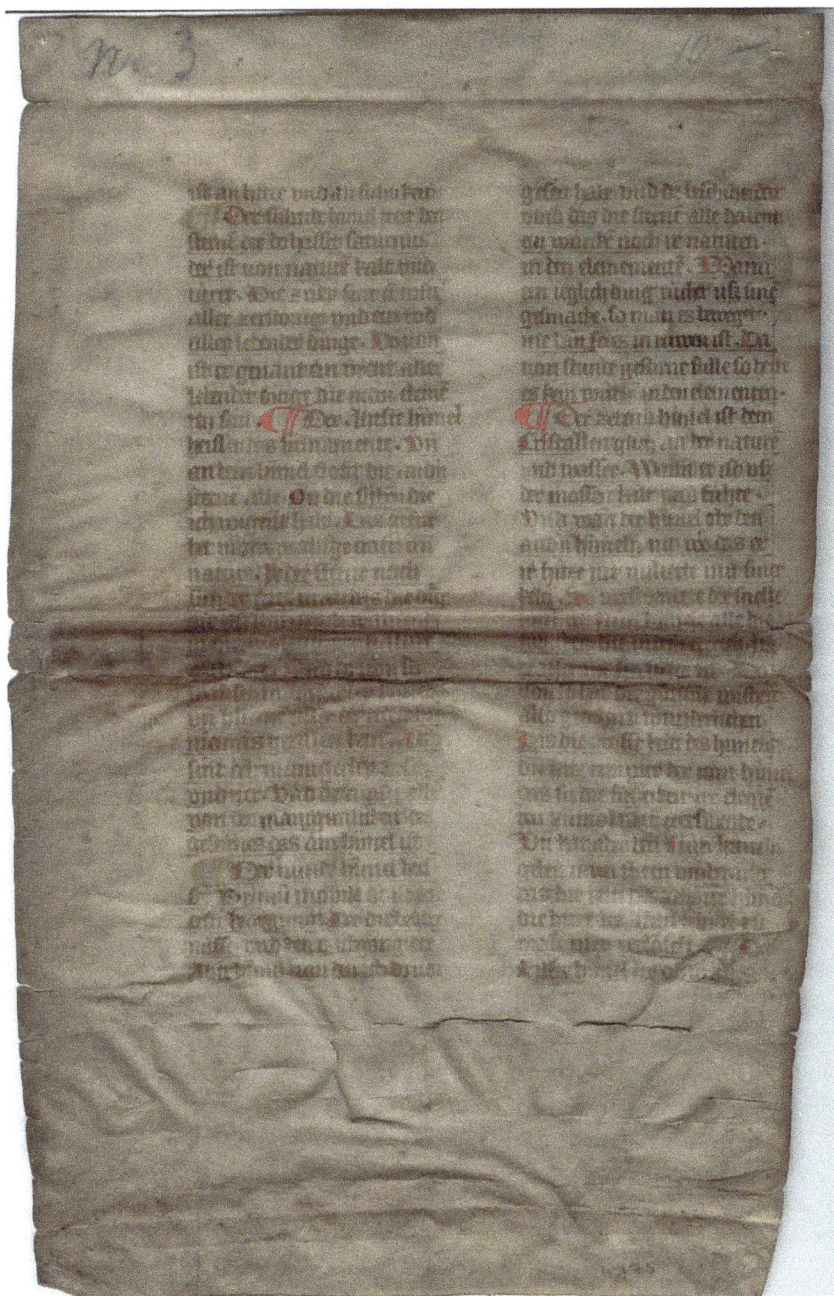

Figure 2: Private Collection Max Schmitz, Luxembourg, no shelfmark, fol. 1v.

ELEVEN HEAVENS

7. [ASTRONOMY.] Von den elf Himmelsphären, in Early New High German; a complete leaf written in a good formal gothic bookhand in double columns of 32 lines, dark brown ink, ruled lightly in ink, EIGHT-LINE INITIAL 'G' (*Gott hiess Abraham das er ansehe den himel*) in burnished gold and blue with elaborate penwork flourishing in red, paragraph marks alternately in red and blue (faded), capitals touched in red; recovered from use in a binding and with consequent wear, fading, soiling and tears, gold largely rubbed away from initial, but mostly legible. 246 x 159 mm (154 x 115 mm)

Germany, 2nd half of 14th century.   £2500

'Von den elf Himmelsphären' is a short anonymous text in Early New High German on the Ptolemaic model of the planetary system. The present fragment, doubtless once part of a larger manuscript containing a number of texts, is early and of notably high quality in terms of script and decoration. It comprises almost half of the text.

Matthias Miller and Karin Zimmermann record six extant manuscripts, most (possibly all) later than the present fragment: Heidelberg, Cod. Pal. germ. 291, ff. 26v–28v (Bavaria, after 1477 but before 1496), and Cod. Pal. germ. 226, ff. 97r–98v (Alsace, 1456–1469); Munich BSB MSS Cgm 349, ff. 51v–55r ('second half of 15th century'), and Cgm 730, ff. 53r–57r ('last quarter of 15th century'); Paris, Bibliothèque nationale, MS Allemand 106, f. 215v ('15th century'); and Strasbourg, Bibliothèque municipal M 711, f. 38v (no date given). See M. Miller and K. Zimmermann, *Die Codices Palatini germanici in der Universitätsbibliothek Heidelberg*, 2005, p. 151.

According to the description, the single parchment leaf, offered starting at £2,500, originated during the second half of the fourteenth century in a German-speaking area. The erstwhile complete codex to which the fragment belonged measured approximately 246×159 mm. Its layout of two columns and 32 lines, as well as its fine decoration (eight-line initial 'G', alternating red and blue paragraph marks), correspond to a manuscript type widely disseminated in the fourteenth century.[2] The Quaritch sales catalog opines that the single folio was "doubtless once part of a larger manuscript containing a number of texts." The full codex, now lost, in all likelihood contained a collection of various technical and instructional texts. It is quite conceivable the codex was an example of a collective manuscript of astronomical and astrological texts, which may well have also contained medical texts and prescriptions. However, such suppositions have to remain speculative in the absence of the original. The codex must have been rendered into binder's waste at a later point in time, which is difficult to determine with any exactness. Binder's waste is produced by dismembering codices in order to utilize the recovered parch-

---

2 A search of corresponding characteristics in the *Handschriftencensus* (handschriftencensus.de) databank (i.e. Codex/fragment, 2 columns, 32–34 lines, 1350–1400) yields 253 hits (1 June 2022). This finding does not contradict estimates. Providing full information as to the details of the search results is beyond the scope of the present chapter.

ment for new purposes.[3] The folio in question was used in its second application as a protective covering for a more modern, perhaps printed book, the measurements of which, based on an analysis of the folio's discoloration and creasing, were approximately 160×115 mm.

The text transmitted by the single folio belongs to an astronomical treatise known among medievalists as *Von den 11 Himmelssphären*, which has received little attention from the research community. The fragment does not transmit the entire text of the treatise. *Der eilfte himel dar ob in allen* is written at the foot of the second column on the verso side (Figure 2). The section about the eleventh heavenly sphere begins with these words. About a third of the entire treatise is missing in the fragment, to judge from a textual comparison with the treatise's easily accessible parallel transmission.[4] This missing last section would presumably have taken up two further columns in the manuscript. Other parchment folia which belonged to this collective manuscript or were connected to the fragmentary specimen in question have yet to be identified. The single folio remains the only evidence for the treatise *Von den 11 Himmelssphären*'s existence as early as the second half of the fourteenth century.

The contents of the treatise have a straightforward structure.[5] A statement recognizing that the heavens should evoke the remembrance of God's omnipotence and that there are eleven heavens, which possess "alle die vier elementen" ["all of the four elements"], follows the opening statement: "Got hiess abraham daz er ane sehe die himel vnd sin gezierde" ["God told Abraham to look at the heavens and its beauty"]. Thereafter, each of the eleven heavenly spheres is named and subsequently described in its own dedicated section. In such manner, a series of several numbered text sections begins with respective formulaic introductions in numerically ascending order: "Der erst hymmel" ["the first heaven"], "Der ander hymmel" ["the other heaven"], "Der drytte hymmel" ["the third heaven"] etc., whereby an inscribed red 'D' lombardic capital, taking up two lines, marks the head of each new heavenly sphere section. The first seven heavens deal with the heavenly bodies known in the Middle Ages: the Moon, Mercury, Venus, the Sun, Mars, Jupiter and Saturn. It is striking that the sections vary greatly in regard to their scope and

---

3 The recovery of binder's waste from medieval parchment manuscripts begins as early as the fifteenth century, especially in the wake of the introduction of the printing press. The demand was high for stable, durable material for book binding and covering, and parchment manuscripts, outdated for whatever reason, presented book binders with a valuable resource not to go unused. For research into binder's waste, cf. Neuheuser (2015), as well as the seminal work of Haebler (1908).

4 I draw for comparison on the manuscript held by the Salzburg university library, the digitization of which is available online: Cod. M III 3, fol. 21ra–21vb (https://handschriftencensus.de/5238).

5 I refer in the following to the already mentioned textual witness held in Salzburg, Cod. M III 3 (cf. Figure 3). All citations are taken from its textual transmission and are, in part, slightly adapted in print to aid in their readability. All translations are my own.

exhaustiveness. Whereas "der vierde hymmel" (the forth heaven, the Sun) and "der sehste hymmel" (the sixth heaven, Jupiter) are addressed extensively with sixteen and ten lines respectively, merely a brief note about "der funften hymmel" (the fifth heaven, Mars) suffices: "[er] treyt den steren der da heysset mars der ist hitzig vnd důrre" ["it contains planet Mars that is hot and dry"]. Following the seventh *hymmel*, Saturn ("von nature kalt vnd durre" ["dry and cold by nature"]), come heavens eight to eleven, which are no longer defined by specific heavenly bodies. The eighth heaven corresponds to the *firmament*, which coalesces all the stars together with the above named heavenly bodies and also exercises influence over the things "wahsent uff dem ertrich" ["grown on earth"]. Last follow the *Primum mobile* (9), the crystal heaven (10), and the final heavenly sphere, "[der] fůryn hymmel in dem got richset mit den ix kôren der engel" ["the fiery heaven in which God rules with the nine choirs of angels"] (11). Attention to the contents focuses one on the significance of the heavens to the human soul with the information that "Dyß sy genuck geseit von den hymmelnn" ["it is enough said about the heavens"], culminating with the admonition: "Ach mentsch da by gedencke war zu du syst geschaffen" ["O Man, consider what you were created for"]. In closing, the twelve signs of the zodiac are mentioned, as is their location in the firmament and the fact that "die meister" ["the masters"] designated these twelve parts with "eynen namen vnd ein gelichnysse eins tieres" ["a name each and an animal as a simile"].

Von den 11 Himmelssphären is as yet unlisted in reference books relevant to German Medieval Studies.[6] Scholarly research papers concerned with the history of the text's tradition are almost completely lacking.[7] The designation of the work's title as used in the research community as well as in the above mentioned sales catalog is based upon reference to particular manuscript repertory catalogs, which among other things, follow medieval titles. Therefore, the difficulty emerges that the treatise has been designated with inhomogeneous titles, for example *Planetenkindertraktat* (Kalning et al. 2016, 159) and *Über die 11 Himmel* (Weimann 1980, 40). Because of the small size of the treatise, it has failed to garner recognition as a stand-alone text and has not been accordingly highlighted by scholarly descriptions (Petzet 1920, 46). Such circumstances make it difficult to get an overview of the actual status of the textual tradition.

---

6 Recently, the *Katalog der illustrierten Handschriften* (Catalog of illustrated manuscripts) highlighted the treatise within the framework of Subject Group 81 (Medicine) as a stand-alone part of the *Iatromathemathisches Hausbuch* collective manuscript. Cf. Freienhagen-Baumgardt et al. (2021), 170–175.

7 Bernard Schnell recently offered the first listing of the *Von den 11 Himmelssphären* treatise's textual witnesses known to him (Schnell 2021, 236) as a companion to the manuscript catalogs, which also document paper manuscripts individually (cf. Miller et al. 2005, 151).

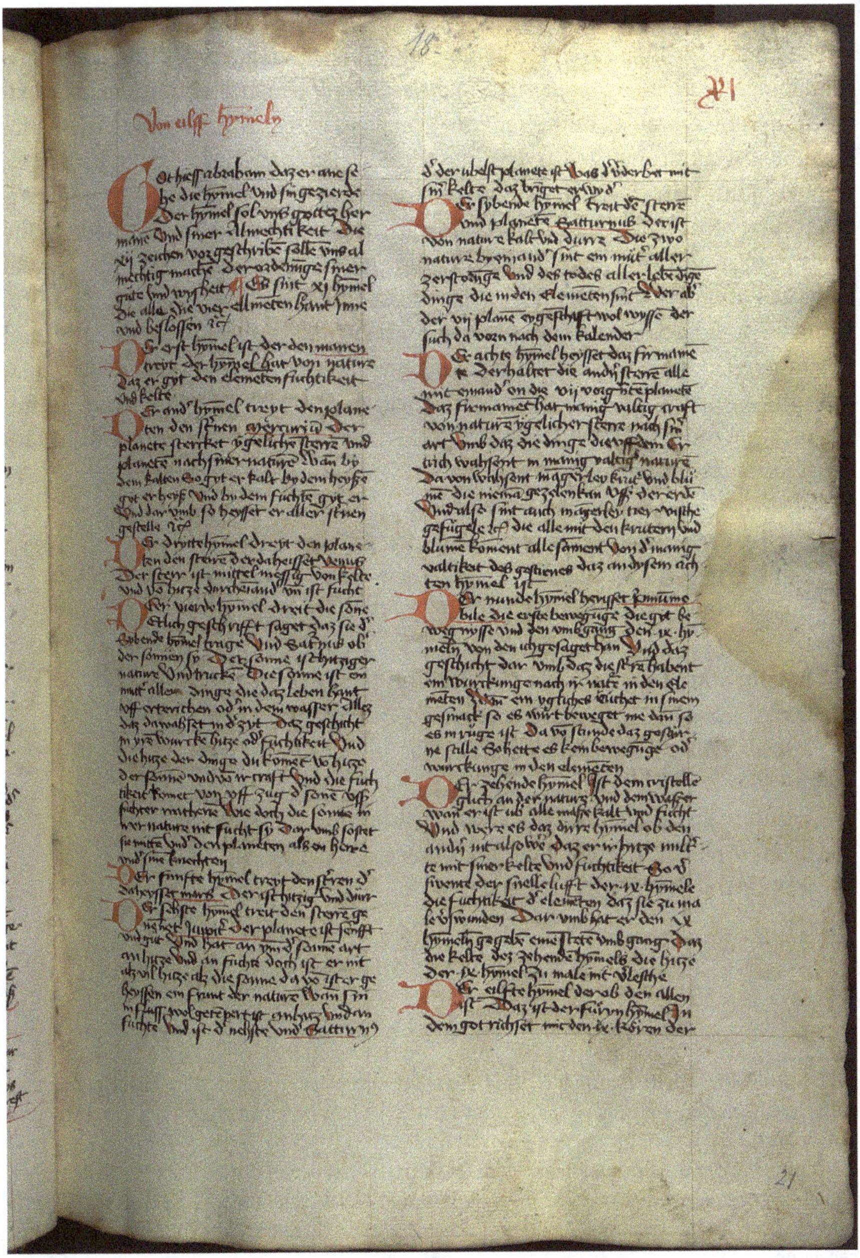

Figure 3: Salzburg, University Libr., Cod. M III 3, fol. 21r. (http://creativecommons.org/licenses/by-nc-nd/4.0/).

The compelling details mentioned in the 2019 Quaritch sales catalog serve as a starting point for the search for additional complete manuscripts and fragments containing the treatise *Von den 11 Himmelssphären*. Its parallel tradition as elucidated therein as well as in the repertory manuscript catalogs already provide indications of additional textual witnesses. Specific searches of incipit-indices, printed manuscript registers and online databases such as *Manuscripta Mediaevalia* (currently being replaced by *Handschriftenportal*)[8] can be conducted via the text's initial words. In such manner, so far, the existence of a total of 29 textual witnesses of the treatise *Von den 11 Himmelssphären* can be established. This provisional list, however, does not likely represent the extent of all existing textual witnesses.[9] On the contrary, forthcoming, methodical examinations and evaluations of the extensive body of astronomical and astrological collective manuscripts dating to the fifteenth century will almost certainly lead to the discovery of yet more textual witnesses.[10]

1. **Berlin, State Libr., Ms. germ. quart. 20, fol. 69v–71v**
   Codex – paper – second half 15th century – illustr. not completed
   https://handschriftencensus.de/11796

2. **Berlin, State Libr., Ms. lat. fol. 262, fol. 18r–18v**
   Codex – paper – second/third quarter 15th century
   https://handschriftencensus.de/26349

3. **Edinburgh, Libr. of the Royal Observatory, Cr. 4.6, fol. 37r–38v**
   Codex – paper – second half 15th century – illustr.
   https://handschriftencensus.de/14955

---

8 Cf. https://handschriftenportal.de/. This site is set to replace http://www.manuscripta-mediae valia.de, but not all features of the previous database have been implemented yet. (Accessed 15 July 2024).

9 I confine myself to information essential to the present paper as drawn from the brief textual witness descriptions. Further, more in-depth information as to the contents as well as to significant research literature can be found on the *Handschriftencensus* website: https://handschriften census.de.

10 The growth in German-language medieval manuscript digitization, consistently undertaken by many libraries, facilitates successful research even when the accompanying metadata provide too little information as to the contents of the shorter texts. The systematic examination by Zinner (1925) is decidedly worthwhile.

4. **Frankfurt a. M., University Libr., Ms. germ. qu. 17, fol. 23ra–24va**
   Codex – paper – first quarter 15th century
   https://handschriftencensus.de/13225

5. **Heidelberg, University Libr., Cod. Pal. germ. 1, fol. 1r–3v**
   Codex – paper – c. 1538
   https://handschriftencensus.de/10691

6. **Heidelberg, University Libr., Cod. Pal. germ. 226, fol. 97r–98v**
   Codex – paper – 1459–1469
   https://handschriftencensus.de/10370

7. **Heidelberg, University Libr., Cod. Pal. germ. 291, fol. 26v–28r**
   Codex – parchment and paper – after 1477/before 1496 – illustr.
   https://handschriftencensus.de/4903

8. **Heidelberg, University Libr., Cod. Pal. germ. 557, fol. 52v–54v**
   Codex – paper – 1468 – illustr. not completed
   https://handschriftencensus.de/4951

9. **Heidelberg, University Libr., Cod. Pal. germ. 718, fol. 1r–4r**
   Codex – paper – before 1500
   https://handschriftencensus.de/10455

10. **Karlsruhe, State Libr., Cod. K 2790, fol. 128v–130v**
    Codex – paper – mid 15th century
    https://handschriftencensus.de/9848

11. **London, British Libr., MS Add. 17987, fol. 81r–84v**
    Codex – paper – mid 15th century – illustr.
    https://handschriftencensus.de/14262

12. **London, University College, MS Germ. 1, fol. 49v–52r**
    Codex – paper – 1471 – illustr. not completed
    https://handschriftencensus.de/5408

13. **Munich, State Libr., Cgm 28, fol. 25r–26r**
    Codex – parchment – mid 15th century – illustr.
    https://handschriftencensus.de/5102

14. **Munich, State Libr., Cgm 349, fol. 51v–54r**
    Codex – paper – second half 15th century – illustr.
    https://handschriftencensus.de/6038

15. **Munich, State Libr., Cgm 596, fol. 24r–26r**
    Codex – paper – c. 1500
    https://handschriftencensus.de/6183

16. **Munich, State Libr., Cgm 730, fol. 53r–56r**
    Codex – paper – fourth quarter 15th century – illustr. not completed
    https://handschriftencensus.de/6267

17. **Munich, University Libr., 2° Cod. ms. 578, fol. 7v–9r**
    Codex – parchment – 1474, 1475 – illustr.
    https://handschriftencensus.de/6436

18. **Nuremberg, State Archive, Rep. 52a (Reichsstadt Nürnberg), Hs. Nr. 426, fol. 23r–24r**
    Codex – paper – c. 1460 – illustr.
    https://handschriftencensus.de/5502

19. **Paris, Bibliothèque nationale, MS Allemand 106, fol. 215v–216v**
    Codex – paper – 1490
    https://handschriftencensus.de/11393

20. **Philadelphia (Pennsylvania), Univ. of Pennsylvania, Rare Book & Manuscript Libr. Collections, LJS 463, fol. 40r–41v**
    Codex – parchment – 1443 – illustr.
    https://handschriftencensus.de/7307

21. **Pürglitz / Křivoklát (Czech Republic), Castle Libr., Cod. I e 7, fol. 77v–81v**
    Codex – parchment – c. 1455 – illustr.
    https://handschriftencensus.de/15171

22. **Saint Louis (Missouri), Concordia Seminary Library, Rare Book Collection, no shelfmark (1), fol. 26v–27v**
    Codex – parchment – 1429 – illustr.
    https://handschriftencensus.de/24273

23. **Salzburg, University Libr., Cod. M III 3, fol. 21ra–21vb**
Codex – parchment and paper – third quarter 15th century
https://handschriftencensus.de/5238

24. **St. Gall, Monastery Libr., Cod. 760, p. 54–58**
Codex – paper – second third 15th century – illustr.
https://handschriftencensus.de/5704

25. **Solothurn, Central Libr., Cod. S 386, fol. 178r–179v**
Codex – paper – 1463–1466
https://handschriftencensus.de/3770

26. **Strasbourg, City Libr., Ms. 258 (olim Ms. 711), fol. 38v–40v**
Codex – paper – second half 15th century – illustr. not completed
https://handschriftencensus.de/26631

27. **Zurich, Central Libr., Ms. C 54, fol. 32v–33v**
Codex – parchment – c. 1469 – illustr.
https://handschriftencensus.de/4014

28. **Private Collection Antiquarian Bookstore Dr. Jörn Günther, Hamburg, Nr. 1997/5,24, fol. 48r[11]**
Codex – paper – c. 1458 – illustr.
https://handschriftencensus.de/15857

29. **Private Collection Max Schmitz, Luxembourg, no shelfmark, fol. 1r–1v**
Fragment – parchment – second half 14th century
https://handschriftencensus.de/26150

Reliable accounts of the treatise's scribal tradition and dissemination can be constructed based on this new and significantly expanded body of knowledge. Already at first glance, the fragmentary single parchment folio dating to the fourteenth century stands markedly apart from the other textual witnesses. It represents the old-

---

11 The current repository of the manuscript sold in 1997 in unknown. The information relayed in the research literature does provide no indication about the version of the text it transmits. The placement of the text within the entire manuscript is also uncertain. Because of the similarity this manuscript has to the Munich manuscript, Cgm 349, and other manuscripts transmitting the short version, it is indicated with some certainty, that this manuscript transmits the short version (cf. Freienhagen-Baumgardt et al. 2021, 205).

est textual witness, as well as the only fragmentary one. As to the remaining body of textual witnesses, the following characteristics and commonalities are to be adduced: in three quarters of the cases we have paper (20 times), or partially paper, manuscripts (2 times), as contrasted with parchment manuscripts. The tradition is focused in the fifteenth century, culminating in the middle and towards the end of that century. Among the complete manuscripts, the Frankfurt Codex (Nr. 4) represents the oldest and the Heidelberg Cod. Pal. germ. 1 (Nr. 5) from the sixteenth century represents the latest example. Linguistically, the manuscripts reveal Upper German tendencies, whereby Bavarian and Alemannic dialects predominate. The vast proportion of manuscripts are painstakingly annotated as well as extensively adorned with rubricated lettering and, in part, colored initials and lombardic capitals, which take up multiple lines. Over half of our manuscripts (18 times) feature illustrations. In five of these cases (Nos. 1, 8, 12, 16, 26), the illustration of the manuscript was discontinued, judging from the dedicated spaces left blank.[12]

The fact that the treatise is typically, and unsurprisingly, included in extensive collective manuscripts, must be attributed to its brevity. Astronomical, astrological and medical subject matter predominates in these collections. A detailed analysis of other text traditions usually accompanying our treatise in these collections is beyond the scope of the present examination. Such analysis, as well as that of the codicological aspects (i.e., dimensions, material, layout, decoration and origin) of these collective manuscripts would be indispensable, however, for future elaborative and in-depth studies of our treatise. An analysis of the text's form could also lead to conclusive evidence as to the interdependence and relationship of the texts found in these collective manuscripts. The present chapter is intended as an introductory essay exploring the potential of further studies expanding on the treatise's wider significance, particularly in light of the expanded body of materials it presents.

What is known about the history of the treatise's tradition gives rise to two justifiable conclusions: Firstly, judging from the observable unity of text and illustration, the representative function of the collective manuscripts played a significant role for their patrons. Secondly, the numerical extent of the surviving textual witnesses as well as their proximity in age point particularly to the treatise's popularity and to that of astronomical texts in general in the late fourteenth and the entire fifteenth century.

---

12 Freienhagen-Baumgardt et al. offer information on the illustrations (2021, 170–174). The illustration titled "Sternseher" (stargazer) is described therein. In most cases, the illustrations show one or more learned people who direct their gaze towards the sky (Figure 5), sometimes with the aid of scientific apparatus.

I was able to view, examine and compare digital representations of nearly all 29 textual witnesses of the treatise.[13] The present cursory examination and comparison primarily takes into account the text form, leaving aside the comparison of word and spelling variables and deviations in various textual witnesses for future in-depth studies. Through the present inquiry, however, it becomes readily apparent that the range of the text tradition is anything but homogeneous. Identical deviations are to be found among about a third of the textual witnesses. These are of such a fundamental nature that they can be used to attribute a particular version of the treatise to said textual witnesses. A result of this initial examination is the determination that there are two disparate versions of the treatise between which must be distinguished. In the following, I present the text of the most commonly passed-down version, corresponding with the manuscript Ms. C 54 in the Zurich Central Library (No. 27; Figures 4 and 5).[14] The printed copy is rendered true to the letter, whereas scribal abbreviations are replaced by the full word, given in italics, and the various forms of the letter 's' are standardized.[15]

[Bl. 32v] <u>Nu saget dis puch | hie nach von den vbrichen koren der himel vnd von | irem lauff vnd natur vnd hebt an an dem der da | haysset das firmament ∵ ~ |</u>
Got hieß abraham das er ansehe den himel vnd | sein gezire der himel sol got ermonen vnd sein al | mechtikayt Di czwelff zaichen sollen andechtig machen | der ordenung seiner güt vnd seiner weishayt Es sind | alf himel mit den siben pla*n*eten di alle di vir eleme*n*ten | haben vmb geben vnd Inbeschlossen vnd hebet also hie nach | von dem achtem himel an zu sagen wann von den siben | planeten das sind siben chȯr der himeln vnd ir natur Ist | vormals geschriben vnd gesagt ~ |

<u>Der acht himel ist das firmament</u> |
Der acht himel hayst das firmame*n*t an dem himel | stend di andere*nn* ster*nn* alle mit ein ander an di siben | steren vorgenant das gestir*nn* hat manigfaltige krafft | von natur yeglicher sterne nach seiner natur vnd art vmb | das di ding di auß dem ertrich wachsent in manigval | tiger natur wachsent vnd davon so wachsent manigerley | kreuter vnd blumen auß der erden di nim*a*nt gezelen kan | [Bl. 33r] Also sind auch manicherley vische vnd tire vnd das kompt | alles sampt von der manigualtikeit des gestirnes das an dem himel ist ~ |

---

13 The only copy which was not accessible to me, because of the reasons mentioned above (cf. note 11), is the copy in Nr. 28. Thanks are due to Pia Rudolph (Munich) and her colleagues, who are affiliated with the *Katalog der illustrierten Handschriften*, for their invaluable help and sharing of information and images.
14 Cf. the facsimile edition Keil, Lenhardt and Weißer 1981–1983.
15 In the printed text, **bold print** indicates lombardic capitals taking up multiple lines. A vertical line (|) indicates line breaks in the manuscript, and <u>underlined text</u> is written in red in the original.

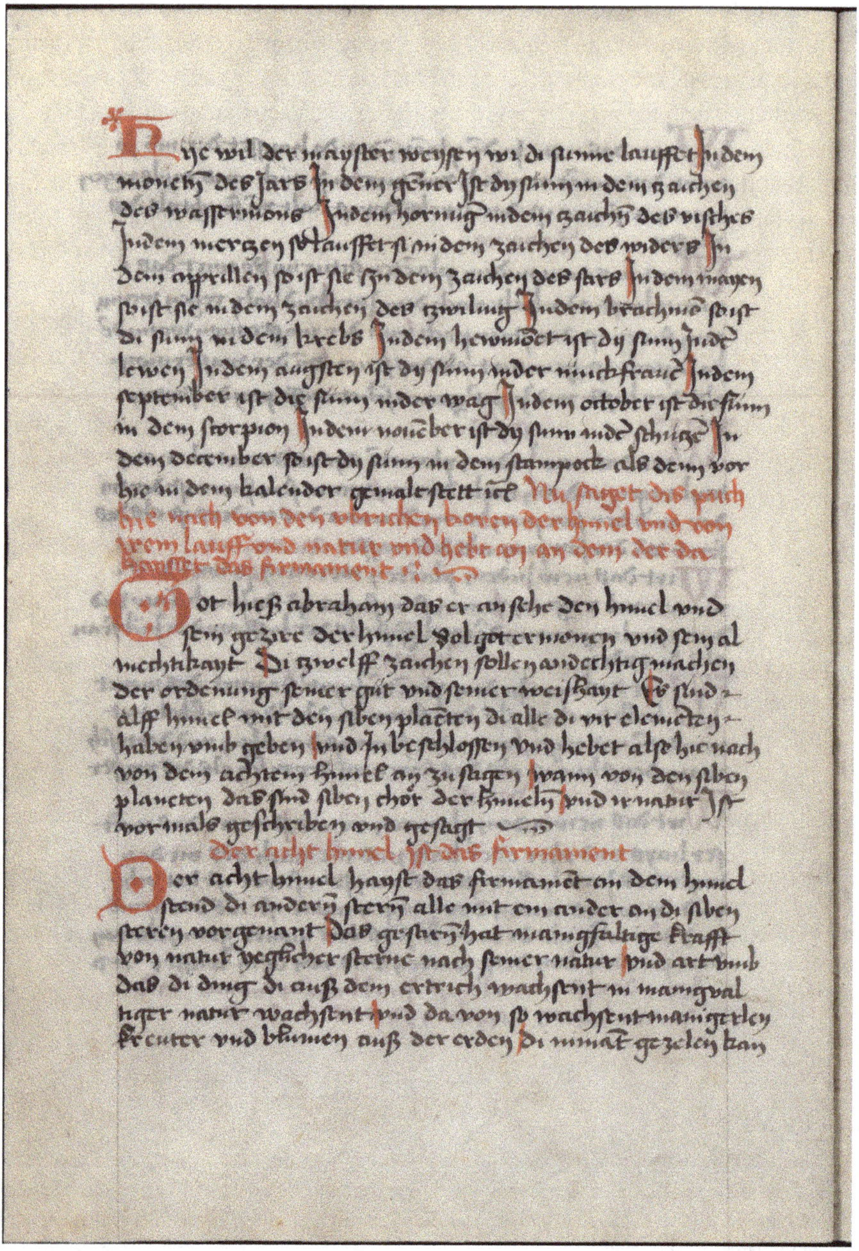

Nu wil der mayster weyssen wie di sternne lauffet in dem
monden des jars in dem gener ist dy sunn in dem zaichen
des wassermans Jn dem hornig in dem zaichen des visches
Jn dem mertzen so lauffet si in dem zaichen des widers Jn
dem aprillen so ist sie jn dem zaichen des stars Jn dem mayen
so ist sie in dem zaichen der zwilng Jn dem brachmon so ist
di sunn in dem krebs Jn dem hewmonet ist dy sunn Jn dem
lewen Jn dem augsten ist dy sunn under die iunckfraw Jn dem
september ist die sunn under wag Jn dem october ist die sunn
in dem scorpion Jn dem nouember ist dy sunn in dem schützen Jn
dem december so ist dy sunn in dem stainpock als denn vor
hie in dem kalender gemalet stett ist Nu saget dis puch
hie nach von den obristen boten der himel und von
irem lauff und natur und hebt an an dem der da
kümet das firmament ⁖

Got hiess abraham das er an sehe den himel und
sein gezirde der himel sol got ermonen und sein al
mechtikayt Di zwelff zaichen sollen andechtig machen
der ordenung seiner gut und seiner weishayt So sind
alss himel mit den siben placten di alle di vir elementen
haben und geben und in beschlossen und hebet alsso hie nach
von dem achten himel an zu sagen wann von den siben
planeten das sind siben chor der himels sind ir nabir ist
vor mals geschriben und gesagt

Der acht himel ist das firmament

Der acht himel hayst das firmament on dem himel
stend di andern stern alle mit ein ander an di siben
stern vor genant Das gesten hat manigfaltige krafft
von natur yeglicher sterne nach seiner natur und art und
das di ding di auss dem ertrich wachsent in manigual
tiger natur wachsent und da von so wachsent manigerley
kreuter und blumen auss der erden di nimat gezelen kan

Figure 4: Zurich, Central Libr., Ms. C 54, fol. 32v.

Figure 5: Zurich, Central Libr., Ms. C 54, fol. 33v.

Der neund himml haisß primum mobile
Der neund himel der haisß primum mobile das ist dy erst | wewegung di gibt webegnus vnd den vmb gang den acht | himel von den ich gesprochen han vnd das geschicht dar | vmb das di sternn all haben ein wrckung nach ir natur In | den elementen wann ein Iglich ding reuchet auß steinen smack | so es wirt wewegt mer denn so es In rwe ist da von stund | das gestirne stille So het es kain wrckung Inden elementen |

Der czehend himel Ist dem cristallen geleich an der na | turen vnd dem wasser wann er ist vber alle masse kalt | vnd feucht vnd warmm das der himel ob dem andernn nit | enwer das er ir hicze miltrote mit seiner kelten So verschwin | dote der schnell lufft der neun himelnn dar vmb das di feuch | tikeit der elementenn zu mal icht verschinde dar vmb so hot | got den neun hi- melnn gegebnn ainen sternn vmb gang das di | kelten des czehenden himels di hicze der neün himelnn zu | mal icht erlesche Der aylfft himel das ist der fewrnn himel

Der âlfft himel das ist der feuren himel vnd ist ob In | allen Indem got regniret mit den neün koren der engelnn | vnd mit allnn menschnn di in seinen gnadnn werdnn sündnn | ¶der himel ist weder hays noch kalt vnd weder feucht noch | kalt warmm wann er ist erhohet vber alle soliche laiplich | aygenschafft wann kayn gebrest hafftiger zu val mag In | nit beruren Er ist da von feu- ren genant wann er aller sampt | leuchtet als das lauter feweren vnd doch nit brennet wann | das er die gayste enhaltet di enczundet sind Inder minne der | warhayt wann erleuchtet Indem himel oder lichte der lau | terkayt Dys sey genüg gesagt von den himelnn nach der | warhayt gesprochen als es inden naturen ist Nach den neun | erstnn himelnn vnd als es in dem glauben nach den Jungsten | zwain himelnn ist vnd In dem sol der mensch gedencken | [Bl. 33v] wi loblichen got Im dar Inne erczayget ist Sider alles das | inden elementen lebt von der krafft der himelnn fleusset | vnd auch geslossen ist | da pey merck das | des menschenn sele di | in dem leibe lebet In | dem oberstnn himelnn | gewrczel ist Siderdas | laub gras kreuter | vnd plümen Als hoch | herkumen ist sind | wer ist denn der gayst | der alle ding gepilden | vnd erkennen mag Aus | so hoher reich- hat ge | flossen wann indem | auß fluß ist so hohe | art das kein creatur | vber dencken Ach mensch da pey gedenck war czu du seist | geschaffnn furbas wissent das Indem firmament czwelff zaich | en sind das sind czwelff taile des selbnn himels mit dem ge | stirnn das an dem selbenn himel stet ¶Des nement di maister | also war vnd sahent das dy sunne vnd die andernn planetnn | In aynem tail des himels het ein ander krafft den Inden | andernn vnd da von so gabent si ydem tail des firmamencz einen | namen vnd ayn gleichnus aynes tirs ~ |

The particular text tradition offered here is doubtlessly representative of the abridged version. The incipit and introduction are laid out identically with the alternative version. However, the text of the Zurich manuscript, instead of de- scribing the first seven heavens, skips over them entirely, beginning its series di- rectly with the eighth. Were these deviations only to surface in one or two manuscripts, one could realistically attribute them to scribal error and/or a faulty text tradition stemming, perhaps, from a flawed prototype. According to this sce- nario, the flawed texts would have to have been passed down in close temporal and spatial proximity to each other, which turns out to not have been the case. The entire tradition of the treatise *Von den 11 Himmelssphären* is listed below, with textual witnesses classified according to the version they represent:

Complete version (11 textual witnesses):

2. Berlin, State Libr., Ms. lat. fol. 262, fol. 18r–18v
4. Frankfurt a. M., University Libr., Ms. germ. qu. 17, fol. 23ra–24va
5. Heidelberg, University Libr., Cod. Pal. germ. 1, fol. 1r–3v
6. Heidelberg, University Libr., Cod. Pal. germ. 226, fol. 97r–98v
9. Heidelberg, University Libr., Cod. Pal. germ. 718, fol. 1r–4r
10. Karlsruhe, State Libr., Cod. K 2790, fol. 128v–130v
15. Munich, State Libr., Cgm 596, fol. 24r–26r
19. Paris, Bibliothèque nationale, MS Allemand 106, fol. 215v–216v
23. Salzburg, University Libr., Cod. M III 3, fol. 21ra–21vb
25. Solothurn, Central Libr., Cod. S 386, fol. 178r–179v
29. Private Collection Max Schmitz, Luxembourg, no shelfmark, fol. 1r–1v

Abridged version (18 textual witnesses):

1. Berlin, State Libr., Ms. germ. quart. 20, fol. 69v–71r
3. Edinburgh, Libr. of the Royal Observatory, Cr. 4.6, fol. 37r–38v
7. Heidelberg, University Libr., Cod. Pal. germ. 291, fol. 26v–28r
8. Heidelberg, University Libr., Cod. Pal. germ. 557, fol. 52v–54v
11. London, British Libr., MS Add. 17987, fol. 81r–84v
12. London, University College, MS Germ. 1, fol. 49v–52r
13. Munich, State Libr., Cgm 28, fol. 25r–26r
14. Munich, State Libr., Cgm 349, fol. 51v–54r
16. Munich, State Libr., Cgm 730, fol. 53r–56r
17. Munich, University Libr., 2° Cod. ms. 578, fol. 7v–9r
18. Nuremberg, State Archive, Rep. 52a (Reichsstadt Nürnberg), Hs. Nr. 426, fol. 23r–24r
20. Philadelphia (Pennsylvania), Univ. of Pennsylvania, Rare Book & Manuscript Libr., fol. 40r–41v
21. Pürglitz / Křivoklát (Czech Republic), Castle Libr., Cod. I e 7, fol. 77v–81v
22. Saint Louis (Missouri), Concordia Seminary Library, Rare Book Collection, no shelfmark (1), fol. 26v–27v
24. St. Gall, Monastery Libr., Cod. 760, p. 54–58
26. Strasbourg, City Libr., Ms. 258 (olim Ms. 711), fol. 38v–40v
27. Zurich, Central Libr., Ms. C 54, fol. 32v–33v
28. Private Collection Antiquarian Bookstore Dr. Jörn Günther, Hamburg, Nr. 1997/5,24, fol. 48r

The complete version, exemplified by our older fragment among other textual witnesses, is represented by eleven manuscripts, whereas the abridged version of the treatise is to be found contained in 18, at almost double the frequency. The consideration of this numerical relationship makes it highly unlikely that the truncation of the text resulted from scribal error and/or the use of a flawed prototype. Even when the existence of the complete version is attested to as early as the second half of the fourteenth century, the abridged version clearly was copied more prolifically. The earliest surviving example of the abridged version is the St. Louis manuscript, which dates to 1429. During the fifteenth century, the text tradition is marked by the coexistence of both versions running parallel to each other. Despite the emergence of the abridged version, the complete version continued to be copied. No tendency of the newer, shorter version displacing the earlier, longer version is evidenced. This suggests that each version was associated independently with certain affiliations, which intentionally required a particular version to suit their own purpose.

It is striking that a survey of the illustration presented in the manuscripts of both versions shows that none of the eleven examples of the complete version are illustrated or, judging from spaces left blank, were intended to have illustrations added to them later. In contrast, eighteen specimens of the abridged version are illustrated, or are at least laid out in such manner as to accommodate later illustration. An illustrated text is, in fact, an obligatory component of the abridged version manuscripts. The interplay of text and image was quite possibly a part of the conceptional design behind the emergence of the abridged version.

A closer look at the relationships among the various text traditions is necessary in order to come to an understanding of the significance of the two divergent text versions, both of which continued in production and coexisted parallel to one another. For what possible reason could the abridged version have been truncated to nearly a third of the original text, while having the manuscript laid out, as a rule, for illustration? Indications can be discerned through an examination of the amalgamation of contents found in the collective manuscripts containing our textual witnesses. Further exploration leads to the recognition of consistent companion traditions or 'communities of texts' present in the respective version manuscripts.

Bernhard Schnell was able to convincingly show in the field of astronomical, astrological and medical texts of the Middle Ages that, time and again, traditions of stable and predictable 'text communities' are observable in collective manuscripts (Schnell 1987; Schnell 2019). Particular texts with related themes were consciously grouped together into standard combinations. Texts compiled in such manner made up fixed groupings, which would be repeatedly reproduced and thus prolifically coalesced into established entities. This principle is not limited solely to text selection, but extends to the structure of the compilations as well. The order of the various texts is arranged according to overarching thematic configurations.

Schnell designated the collective manuscript: the *Iatromathematisches Hausbuch*[16] as a compelling representative of just such a 'community of related texts' compiled in such a way as to acquire the character of an autonomous work. He

---

[16] Cf. the *Handschriftencensus* entry (https://handschriftencensus.de/werke/2939) as well as the primary work: Lenhardt and Keil 1983/2004. The most up-to-date research is offered by Freienhagen-Baumgardt et al. 2021, 166–214 with Fig. 45–65.

postulates that an unnamed compiler/editor must have emerged who seized upon extensive fixed works out of the astronomical, astrological and medical traditions, adopted them nearly unchanged, and then deliberately revised and arranged them (Schnell 2019, 231–232). He goes on to posit that this compiler/editor even modified the included texts according to his or her purpose (Schnell 2019, 243). He estimates that the origin of the *Iatromathematisches Hausbuch* goes back to the beginning of the fifteenth century (Schnell 2019, 231–232).

To aid in better recognizing these 'text communities', Schnell formulated the *Schürstab* version as a model of the *Iatromathematisches Hausbuch* based upon manuscript C 54 held by the Zurich Central Library (Schnell 2019, 234–242). In so doing, he identifies three overarching thematic sections in the collective manuscripts, which are composites of various stand-alone texts of similar content-related themes: the first section deals with texts about human health in the course of the seasons; the second section presents texts concerning the influence of celestial bodies on human health; and the final, third section consists of texts instructive about healthy lifestyle, especially by way of purging harmful bodily fluids.

The treatise *Von den 11 Himmelssphären* is found in the abridged version towards the end of the second thematic section of the *Hausbuch* about the influence of celestial bodies. Because we know the earlier complete version was already in existence in the late fourteenth century, it could have well been a work which the compiler/editor came upon among texts circulated at the time, and then adapted and included in his compilation. An examination of all manuscripts of the abridged version confirms that the noted connection between the abridged version and the *Iatromathematisches Hausbuch* is in no way an isolated case. On the contrary, all known copies of the abridged version are found in the context of the tradition of the *Iatromathematisches Hausbuch*. On the other hand, the complete version may also surface in collective manuscripts with astrological, astronomical and medical themes. However, it is yet to be evidenced in the Schürstab version, which has been asserted as the standard of the *Iatromathematisches Hausbuch*.

The abridged version of the treatise is passed down without exception in specimens of the *Iatromathematisches Hausbuch*. It raises the question: why did the compiler/editor see fit to truncate the original text, excluding the first seven heavenly spheres? By comparing both versions, it becomes readily apparent that the changes already occur at the very beginning of the text. Two passages are particularly striking in this regard. Even before the body of the text begins, the compiler/editor placed an introductory inscription in advance of the incipit, "Got hieß abraham [. . .]" ["God asked Abraham . . ."], which reads as follows: "Nu saget dis puch hie nach von den vbrichen koren der himel vnd von irem lauff vnd natur vnd hebt an an dem der da haysset das firmament" ["This book tells of the remaining heavenly spheres and of their course and nature, and it begins with the firmament"]. This entry highlights that the text presented shall only deal with the remaining heavenly spheres continuing after the first seven, the "vbrichen koren der himel" ["remaining heavenly spheres"].

This deliberate curtailment of the text gives way to the hypothesis that the first seven heavenly spheres were already dealt with in other passages contained within the collective manuscript and thus need not be repeated. The abridged version native to the *Hausbuch* begins with the eighth heavenly sphere for exactly this reason. In a later passage the compiler/editor inserted yet another textual change directly before the section on the eighth heavenly sphere. He writes: "vnd hebet also hie nach von dem achtem himel an zu sagen wann von den siben planeten das sind siben choer der himeln vnd ir natur Ist vormals geschriben vnd gesagt" ["and (the book) now tells of the eighth heaven, for the seven planets, which are the seven heavenly spheres, and their constitution have already been written about"]. Thus, he again points out that the first seven planets have already been dealt with in the *Hausbuch*.

Both of these textual interventions, not present in the older, complete version, serve to explain the extensive truncation of the text by way of editorial commentary. Had the compiler/editor instead simply submitted the text's complete version unchanged, then unnecessary content-related redundancies regarding the seven heavenly spheres and the planets associated with them would have resulted. Against this backdrop, the observed editorial intervention is plausible and even to be anticipated because it constitutes a qualitative upgrade to the text in the context of the *Hausbuch*. This conclusion arises through consideration of the structure of the *Iatromathematisches Hausbuch's* second section as modelled by Schnell (2019, 235–236), whereby the placement of the passage concerning the seven heavenly spheres within the collective manuscript is shifted:

2. Texts about the influence of celestial bodies on human health

2.1. About the 12 constellations; illustrated

2.2. About the 7 planets; illustrated

2.3. About the course of the planets

2.4. Weather forecasting according to the planets

2.5. *Von den 11 Himmelssphären*, with one illustration

The text about the seven planets to which the revised version of our treatise refers is found under item 2.2 and often only a few pages before *Von den 11 Himmelssphären* in the manuscripts. *Von den 11 Himmelssphären* begins on folio 32v in the specific case of the Zurich manuscript, while the text about the seven planets takes up folios 24v–31v. Both texts stand in close proximity to another. It was necessary to avoid the redundancy of presenting two texts about the seven heavenly spheres in this specialized tradition grouping. Therefore, the conscious decision of the compiler/editor to conceptionally tailor the second section of the *Hausbuch* superseded concerns about preserving the integrity of a single text. He or she was acquainted with the textual contents of the compilation and viewed this amalgamation of single texts as a cohesive unity.

The abridged version exists exclusively within the *Hausbuch* tradition because the contents of a significant part of the complete heavenly sphere treatise was already covered by another text included in the collective manuscript. The abridged version is yet to be evidenced outside of this particular tradition's context, which leads one to surmise the text was purposefully revised to be incorporated into the *Hausbuch* compilation. Along with the treatise's adaption followed the addition of accompanying illustration, which conveyed elements of the contents through images. The compiler/editor introduced this newly fashioned version wonderfully into the concept of the comprehensive illustration of his or her compilation, which, according to Schnell makes up an important hallmark of the *Iatromathematisches Hausbuch* (Schnell 2019, 233).

Examinations such as the present one show how productive such explorative soundings can be. They can pave the way for further insights into the tradition history and aid in the understanding of the contemporary significance and function of such treatises as *Von den 11 Himmelssphären*. The consideration of information stemming from the entire tradition makes it possible to resolve apparent contradictions and irregularities, which otherwise would be difficult to explain. As a rather unusual textual witness, the fragment of *Von den 11 Himmelssphären* served as a starting point for future investigations. The parallel examination of both versions of the text shows that the respective traditions follow explicable patterns and the existence of the abridged version need not be ascribed to supposed flawed prototypes or scribal error. This version appears only in the context of the *Iatromathematisches Hausbuch* for which, in all likelihood, it was tailored. The dissemination of the abridged version is much more prolific than its older and lengthier counterpart's, which is owed to the great popularity of the *Iatromathematisches Hausbuch*. Our treatise profited enormously from its inclusion in this widespread collective manuscript. The present examination could well add to nascent theories as to the emergence of the *Iatromathematisches Hausbuch* (Schnell 2019). The example of this text might be consulted as a resource in exploring the additive method of compilation employed by the *Hausbuch* compiler/editor.

Two distinct versions of the treatise *Von den 11 Himmelssphären* exist according to our present understanding of its tradition. The abridged version takes on great significance as well as a special function for the present examination. While, according to findings of the research community, the earlier, complete version is associated with astronomical and astrological collective manuscripts in general, the later, abridged version clearly belongs specifically to the tradition complex of the *Iatromathematisches Hausbuch,* so much so that its presence can be used as an indicator helpful in the identification of the *Schürstab* version of the *Iatromathematisches Hausbuch.* This identification can be reliably verified by checking whether a particular fifteenth century collective manuscript of astronomical, astrological and medical texts contains the abridged version of our treatise, as well as whether it finds itself copied in close proximity to a distinct text about the seven heavenly bodies. In such a case, one has, in all likelihood, a textual witness of the *Iatromathematisches Hausbuch* at hand.

While looking more closely at the interrelations of textual traditions, it is to be anticipated that further findings about fixed text and tradition communities as embodied in collective manuscripts are to follow. The present chapter provides an initial stage which, when added to the mosaic of medieval astronomical, astrological and medical literature, makes way for much-needed further investigation into the subject.[17]

Bibliography

Bernard Quaritch Ltd., *Catalogue 1439, Medieval & Renaissance Manuscripts.* London: Quaritch, 2019.

Freienhagen-Baumgardt, Kristina, Polina Gedova, Nino Nanobashvili, Pia Rudolph, and Nicola Zotz. *Katalog der deutschsprachigen illustrierten Handschriften des Mittelalters, begonnen von Hella Frühmorgen-Voss und Norbert H. Ott.* Vol. 9,2/3 *(Medizin – Albertanus von Brescia, 'Melibeus und Prudentia').* Munich: C. H. Beck, 2021.

Haebler, Konrad. "Makulaturforschung." *Zentralblatt für Bibliothekswesen* 25 (1908): 535–544.

Handschriftencensus. An inventory of the manuscript tradition of medieval German-language texts. https://handschriftencensus.de.

Handschriftenportal. https://handschriftenportal.de. (Accessed 15 July 2024)

Kalning, Pamela, Matthias Miller, and Karin Zimmermann. *Die Codices Palatini germanici in der Universitätsbibliothek Heidelberg (Cod. Pal. germ. 671–848).* Kataloge der Universitätsbibliothek Heidelberg XII. Wiesbaden: Harrassowitz, 2016.

Keil, Gundolf, Friedrich Lenhardt, and Christoph Weißer. *Vom Einfluss der Gestirne auf die Gesundheit und den Charakter des Menschen, Bd. 1: Faksimile, Bd. 2: Kommentar zur Faksimile-Ausgabe des Manuskripts C 54 der Zentralbibliotek Zürich (Nürnberger Kodex Schürstab).* Lucerne: Faksimile, 1981–1983.

Lenhardt, Friedrich and Gundolf Keil. "'Iatromathematisches Hausbuch'." *Die deutsche Literatur des Mittelalters. Verfasserlexikon* 4 (1983): 347–351.

Manuscripta Mediaevalia. http://www.manuscripta-mediaevalia.de.

Miller, Matthias, and Karin Zimmermann. *Die Codices Palatini germanici in der Universitätsbibliothek Heidelberg (Cod. Pal. germ. 182–303).* Kataloge der Universitätsbibliothek Heidelberg VII. Wiesbaden: Harrassowitz, 2005.

---

17 Translated by Robert Whitley (Houston).

Neuheuser, Hanns Peter. "Zu den Perspektiven der Fragmentforschung." *Fragment und Makulatur: Überlieferungsstörungen und Forschungsbedarf bei Kulturgut in Archiven und Bibliotheken*. Eds. Hanns Peter Neuheuser and Wolfgang Schmitz. Wiesbaden: Harrassowitz, 2015. 1–14.

Petzet, Erich. *Die deutschen Pergament-Handschriften Nr. 1–200 der Staatsbibliothek in München*. Catalogus codicum manu scriptorum Bibliothecae Monacensis V,1. Munich: Harrassowitz, 1920.

Schnell, Bernhard. "Ein Würzburger Fragment des 'Iatromathematischen Hausbuchs'. Ein Beitrag zu dessen Überlieferungsgeschichte." *Würzburger medizinhistorische Mitteilungen* 5 (1987): 123–141.

Schnell, Bernhard. "'Iatromathematisches Hausbuch (Schürstab-Fassung)', früher 'Deutscher Kalender'. Ein kritischer Forschungsbericht." Bernhard Schnell, *Arzneibücher, Kräuterbücher, Wörterbücher. Kleine Schriften zur Text- und Überlieferungsgeschichte mittelalterlicher Gebrauchsliteratur*. Ed. Dorothea Klein. Würzburg: Königshausen & Neumann, 2019. 223–254.

Weimann, Birgitt. *Die mittelalterlichen Handschriften der Gruppe Manuscripta Germanica*. Kataloge der Stadt- und Universitätsbibliothek Frankfurt am Mainz 5,IV. Frankfurt am Main: Klostermann, 1980.

Zinner, Ernst. *Verzeichnis der astronomischen Handschriften des deutschen Kulturgebietes*. Munich: C. H. Beck, 1925.

II The Sixteenth and Seventeenth Centuries

Helge Perplies

Heavenly Theater: Writing about Astronomy and Astrology in Jean Bodin's *Démonomanie des sorciers*

Abstract: In his *Démonomanie des sorciers* (1580), Jean Bodin presents astrology as a natural form of divination within clear limits, set by his understanding of the divine forces at work in the celestial realm. He uses two distinct manners of writing about these topics, where his assertive use of astronomical and astrological terminology is in contrast with the image of the heavens as a theater, used to showcase God's creation and to instill his readers with a sense of wonder and humility.

1 Introduction

Astrology has a curious place in the work of Jean Bodin,[1] being one of only a few acceptable – natural – means of learning about the future, whereas most other methods are presented as the work of the devil. The devil is, of course, a constant protagonist in Bodin's *Démonomanie des sorciers*, his book on magic, witchcraft and, quite literally, the *demon-mania* of witches. Published in Paris in 1580, this work attempts a systematization of all the various forms of magic and sorcery in the past and the present.[2] For this, Bodin ventures into the fields of religion, medicine, philosophy, natural history and, of course, jurisprudence. The prosecution – and, one could argue, persecution – of alleged witches and sorcerers is the beginning and end of Bodin's argument, the last parts of the work being explicitly dedicated to the legal process.[3] He was, after all, a jurist and professor of law in

---

[1] Jean Bodin was born around 1530 near Angers. From 1549, he studied at the university in Paris as well as at the Collège des Quatre Langues. In the 1550s, he studied and later taught Roman law at the University of Toulouse; in 1561 he was called to the bar in Paris. He entered the world of high politics when he became secretary of commands in the household of François, Duc d'Alençon, in 1571. Bodin published books on history, economics, political theory, demonology, natural philosophy, and religion. He died in Laon in 1596. Cf. also the synopsis of his biography by Campion (1995, 95–98).

[2] All quotations are from the edition prepared by Virginia Krause, Christian Martin, and Eric MacPhail (Bodin 2016).

[3] On the legal-historical context, cf. Lattmann (2019).

Toulouse, even if he is best known as a political philosopher. But in the *Démono-manie*, his political and legal arguments revolve around the ubiquitous influence of witches – and in consequence the influence of the devil – that he sees at work everywhere. In this light it is all the more surprising to see him explicitly embracing astrology – or at least parts of astrology.

I will try to explain this curious fact by briefly going into Bodin's understanding of natural divination and the limits this sets for astrology, before I present in more detail the way Bodin puts forth his knowledge of celestial matters. Bodin, while certainly versed in these topics, writes largely for a lay audience, and I will argue that he uses two distinct manners of writing the heavens to achieve his goals. In this context, I will also take into consideration the German edition of Bodin's work, translated by the prolific author Johann Fischart[4] and published in Strasbourg, first in 1581 – shortly after the French original – and then again in 1586 in an expanded version.[5] Here I will show how Fischart amplifies the celestial descriptions he finds in Bodin's text and expands upon them to further his own linguistic agenda.

2 Bodin's understanding of astrology and its place within the framework of the *Démonomanie*

The *Démonomanie* opens with a discussion of the place that humankind occupies in the larger scheme of beings and forces, between God, the Devil, angels, demons, spirits et cetera. Bodin suggests possibilities to influence the various entities as well as communicating with them – for example via dreams and prophetic speech. This is followed by the chapter I will discuss here, which focusses on nat-

---

4 Johann Fischart was born in 1546 or 1547 in Strasbourg. In 1574, he finished his studies with the degree of *doctor juris* at Basel University. He worked in the Strasbourg-based print shop of Bernhard Jobin, his brother-in-law, who published most of Fischart's books. In the early 1580s, Fischart was an advocate to the imperial court of appeal at Speyer; in 1583 he was appointed magistrate at Forbach, where he died in 1591. His literary production ranges from polemical and satirical to moral and philosophical works, including many translations, such as *Geschichtsklitterung* (1575), a free adaptation of Rabelais' *Gargantua*.

5 Bodin and Fischart (1581) and Bodin and Fischart (1586). In the following, I quote from the historical-critical edition, prepared by Tobias Bulang, Nicolai Dollt, and Joana van de Löcht (Bodin and Fischart 2024). Without marking this individually, in several instances I make use of the preliminary work for the unreleased commentary volume to the above-mentioned edition, prepared by Tobias Bulang, Raffaela Kessel, Isabella Managò, Helge Perplies, Joana van de Löcht, and Katharina Worms.

ural means to learn hidden things – thus the chapter heading –, and which is in turn followed by two chapters on unnatural or unlawful means.

Bodin argues that the celestial bodies are moved by powers or forces that he calls *angels*. These forces are subject to God, who could, as Bodin writes, of course move them directly, without any instrument, "mais il est plus seant à la Majesté divine d'user de ses creatures" (Bodin 2016, 140) ["but it is more fitting for the divine Majesty to use His creatures"].[6] This ties in neatly with Bodin's understanding of communication with the divine, which is also conducted through angels. In fact, as Bodin points out, God does not interfere directly with physical bodies, but uses the influence of heavenly bodies instead. Therefore, Bodin argues, it is and always has been lawful to inquire about the virtues and properties of the celestial lights. Their enormous powers do not detract from God's own might, on the contrary, they elevate it.

However, even this amazing power is limited, inasmuch as it only concerns the material world. The celestial bodies have influence over the human body because they can change the disposition and the humors in the same way they change the ebb and flow of the tides. Bodin lists, for example, the influences of the moon on the course of diseases, referring to Galen and the concept of Critical Days, even though, as he points out, Galen would have been surprised had he known about the influences of all the other heavenly bodies besides the moon (Bodin 2016, 143–144).[7] In the Arab world, Bodin claims, this knowledge is so widespread, that all doctors know about the celestial influences on the body and are therefore called *Iathromathematicians* (Bodin 2016, 144).[8] The important lesson of this is, however, that the influence of heavenly bodies on human bodies cannot just be established post facto, but also be predicted. Thus, astrologers can use horoscopes – including natal charts – to predict the future development of a person's humors, temperaments and physiognomy. This, according to Bodin, is a perfectly acceptable means of divination, that is, of learning hidden things.

But, and this is an important *caveat*, "[m]ais il ne faut pas que les Astrologues se meslent de juger des ames, des esprits, des vices, des vertus, des dignitez, des supplices, et beaucoup moins de la religion, comm plusiers ont faict" (Bodin 2016,

---

6 The English translations of French and German quotations are mostly my own, sometimes based on the (partial) translation of the *Démonomanie* by Randy A. Scott (Bodin 2001).

7 The concept of Critical Days, which indicate the course and chances of recovery from disease, goes back to Hippocrates. The second-century Greek physician and philosopher Galen explicates the connection between the course of diseases and the position of the celestial bodies in his *De diebus decretoriis* (here especially Galen 2011, 911.14–913.15).

8 The field of iatromathematics applies astrological knowledge to medical practice in the framework of humoral pathology. Ptolemy attributes the origin of iatromathematics to the Egyptians (Ptolemy 1940, 1,3,16).

144) ["astrologers must not get involved with making judgments on souls, spirits, vices, virtues, honors, punishments, and much less on religion, as many do"]. They must not, Bodin adds, turn to "choses qui ne touchent en rien le corps, à sçavoir, aux mariages, aux dignitez, voyages, richesses, et autres choses semblables, où les astres n'ont ny force ny puissance" (Bodin 2016, 144) ["things that do not concern the body at all, such as marriages, honours, voyages, riches, and other similar matters, where the stars have neither force nor power"].

Especially the topic of astrology in religious matters is a thorn in Bodin's side. He criticizes the fourth-century astronomer Julius Firmicus Maternus, who claimed that people born under a certain constellation will go directly to heaven after their death,[9] and ninth-century Abu Ma'shar for his belief that prayers made under a certain constellation will inevitably be answered (Bodin 2016, 145).[10] Not only do these examples not concern the body, they also subject the divine will to the power of the heavens. But God has dominion over everything, even the celestial bodies, Bodin writes, pointing to the prophet Joshua, on whose request God stopped the course of sun and moon (Bodin 2016, 148; cf. Jos. 10:12–13).

Another point of contention is the idea of a celestial influence not just on the individual believer, but on the rise and fall of whole religions, as formulated amongst others by Abu Ma'shar,[11] Pierre d'Ailly[12] and Cyprian Leowitz.[13] In a sim-

---

9 Firmicus Maternus wrote his *Matheseos libri VIII*, the most comprehensive astrological manual in Latin, between 334 and 337. The passage in question is in the fifth book (Firmicus Maternus 1968, 5,3,22). Interestingly, Fischart adds the phrase "im Buch vom Herrn der Genitur" ["in the book of the lord of the moment of birth"] (Bodin and Fischart 2024, 129), referring to the term *dominus geniturae* used by Firmicus for the dominant planet of a horoscope (cf. for details Firmicus Maternus 1968, 4,19,1–40). This shows that Fischart himself was familiar with astrological terminology and probably with the *Matheseos libri VIII* as well.

10 In fact, the passage cannot be found in the over thirty works attributed to the Persian astronomer and astrologer Abu Ma'shar. Instead, it is recorded in the *Excerpta de secretis Albumasaris* (also known as *Albumasar in Sadan*), a collection of astrological instructions and anecdotes, composed by Abu Ma'shar's pupil Sadan in form of a dialogue between the two men.

11 Abu Ma'shar included in his description of the stellar influence on worldly dynasties also a sequence of six dominant religions, which would replace each other in the rhythm of ten Saturn orbits (about 300 years) (Abu Ma'shar 2000, 2,8,33f.).

12 Pierre d'Ailly, French theologian and cardinal, wrote a series of treatises between 1410 and 1414 linking historiography and astronomy. There he describes, among other things, the succession of a total of six religions (especially in *De legibus et sectis contra superstitiosos astronomos*, printed *c.* 1490).

13 The sixteenth-century court astronomer Cyprian Leowitz wrote several works with astronomical calculations, especially on solar and lunar eclipses as well as on planetary conjunctions. In his *De Coniunctionibus Magnis Insignioribus Superiorum planetarum* (1564), he predicted that the conjunction of Jupiter and Saturn in 1583/1584 would mark the return of Christ.

ilar vein, Bodin criticizes astrology-based predictions about the coming of the antichrist, namely by Arnau de Vilanova[14] (Bodin 2016, 145). For this criticism, Bodin relies heavily on the *Disputationes adversus astrologiam divinatricem* by Giovanni Pico della Mirandola,[15] quoting him almost *verbatim* (Pico della Mirandola 1969, 1,550f.). This is all the more surprising in the face of the damning invectives that Bodin launches against another, more famous work by the Count of Mirandola, the *900 Theses* (e.g., Bodin 2016, 151–152).

It must be added, however, that Bodin is not averse to the idea that large cycles of history are influenced by the heavens, especially by the Great Conjunctions of Jupiter and Saturn. In his work on political theory, *Les Six livres de la République* (Bodin 1576), as well as in his treatise on history, *Methodus ad facilem historiarum cognitionem* (Bodin 1566), Bodin writes extensively about the societal changes that occur in the rhythm of the planets as they move through the zodiac. It is the idea of celestial influence on *religious* matters that he abhors.[16]

Interestingly, his critique of Cyprian Leowitz and others on this topic is twofold: not only do they promote impious ideas, they also prove to be bad astronomers. Pierre d'Ailly, for example, based his horoscope for the creation of the earth on the assumption that the sun was in the sign of Aries, when, according to Bodin, the Bible clearly shows that the sun was in the sign of Libra (Bodin 2016, 145–146). By pointing out their errors – or what he claims to be errors –, Bodin emphasizes his own expertise in astronomical, astrological and, in this case, theological matters.

However, Bodin also presents two important caveats as to the reliability of astrology: one, and this is especially important with regards to the large-scale societal changes, there are just not enough datapoints from the mere 3,000 years of astronomical observations to predict the future with any certainty. Two, and this is important for all astrological predictions from global events to a person's disposition, the celestial influence is merely a natural inclination, not a necessity (Bodin 2016, 147). Bodin has no doubt "que l'homme qui se fie en Dieu ne soit plus fort, et plus

---

14 The thirteenth-century Spanish physician and alchemist Arnau de Vilanova, who was convicted of heresy by the Inquisition but pardoned by Boniface VIII. In his *Tractatus de tempore adventus Antichristi* and the *Expositio super Apocalipsim* – attributed to, but most likely not actually written by Arnau – he calculated the time of the arrival of the Antichrist on the basis of biblical passages.

15 Giovanni Pico della Mirandola, fifteenth-century *wunderkind* philosopher, wrote his *900 Theses* on natural philosophy, religion, and magic at the age of 23. After they had been banned by the Church and Pico had fled to France, he was allowed to return to Florence in 1488, where he wrote the *Disputationes adversus astrologiam divinicatrium*, condemning a deterministic astrology that stood in conflict with the Christian notion of free will. The work was posthumously published in 1493.

16 For more on this point, cf. Halbronn (1987, 207–209).

puissant, que toutes les influences celestes" (Bodin 2016, 148) ["that the man who trusts in God is stronger and more powerful than all celestial influences"].

To summarize, Bodin's understanding plants astrology firmly within the means of natural divination. The heavens are created by God and moved by his angels, and they can influence the natural world, from tides to weather phenomena to human temperaments and dispositions. However, they are limited to this material world and do not have any influence on souls, spirits or religious matters. Also, the influence on the body is not an inevitable force but manifests only in certain inclinations.

3 Bodin's use of astronomical and astrological vocabulary

Next, I turn to the question of *how* Bodin presents his understanding of astrology to his readers; readers who, given the jurisprudential nature of the *Démonomanie*, would for the most part not have been experts on the topic. However, since astronomy and astrology were taught at universities as part of mathematics,[17] he could expect at least a general understanding of the matter and a familiarity with the basic technical terms. When, for example, he introduces Galen's concept of Critical Days, Bodin writes about patients' horoscopes and notes "que l'opposition ou quartier de la Lune au Soleil donne un changement notable aux malades, et quand la Lune attainct l'opposition ou quartier du lieu où elle est partie, quand la maladie a commencé" (Bodin 2016, 143) ["that the opposition or quarter of the moon in the sun produces a notable change in patients, as well as when the moon reaches the opposition or quarter of the place from where it set out when the illness began"]. He doesn't explain the terms *opposition* or *quarter*, expecting his readers to have a basic knowledge of horoscopes and celestial mechanics. Similarly, when Bodin points out Galen's limited knowledge, he talks about the effects of the planets and their conjunctions, both in relation to each other and to the fixed stars (Bodin 2016, 144), again without explaining what exactly conjunctions are.

The examples that follow are not much more forthcoming: "Car les anciens ont remarque pour maxims, et par experience de plusieurs siecles, que Saturne et Mercure estant opposites en un signe brutal, l'homme ordinairement, qui naist alors, est begue ou muet" (Bodin 2016, 144) ["For the ancients have noted as max-

---

17 For a discussion on this subject in the context of Kepler's horoscopes, cf. Boockmann (2010, 5–7).

ims, and by experience of many centuries, that when Saturn and Mercury are opposite to each other in a violent sign, the man who is born then is usually tongue-
tied or mute"]. Even with astrological knowledge, both sixteenth-century readers
and twenty-first-century scholars are left to speculate what Bodin exactly means
with "violent sign"; presumably, he refers to those astrological signs whose planet
rulers are associated with negative effects, knowledge he and his readers could
find in Ptolemy's *Tetrabiblos* (Ptolemy 1940, 3, 12, 150). Other examples use similar
terms to determine the celestial situation at the time of birth, such to as "la Lune
estant au Levant" ["the moon is in the ascendant"], "naist en la conjonction de la
Lune" ["born in the conjunction of the moon"] or simply "en l'eclypse" ["during
the eclipse"] (all quotations Bodin 2016, 144).

When Bodin criticizes Firmicus Maternus for his belief that people born in a
certain constellation will go directly to heaven after their death, he uses a kind of
shorthand: "celuy qui a Saturne au Leon" (Bodin 2016, 145) ["one who has Saturn
in Leo"], again indicating both a familiarity with astrological terminology and the
presumption of the same familiarity in his readership. A bit more complicated is
the case of Abu Ma'shar's belief that prayers made under a certain constellation
will inevitably be answered: Bodin writes that this has to happen – according to
Abu Ma'shar – "estant la Lune conjoincte à une autre Planette, que je ne mettray
point, et tous deux au chef du Dragon" (Bodin 2016, 145) ["while the moon is in
conjunction with another planet, which I will not indicate, and both are at the
head of the Dragon"]. This is one of only a few times where Bodin actually adds
some discussion of an astronomical term: "le chef, et queuë du Dragon ne sont
rien que deux poincts d'une intersection imaginaire, et de deux cercles imagi-
naires, et qui n'ont ny estoille ny planette, et variables à tous momens" (Bodin
2016, 145) ["the head and tail of the Dragon are only two points of an imaginary
intersection, and of two imaginary circles, which have neither star nor planet,
and are variable at all times"]. This description of the lunar nodes, the intersections of the lunar orbit and the ecliptic, is not really an explanation, since he mentions neither moon nor earth. Building on a presumed knowledge of the term
'head of the Dragon', his point seems to be that the lunar nodes themselves are
not relevant for the influence of the celestial bodies. However, it must be mentioned that Bodin does not need his readers to understand the term; he is less
interested in explaining the details than he is in criticizing Abu Ma'shar. It is difficult to discern the extant of the knowledge he expects in his readership, but he
seems to be perfectly content with keeping them in the dark from time to time.

This relates to the curious censure of the planet in question. There are various similar instances in the *Démonomanie* where Bodin explicitly refuses to include some detail, usually the wording of magic spells or ingredients for potions
(e.g., Bodin 2016, 177–178). His motivation for doing so is presumably to prevent

his readers from attempting to use the dangerous knowledge contained in his book. Here he even adds the case of Pietro d'Abano – "maistre Sorcier, s'il en fut oncques" (Bodin 2016, 145) ["a Master Sorcerer if ever there was one"] – who not only practiced this method to have his prayer answered, but also enticed other people to try it. It seems likely that Bodin's reluctance to explicate astrological details is connected to his attempt to keep control over the information.[18]

A special nemesis for Bodin is his contemporary Cyprian Leowitz and his belief that both the Christian religion and the world at large are destined to end in 1583/1584, "pour la grande conjonction en la triplicité aquatique de Jesus Christ" (Bodin 2016, 146) ["due to the Great Conjunction in the watery triplicity of Jesus Christ"]. Bodin sharply criticized Leowitz and his scientific methods in the *Six Livres de la République*, mocking the wrong predictions made by him (Bodin 1986, ch. 4.2). Here he again uses a double-edged critique, claiming that the prediction regarding the end of the Christian religion is both "une incongruité notable en Astrologie, et impieté en termes de religion" (Bodin 2016, 146) ["a notable incongruity in astrology and an impiety in terms of religion"]. He goes on: "car jamais Planette ne ruina son signe ny sa maison, et Juppiter est conjoinct aux poissons, en la conjunction qu'il craint si fort, qui est le signe de Juppiter conjoinct avec Saturne, qui est son amy" (Bodin 2016, 146) ["for never has a planet ruined its sign nor its house, and Jupiter is conjoined to Pisces, in the conjunction that he (i.e., Leowitz) fears so much, which is the sign of Jupiter conjoined with Saturn, which is his friend"]. While he meticulously points out Leowitz's errors, Bodin does not bother to explain any of the terminology he uses, from the 'Great Conjunction' to the 'watery triplicity' (the zodiac signs of Cancer, Scorpio and Pisces) to the 'friendship' of Jupiter and Saturn. As if he anticipated that not all of his readers might understand the intricacies of his critique, Bodin adds that Leowitz has also published printed ephemerides (tables to compute the positions of celestial objects) that go far beyond his proposed end of the world, indicating that Leowitz may not believe in his own predictions – a point of contention much easier to grasp (Bodin 2016, 146).

Probably the most detailed astrological description is also part of a refutation: Bodin criticizes as inept the judgement of sixteenth-century polymath Girolamo Cardano, who created a natal chart of Jesus Christ that was printed and widely distributed throughout Europe:[19] "disant que Saturne en la neufieme maison sig-

---

18 The planet in question, by the way, is Jupiter, in case readers want to try this at home.
19 Like others before and after him, Girolamo Cardano produced a natal chart of Christ, which he included in his commentary on Ptolemy's *Tetrabiblos* (1554). Although Cardano stressed that the stars only indicated coming events and did not cause Christ's power, he was nevertheless heavily criticized (Grafton 1999, 151–155).

nifioit la desertion de sa religion, et Mars avec la Lune en la septieme, monstroit le genre de mort, chose ridicule, attendu que Mars estoit en son propre signe, qui est ignee" (Bodin 2016, 146) ["saying that Saturn in the ninth house signified the desertion of his religion, and Mars with the Moon in the seventh showed the manner of death, a ridiculous thing, seeing that Mars was in his own sign, which is fiery"].[20] While the astrological critique is made explicit by Bodin, the accusation of impiety is only implied, because he goes on to say: "Mais l'impieté est beaucoup plus grande de vouloir asservir la religion aux Astres, comme aussi a faict Abenesra, qui avoit predict qu'il naistroit un grand Capitaine, pour afranchir les Juifs, qu'il appelloit Messie, l'an M.CCCC.LXIIII, ce qui n'est poinct advenu" (Bodin 2016, 146) ["But it is a much greater impiety to try to subjugate religion to the stars, as did also Abenesra who had predicted that a great captain would be born to free the Jews, whom he called Messiah, in the year 1464, which did not happen"].[21] Here Bodin criticizes both the wrong prediction – although he does not explicate the underlying astrological error – and the suggestion that the advent of the messiah could be predicted from the stars.

In summary, Bodin is using astronomical and astrological terms in an assertive way that demonstrates his deep understanding of the matter. While many of the terms would have been familiar to his readers, the *Démonomanie* – and the same can be said for his *Six Livres de la République* – is not directed at experts in the fields of astronomy and astrology. The lack of explanations regarding more complicated matters seems therefore to have a double purpose: It forms a barrier for those readers who are uninitiated in the topic, keeping potentially dangerous knowledge from them. At the same time, his assertiveness lends his propositions a certain authoritative nature that belies the fact that Bodin himself takes a lot of his arguments from other sources, such as the before-mentioned Pico della Mirandola, whom he quotes (mostly without attributions) throughout the chapter. However, this factual, almost technical discourse is at odds with another manner of writing about the heavens: the heavens as a theater.

---

20 Fischart's opinion of this matter seems to differ from Bodin's: when he discusses Christ's natal chart for the first time, he adds a marginal note, saying that "Christo sein Natiuitet stellen ist nit Astronomisch" (Bodin and Fischart 2024, 129) ["it is not astronomical to create Christ's natal chart"]. Whereas Bodin criticizes Cardano's *faulty* astrology, Fischart seems to reject *any* use of astrology.

21 There is no such prediction recorded by the twelve-century scholar and poet Abraham Ben Meïr Ibn Ezra. This is probably a mistake by Bodin, who confused him with the older mathematician and philosopher Abraham Bar Hiyya – both are known as *Abraham Judaeus*. Bodin most likely quotes Pico della Mirandola, who refers to the prediction by *"Abraam Iudæus"* (Pico 1969, 1,550) about the appearance of the Jewish messiah in 1464 (Sarachek 1968, 323–326; Töyrylä 2014).

4 The heavens as evidence of God's glory

In Bodin's view, the heavens and especially the movement of the celestial bodies give testament to God's power. He uses the image of a theater, showcasing God's creation in all its glory: "Et à dire vray, le Ciel est un tresbeau theatre de la louange de Dieu, et plus on cognoist les effects de ces lumieres celestes, plus on est ravi à louër Dieu" (Bodin 2016, 143) ["And truly, the Heavens are a most beautiful theater for the praise of God, and the more one knows the effects of these heavenly lights, the more one is inspired to praise God"]. This metaphor, probably taken from Philo of Alexandria (Philo 1929, XVII,54), is further elaborated in a later work of Bodin's, in the *Théâtre de la nature universelle* (1597).[22] He also quotes Psalm 8, using a rendition by French poet Clément Marot:

> Mais quand je voy, et contemple en Courage
> Les Cieux, qui sont de tes doigt haut ouvrage,
> Estoilles, Lune et Signes differens,
> Que tu as faicts, et assis en leurs rancs:
> Adonc je dy à part moy ainsi, comme
> Tout esbahi, "et qu'est-ce que de l'homme?"
> (Bodin 2016, 142)

> [But when I see and contemplate in courage the heavens, which are your finger's great work, stars, moon and different signs, that you have made, and sitting in their ranks: therefore, I say to myself thus, as if all amazed, 'and what is man?']

Here Bodin uses the astronomical images to instil his readers with a sense of wonder and humility in the face of God's creation, which, at the same time, reaffirms the idea that God alone is in control of the celestial bodies. This ties in nicely with a short passage, inserted between the cosmological introduction and the musing on celestial influence. Bodin reflects on the story of Job, especially the colloquy at the end, where God puts a number of questions to Job to show him the limits of his power: "Pourras tu dict-il, lier les Pleiades, ou desjoindre les estoilles de la grand' Ourse? Produiras tu les hyades, et si tu pourras gouverner les estoilles d'Arcturus?" (Bodin 2016, 142) ["Will you be able to say: bind the Pleiades or join the stars of the Great Bear? Will you bring forth the Hyades, and are you able to govern the stars of Arcturus?"].[23] He inserts this passage in an attempt, I would

---

22 For more on the metaphor of the theater, cf. Blair (1997, 153–179).
23 Job 38:31–32. As is often the case in the different biblical traditions and translations, the stars, planets and constellations that are mentioned here vary a lot; I have not been able to identify the direct source for Bodin's version.

argue, not only to establish God's dominion over the heavens, but also to emphasize their beauty and grandeur.

The German translator of the *Démonomanie* at least seems to have understood
it in this way. One striking feature of Fischart's translation is his amplification of
the source material, in many cases to show off the versatility of the German language and its suitability for scientific discourse.[24] In this passage we see the same
tendency; however, I think we can see something more: Fischart's version not only
offers a variety of different designations for the celestial bodies, showing the various possibilities the German language has to offer; it specifically presents a vivid
imagery that fits well with Bodin's idea of the heavens as a theater.

> Kanst du auch die Zwitzerende Gluckhenne oder Glåntzige *Pleiadas* binden/ oder jhre Hůnlin
> zusamen bringen? Kanst du auch den vmbschweiff des hellen Sterns des Wagenmanns vnnd
> den Schwantz der grossen Bårin oder *Arcturi* von den anderen absönderen? Oder magstu den
> Zeug des Heerwagens zertrennen? Oder kanstu daß Rågenlich Sibengestirn im Kopff des
> Stiers oder die *Hyadas* herfůr locken? Kanstu den Morgen vnnd Abendstern zu bestimpter
> zeit vber die Kinder der Erden außfůhren/ daß du sie widerumb zu rechter zeit heimfůhrest?

> [Can you bind the glittering clucking hen or the glistening Pleiades, or gather her chicks?
> Also, can you separate from the others the orbit of the coachman's bright star and of the
> Great Bear's tail or of Arcturus? Or can you disjoint the harness of the chariot? Or can you
> coax out the rainy seven-stars in Taurus's head or the Hyades? Can you lead forth the morn-
> ing and evening star over the children of the earth at a certain time so that you can lead
> them home at the right time?] (Bodin and Fischart 2024, 124)

As is the case in Bodin's version, Fischart here refers by name to the open star
cluster of the *Pleiades*, but he adds the popular image of the clucking hen and her
chicks, common already in antiquity. *Zwitzernd* can be translated here either as
'glittering' or as 'chirping', showing right away the subtility of Fischart's writing.
He also includes the names of other stars and/or constellations: The 'coachman'
can either refer to the constellation Auriga – literally the coachman or chariot-
eer –, in this case the bright star would be Capella, or to the (albeit not very
bright) star above the shaft of the asterism of the Wagon – also called *Chariot*.
This asterism forms the 'Great Bear's tail' mentioned in the passage, so there is
some plausibility to the assumption. *Arcturus* seems out of place here in an astronomical sense, being the brightest star in the constellation Boötes, but mythologi-
cally it is tied to Arcas, the son of Zeus and Callisto, whose mother had been
transformed into a bear and later placed among the stars as Ursa Major. It is plausible that Fischart adds Arcturus because of the mythological context, but it
should be noted that Arcturus is mentioned in the Vulgate version of this pas-

---

24 On Fischart's inserts and additions, cf. Schüz (2011, 174–241).

sage.[25] The *Hyades*, the open star cluster in the head of the constellation Taurus, are commonly associated with rain, which, as well as their number, refers to Greek mythology. Fischarts adds these details, tying the astronomical phenomenon to the lived experience of his readers and emphasizing the close connection between the heavens and the physical world established in this chapter. The same can be said for the last phrase, where Fischart has God asking Job if he can "lead forth the morning and evening star over the children of the earth at a certain time so that you can lead them home at the right time?" 'Them' seems to refer to the morning and evening star – that is to say, Venus – being lead home at the right time, but it could also apply to the children of the earth and their dependence on the stars for daily routines. In fact, Bodin goes on to discuss briefly the role of the celestial lights as markers of time as described in Genesis 1:14 (Bodin 2016, 142), so it is possible that Fischart anticipates this argument here. It is notable that Fischart uses a variety of different designations, from astronomical terms in Latin and Greek to popular imagery in German. He evokes a vividness and immediacy that is in line with Bodin's conception of the heavens as a stage on which God presents both the wonders of his creation and the immense powers associated with them.

This sense of divine forces at work combined with the assertive use of astronomical and astrological theory and terminology provides the framework for Bodin's views on the stars, planets and other celestial bodies. Their influence on the material world can be measured, calculated and computed – and he gleefully points out any errors in the process –, but the real danger is a lack of humility in the face of God's glory. In the end, astrology may be a natural kind of divination, an admissible way of interpreting celestial signs to learn about the future, but Bodin tirelessly points to the limits of astrology specifically by highlighting the limitless power of its ultimate source.

Bibliography

Abu Ma'shar. *On Historical Astrology*, vol. 2: *The Latin versions*. Eds. and transl. Keiji Yamamoto and Charles Burnett. Leiden, Boston, and Cologne: Brill, 2000.

Arnau de Vilanova. *Expositio super apocalypse*. Ed. Joachim Carreras i Artau. Barcelona: Institut d'Estudis Catalans, 1971.

Arnau de Vilanova. *Tractatus de tempore adventus Antichristi: ipsius et aliorum scripta coaeva*. Ed. Josep Perarnau. Barcelona: Institut d'Estudis Catalans, 2014. 169–276.

---

25 "numquid coniungere valebis micantes stellas Pliadis aut gyrum Arcturi poteris dissipare | numquid producis luciferum in tempore suo et vesperum super ilios terrae consurgere facis" (Job 38:31–32).

Blair, Ann. *The Theater of Nature. Jean Bodin and Renaissance Science*. Princeton: Princeton University Press, 1997.

Boockmann, Friederike. "Johannes Kepler's Horoscope Collection." *Culture and Cosmos*, 14.1/2 (2010): 1–32.

Bodin, Jean. *Les Six Livres de la République*. Ed. Christiane Frémont. Paris: Fayard, 1986.

Bodin, Jean. *On the Demon-Mania of Witches*. Transl. Randy A. Scott. Toronto: Centre for Reformation and Renaissance Studies, 2001.

Bodin, Jean. *De la Démonomanie des Sorciers*. Eds. Virginia Krause, Christian Martin, and Eric MacPhail. Geneva: Droz, 2016.

Bodin, Jean, and Johann Fischart. *De Daemonomania Magorvm. Vom Außgelaßnen Wütigen Teuffelsheer* [. . .]. Strasbourg: Bernhard Jobin, 1581.

Bodin, Jean, and Johann Fischart. *De Magorvm Daemonomania. Vom Außgelaßnen Wütigen Teuffelsheer* [. . .]. Strasbourg: Bernhard Jobin, 1586.

Bodin, Jean, and Johann Fischart. *De Magorum Daemonomania. Edition von Johann Fischarts Übersetzung der Démonomanie des Sorciers Jean Bodins*. Eds. Tobias Bulang, Nicolai Dollt, and Joana van de Löcht. Stuttgart-Bad Cannstatt: frommann-holzboog, 2024.

Campion, Nicholas. "Astrological Historiography in the Renaissance. The Work of Jean Bodin and Louis Le Roy." *History and Astrology. Clio and Urania Confer*. Ed. Annabella Kitson. London: Mnemosyne Press, 1995. 89–136.

Firmicus Maternus, Julius. *Iulii Firmici Materni Matheseos libri VIII*. Eds. Wilhelm Kroll and Franz Skutsch. 2 vols. Stuttgart: Teubner, 1968.

Galen. *De Diebus Decretoriis, from Greek into Arabic: A critical edition, with translation and commentary of Ḥunayn ibn Isḥāq, Kitāb ayyān al-buḥrān*. Ed. Glen M. Cooper. Farnham and Burlington: Ashgate, 2011.

Grafton, Anthony. *Cardano's Cosmos. The Worlds and Works of a Renaissance Astrologer*. Cambridge, MA: Harvard University Press, 1999.

Halbronn, Jacques E. "The Revealing Process of Translation and Criticism in the History of Astrology." *Astrology, Science and Society. Historical Essays*. Ed. Patrick Curry. Woodbridge and Wolfeboro: Boydell, 1987. 197–217.

Lattmann, Christopher. *Der Teufel, die Hexe und der Rechtsgelehrte: Crimen magiae und Hexenprozess in Jean Bodins De la Démonomanie des Sorciers*. Frankfurt am Main: Vittorio Klostermann, 2019.

Leowitz, Cyprian. *De Coniunctionibus Magnis Insignioribus Superiorum planetarum* [. . .]. Lauingen: Saltzer, 1564.

Philo. *On the Creation: Allegorical Interpretation of Genesis 2 and 3*. Transl. F. H. Colson and G. H. Whitaker. Cambridge, MA: Harvard University Press, 1929.

Pico della Mirandola, Giovanni. *Opera Omnia* (1557–1573). Hildesheim: Olms, 1969.

Pierre d'Ailly. *De legibus et sectis contra superstitiosos astronomos* [. . .]. Rouen: Guillaume Le Talleur, c. 1490.

Ptolemy. *Tetrabiblos*. Transl. Frank Egleston Robbins. Cambridge, MA: Harvard University Press, 1940.

Sarachek, Joseph. *The Doctrine of the Messiah in Medieval Jewish Literature*. 2nd ed. New York: Hermon Press, 1968.

Schüz, Jonathan. *Johann Fischarts Dämonomanie. Übertragungs- und Argumentationsstrategien im dämonologischen Diskurs des späten 16. Jahrhunderts*. Dissertation submitted 2011. http://www.diss.fu-berlin.de/diss/receive/FUDISS_thesis_000000021283 (29 April 2022).

Thorndike, Lynn. "Albumasar in Sadan." *Isis* 45.1 (1954): 22–32.

Agata Starownik

Astronomy for the Public
The 1580 Warsaw Parade of Planets in Martin Gruneweg's Relation

Abstract: The article discusses the parade of the planets, which took place on 15 February 1580 in Warsaw and was described by Martin Gruneweg in the early seventeenth century. It was based on the iconography of the seven planets associated with the signs of the zodiac – their planetary houses. The main theme of the show was the celestial bodies, which distinguishes it from other shows in the Polish-Lithuanian Commonwealth and links it to European spectacles. The parade was a kind of urban carnival procession but also referred to Renaissance triumphs. It was an entertainment for both the townspeople and the royal court of Stephen Báthory and his wife Anna Jagiellon. It was probably created through the collaboration of these two circles. The authors used their astronomical or astrological knowledge, which was based on a geocentric model of the cosmos. They adapted this knowledge to the needs of a wide, diversely educated audience. The parade can therefore be seen as an example of the reception of nature studies in a possibly attractive performative form. It mixed cosmology, astrology and mythology, as well as a carnivalesque ludic element. To some extent, the parade reflects the astronomical consciousness of Warsaw's inhabitants – from the royal couple and magnates to the bourgeoisie and commoners. Moreover, the astronomical meanings of the spectacle are filtered through the complicated worldview and mentality of the author of the account – a Lutheran merchant who travelled a lot and eventually converted to Catholicism and became a Dominican. He interpreted the procession in the context of Christian spirituality and townspeople's virtues.

1 Introduction

On 15 February 1580 a carnival parade of the seven planets took place in Warsaw. This spectacle can be considered an interesting example of the reception of astronomical/astrological[1] knowledge in the Early Modern multicultural culture in the

---

[1] Two fields of knowledge about the cosmos – astrology and astronomy, which began to be slowly distinguished during the Renaissance – will be treated jointly. Using the terms explained by Eugenio Garin (1982, 3–17), the tradition of prophetic astrology (closer to astrology in contemporary meaning) will prevail, not the mathematical (closer to astronomy). It results from the

Polish-Lithuanian Commonwealth. The procession combined elements of court and city carnival culture. The event was witnessed by Martin Gruneweg who left a detailed description of the performance in his notes (2008, 602–604).

The main part of the procession were actors dressed up as planetary deities with their characteristic attributes (known from the iconography) and the planetary houses, it means the zodiac signs assigned to them according to the astrological tradition. The parade was opened by three pairs of musicians in antique fashioned costumes and closed by figures typical of the carnival: Bacchus, an Ottoman, and a cart-ship. These external components were distinguished by their mode of transport: they walked on foot, unlike the planets, which moved on horseback. Only Bacchus had a mount, but it was smaller and pretended to be a goat. Anyway, as a Greco-Roman deity, like the planets, it befits him to ride a horse rather than walk.

The aim of this article is to describe the significance of the astronomical motifs in the parade. In order to do so, various aspects of the event must be taken into account, notably specificity of the source for the pageant, astronomical knowledge of the likely organisers and recipients of the march, its functions in relation to the creators' and viewers' state of knowledge about nature and the culture, the basic theatre traditions from which the spectacle grows, as well as astrological contexts of its structure and iconography of the procession. All these layers will be discussed, but it should be noted that the conclusions are to some extent hypothetical, so they need to be deepened through solid background research and extended with source material.

2 The source and its author

The parade of planets was watched by Martin Gruneweg, whose varied experiences, complex identity and eventful life have influenced his account. He was born in 1562, in Gdańsk, in a middle-class, Lutheran, German-speaking family of traders. He apprenticed with the Warsaw merchant Georg Kersten from Nuremberg, and during his apprenticeship witnessed the parade. When the master went bankrupt, Gruneweg moved to Lwów, where he worked together with Armenian merchants, with whom he travelled a lot (e.g., to the Ottoman Empire and Mos-

---

specificity of the research subject: the planets pageant is based on the connection of celestial objects with ancient deities and their iconography. It must not be forgotten, however, that in Gruneweg's time even prophetic astrology was based on an idea generally believed to be credible – on the assumption of an analogy between the macrocosm and the microcosm.

cow). In that time, he polonised, converted to Roman Catholicism, and joined the Dominican Order. In the monastery, he was in charge of pastoral and administrative matters. Because of conflicts within the community he started a pilgrimage to Rome. Upon return to the Commonwealth, he lived for longer time in Cracow, Płock and finally in Warsaw, where he probably died after 1615.

Gruneweg described the planets march in notes which he kept for about five years, in 1601/1602–1606, in Lower German. The manuscript has almost 2,000 pages and is stored in the Gdańsk Library of the Polish Academy of Sciences (MS 1300, old number I E.f.77). It has been published by Almut Bues in 2008, in four volumes. The text includes memoirs, travel accounts, elements of chronicle and apologia, religious reflections, extracts from the Bible, ancient and theological texts, sermons, as well as numerous illustrations in the margins. One of them is connected to the Warsaw procession and shows a triumphal gate in the Market Square, through which the participants of the planets parade passed.

Generally, Gruneweg's report can be considered credible and valuable. The author was a sharp, attentive, open-minded observer with "extraordinary visual sensitivity" (Crăciun 2009, 246) and attention to detail (Walczak 1960, 67–68). He had an excellent memory, perhaps supported by a diary he kept up to date and used while working on the final work. Although he was aware of the limitations of human remembering, his sensitivity to the colourful details of everyday life allows us to trust his account (Hapanowicz 2014, 40–41).

Gruneweg's plasticity of expression was expressed not only in the illustration, but also in the allusion to traditional planetary iconography in the description of the Sol figure: "Das angesichtte war eittel goltt unde mit langen stremen, wie man sie moelet" (Gruneweg 2008, 603) ["The face was like burning gold, with long rays, as it is usually painted"[2]]. In the beginning, his report on the march mentions that the planets' costumes corresponded to how these figures were usually depicted ("wie man die Planetenn tzumaelen pflegett"; Gruneweg 2008, 603).

The author became acquainted with a variety of communities: the Protestant German-speaking merchants from whom he originated, circles of Armenian traders with whom he worked, the numerous countries he visited, including Muslim and Orthodox regions, Catholic clergy of which he became a part, and the Commonwealth royal court with which he collaborated as a trader and monk. Gruneweg was a Lutheran when he watched the planets parade, but he described it from the Catholic point of view that was closest to his own when he kept his notes. These two perspectives intermingle in his interpretation of the spectacle, and have a common Christian basis, contrasted with a showy but deceptive my-

---

2 All translations from Gruneweg's notes are mine.

thology. However, the elements of commentary can be easily separated from the layer of facts.

Gruneweg received a solid general education but did not complete university studies. His astronomical and cultural awareness was quite typical of the more intellectually educated townspeople and nobility. Although he did not belong to the intellectual elite and did not know the secrets of mathematical or prophetic astrology, he accurately reflected the details of the event, and supplemented any gaps in knowledge with keen observation. His purpose was not a profound interpretation, but providing a picturesque and accurate story, exhaustive and persuasive.

3 Place, audience, and organisers of the parade

The planets parade took place in Warsaw, which in 1580 was a relatively small town in Mazovia. It had no university or well-developed education, and its glory days as political and cultural centre were just beginning. However, the parliament was meeting and the king was elected there. Moreover, Queen Anna Jagiellon, who resided in the city, supported it. Therefore, Warsaw became a field of influence for the universal court culture. The city was also not a centre of astronomy, unlike, e.g., Cracow. Besides, astronomical research did not flourish at all in the second half of the sixteenth century in the Commonwealth. The astronomers were more interested in prophetic and medical astrology. Copernicus' discoveries were not widely popularised, even though the data collected and calculated by him was used (Dobrzycki et al. 1975, 185–213).

The route of the procession may have been as follows: Market Square (the actors passed through a triumphal gate there) – Grodzka Street (now Świętojańska) – the Royal Castle. The planets probably followed the main street of Old Warsaw, and at the same time the shortest route between the starting and finishing points mentioned by Gruneweg. When the artists reached the castle, "rietten sie etliche mal ume unde prengten vor des Königes augen" ["they went around it a few times, presenting themselves to the king's eyes"] (Gruneweg 2008, 604).[3] This was the end of the main part of the parade. Unofficially, and without costumes, the participants returned to the Market, probably along the same route.

The most important addressee of the procession was King Stephen Báthory, as its route and the narrative indicate. Gruneweg emphasises the connection of the parade with the seat of the monarch: it was the planets' destination, where

---

3 It is not clear exactly how they did this. Perhaps Gruneweg had the courtyard in front of the building in mind.

they presented themselves to the king during several laps. He was a viewer all the more valuable as he rarely visited Warsaw, avoiding his unloved wife Anna Jagiellon who lived in Warsaw permanently. In winter 1579/1580, Báthory visited in connection with a meeting of Parliament; he departed 18 February (Besala 1992, 290–292; Bogucka 2009, 141–145, 156; Wrede 2010, 102).

The second most important viewer was the Queen. She could observe the event from a house in the Market Square – according to Gruneweg, the planets walked through the Market next the Queen's chambers. Moreover, it can be assumed that the initiative to organise the show came from Queen Anna's court, although Gruneweg does not mention it explicitly. The king was at that time occupied with the war with Moscow and generally not so closely associated with Warsaw as his wife. The queen, who was quite well educated, willingly cooperated with the townspeople to develop the city and her seat at Jazdów (Bogucka 2009, 151–153). She also took care of her court, e.g., she arranged games and her courtiers' wedding festivities. Gruneweg writes that the parade took place the day after such a wedding at the Castle. Maybe the two events were linked. Additionally, by organizing performances, Anna supposedly tried to win her husband's heart.

The idea, which came from the queen's circle, was implemented with the participation of the townspeople. Collective undertakings were facilitated by the mixing of the circles of the city and the court, whose members sometimes lived with the townspeople. It can be assumed that the city was responsible for, *inter alia*, the actors, who joined the general city play after the parade. The townspeople also may have built the triumphal gate, located in the city's main square. The court chose the iconographic concept, using similar motifs to other court performances (more on this later). This group probably also provided the costumes – in another part of the notes, Gruneweg mentions that the townspeople and courtiers organized a masquerade together and borrowed costumes and jewels from the Castle for it (Gruneweg 2008: 2, 616).

Although it is not difficult to guess who the most important spectators were, the group of addressees was wider and very diverse. The audience connected people with extremely different cultural competences and education: illiterate plebeians, average educated middle townspeople (including Gruneweg) and courtiers originating from the middle nobility, as well as the elite members, who were educated, distinguished politically and economically, but rather not scientifically.

Unlearned viewers could enjoy the richness and technical effects. However, the performance could be fully appreciated just by a person who was at least a bit familiar with mythology and astrology. Therefore, the content of the pageant seems to be directed at the recipients in the middle, who had received an education but were not scholars. They would know the basics of iconography and the connections between the planets and the signs of the zodiac. However, they did not need to be specialists, neither in the humanities nor in science. They were not required to be able to calculate the positions of stars, to have detailed knowledge of their influence on Earth, or to be familiar with the symbolic and literary meanings of celestial bodies.

The spectacle was based on cultural rather than mathematical aspects of astrological knowledge, and its basic connection to natural philosophy, according to which the planets were moved by celestial spirits, identified with mythological deities or, in the Christianised version, angels. It was rooted in the iconography of planetary deities, identified with the Greco-Roman. On one hand, they are connected with the Renaissance fascination with antiquity, its knowledge and research, on the other hand, with hermetic thought, magical practices, and the popularity of divination. Therefore, the basis of astrology, which was known to an average educated nobleman or townsman, was connected with this knowledge of nature (celestial bodies) and culture (mythology and iconography), as well as spirituality, which had many faces during the Renaissance: from Christian belief in providence, through subtle Neoplatonic concepts, to popular beliefs in horoscopes, elements of magic and superstition. All these areas, nowadays separated, permeated each other and formed the backdrop to a carefree carnival parade.

4 Objectives of the procession

The first goal of the Warsaw parade was honouring the King and Queen. However, the spectacle did not seem to be aimed at ostentation of power or political ambitions, at least as far as the rulers of the state are concerned. Contrary to many similar court processions in Europe and the Commonwealth, the show was rather not ceremonial. Although it might have involved a wedding at court, it did not serve to celebrate the greatest people in the state, like the famous later ceremonies accompanying the wedding ceremony of Jan Zamoyski and Griselda Báthory (1583) or Sigismund III Vasa and Anne of Austria (1592). The planets pageant was rather a manifestation of Varsavians' capabilities, their hospitality and kindness. The inhabitants may have wanted to show their possibilities and celebrate the city's most important citizens: Báthory, a rare but most generous guest, and Warsaw's benefactress, Anna. Thanks to this play, the relationship between the court and the city was strengthened.

At the same time, the parade was simply meant to entertain. It had features of a joyful wedding and carnival party of a public nature. The goal was carefree, possibly egalitarian play, based on respectable, but above all fashionable, attractive and relatively well-known mythological themes. Recognising them provided intellectual pleasure, watching their splendour, aesthetic delight.

The spectacle probably did not contain any hidden allegorical meanings that could be noticed by a narrow circle of experts. This conjecture results not so much from Gruneweg's description, as from the specifics of the happy, dynamic carnival procession. Its main goal was to celebrate carnival joy, to dazzle the audience, to give them the pleasure of watching a spectacular show – not to contemplate very complex or hard to perceive significations. It was not of an academic nature, especially as it was not watched by scholars. Although some potential viewers from among the courtiers (like Jan Zamoyski or Andrzej Patrycy Nidecki) can be counted among them, the King, who was interested rather in practical activities and military tactics, or the Queen, who was not considered a *connoisseur* of literature, art or scientific curiosities, can hardly be numbered among them.

5 The parade against the background of theatrical traditions

The Warsaw procession was derived primarily from two types of performances: city carnival processions and court performances such as *trionfi* or *maszkary*, also often connected with the carnival. Situating them in the context of these two performance traditions allows us to see their complex interrelationships and sheds new light on the specificities of these circles. More visible is the ludic aspect of classical motifs and official ceremonies, as well as the cultural entanglements of seemingly simple and purely entertaining carnival games.

Greco-Roman heritage plays a double role here. Firstly, it is a treasury of attractive artistic motifs, especially planetary deities. On the one hand, they are allegorical figures – personified planets, rooted in the allegorical-cosmological Renaissance hermeneutics of myth. On the other hand, they belong to the already mentioned tradition according to which wandering stars are animated or closely related to the spirits that move them. Secondly, ancient literature, most fundamentally Ptolemy's *Almagest* and *Tetrabiblios*, provided astronomical knowledge, describing the movements of the planets and their interaction with the Earth.

In addition to the seven planets, the parade also featured other ancient elements belonging primarily to city carnival culture. They were the framework for the actual astronomical part of the show. It opened with three pairs of musicians playing on cornets "ein Heydnisch thönlein" (Gruneweg 2008, 603) ["a pagan melody"]. Their costumes, according to the author, were classically fashioned, extremely scanty, sleeveless, knee-length, green, decorated with shiny gold. They wore laurel wreaths on their heads and in the costume of Greco-Roman antiquity, they sustained the atmosphere of the main part. At the same time, they introduced a joyful, carefree mood, which corresponded especially with the final elements of the show.

The parade was closed by Bacchus, a figure associated with both carnival topics and the Greco-Roman tradition, as were the musicians at the beginning. He was an intermediate element between the planets, which belonged to the venerable classical repertoire, and the carnival play which the parade turned into. He was the ancient deity of wine and also a kind of personification of the carnival, known from the old Polish carnival comedies (a figure called *Mięsopust*, which was also the name of the last days of carnival) or Bruegel's *The Fight Between Carnival and Lent* (1559). Therefore, it is obvious that from the two types existing in the iconography, the creators of the parade chose not a beautiful youth but a dishevelled-looking overweight man. Bacchus was accompanied by his classical attributes: a wreath of vines, jugs, a goat (i.e., a small, suitably dressed horse). Behind him ran a herd of pigs, maybe less obvious but linked to Bacchus in the influential mythographic work written in the twelfth century and transcribed by Albericus (1681, 321). They can also be related to the god of the vine as a chthonic deity, and they are the opposite of seriousness and order, so they fit in with the Bacchic and Carnival symbolism.

When returning to the Market, Bacchus was followed by two elements typical for the carnival. These were a man dressed richly in Turkish – i.e., an oriental – style, analogous to, e.g., the Venetian *Flight of the Turk*, and the cart-ship, which frightened the audience with pyrotechnic effects ("vertrieb die gest mitt seinem schiessen unde rakett werffen"; Gruneweg 2008, 604). It probably rolled on wheels and derived from the medieval carnival ships-wagons, which were supposedly derived from the wagons of Dionysus, called *carrus navalis* (Dudzik 2005, 14–17). Shots and rockets can be related to fireworks display, which in the sixteenth century caused a lot of excitement as a new phenomenon (Dudzik 2005, 86).

In the perspective of the research on the astronomical topics in the planets parade, the reflection on the carnival forms serves rather to describe the function and generic context of the spectacle. The carnival motifs are less important, more the circumstances of the season. This is the direct context of the performance, which in its basic layer was an urban carnival play, watched by the Warsaw people. Moreover, one can appreciate the importance of the medieval and early Renaissance reception of antiquity – both the carnival elements and the Greco-Roman vision of the universe were filtered through it. Albericus' treatise influenced not only the costume of Bacchus, but also the appearance of the planets (more on this later).

The carnival was a typical time for various games, including those referring to the celestial bodies. It was no coincidence that the parade took place on the last Monday of the carnival,[4] a day after the wedding feast ("Aufn morgen nach essens"; Gruneweg 2008, 602) on Sunday, 14 February 1580 ("am Sontage tzu Fastnachten"; Gruneweg 2008, 602). The last days before Lent were a time of the most intense fun, and the parade of planets was probably part of it. Moreover, the wedding that Gruneweg mentions is also in keeping with the specifics of the carnival, when such celebrations were often organised (Dudzik 2005, 120; Zadrożyńska 2002, 74–75). Furthermore, the venue of the procession partly links the event to the carnival. It began and ended in the city's main square, a typical setting for carnival entertainment as well as for most public events.

Typical for the carnival is, of course, the very idea of a costume parade, very popular in early modern Polish culture. Specific to it was the lesser importance of the idea of the world turned upside down in the carnival, and the greater role of celebrating the joys of life and enjoying its charms during feasts, processions and dressing-up games (Bogucka 1994, 142–144; Zadrożyńska 2002, 75–77).

The carnival and situation of wedding were also typical circumstances of the court spectacles, typical in the Italian Renaissance culture: especially the allegorical scenes and triumphal processions (*trionfi*). Such spectacles were adapted in their own way and on a smaller scale in the Commonwealth. They originated from the oldest form of this type, Roman triumphs and later medieval processions and royal entrances (Komza 1995, 43–56). These shows added more splendour to the celebrations of important events, e.g., introductions of cult objects, funerals, birthdays, weddings, coronations, or visits by important guests. Although they were usually organised by the wealthy and staged in their honour, they took place in public spaces, thanks to which they were generally accessible – like the Warsaw parade. One obvious correspondence is the allegorical character of the spectacle, using astronomical/mythological motifs. The Italian scenes were often accompanied by a verbal commentary, which Gruneweg, however, does not mention.

A more distant analogy was the way the planets moved. In Warsaw, they were riding on horseback. It can be considered similar to Italian shows, where the figures moved on foot and on horseback and rode in carts. Although the form

---

4 According to further notes, 27 February was *Reminiscere* Sunday, the second Sunday in Lent (Gruneweg 2008, 604). Therefore, two Sundays earlier fell the last Sunday of the carnival. Moreover, Gruneweg's *Fastnachten* can mean Shrovetide, the last days before the Ash Wednesday (this is the reading proposed by the Polish translator of a fragment of the notes; Bues and Borg 2007, 165). However, it is not possible that 14 and 27 February are Sundays in the same year – one day is missing. Therefore, the topic of Gruneweg's chronology requires more research.

is quite modest in the Polish show, the visual recognition of main actors is clear – they tower over the audience.

Another element linking the planets procession with *trionfi* was the aforementioned triumphal gate in the Market Square. This form might have arisen from the architectural setting of Italian performances. According to Gruneweg (2008, 602), it was made of fir tree branches ("dannen lob"). It had the shape of an ogee arch. On top was a ball, and near the centre a second ball was placed. In the context of the astronomical theme of the procession, these details can relate to the spherical shape of the celestial spheres or planets. The evergreen branches, in turn, seem to be the northern equivalent of the Mediterranean laurel crowning the winners. The masqueraders ran to it from the Castle after the parade, after partially removing their costumes. However, it can be assumed that they had already passed through it earlier, when they "officially" walked from the Market Square to the Castle.

Moreover, the gate was located "vor des Königes fenster" (Gruneweg 2008, 602) ["before the royal window"], so it honoured the queen who was watching the parade from one of the houses in the Market. Then, both the starting and the arrival point of the planets were connected with the rulers: the first with Anna Jagiellon, the second with Stephen Báthory. As has been said, the royal couple were the most important spectators and the procession was in a sense to honour them, even if it was not ostentatiously representing their power. Moreover, Anna was a daughter of Bona Sforza, an Italian aristocrat, whom she followed in her ambition to be a patron of the city and court culture. The Queen can be seen as a direct link between the Warsaw parade and Italian culture.

Heavenly bodies or deities appeared quite often in courtly spectacles of a panegyric and allegorical nature, not only processions. It is worth mentioning examples of the Italian *La festa del paradiso*, which took place in 1490 (cf. Garai 2014), and the intermedio *L'armonia delle sfere* from the staging of the play *La pellegrina*, which took place in 1589 (cf. Osiecka-Samsonowicz 2006, 208–210), as well as aforesaid weddings in the Commonwealth. There may be dependencies between the Warsaw pageant and earlier performances. It may be an echo of *La festa del paradiso*, staged on 13 January (i.e., during the Carnival) 1490 in Milan, at the court of Ludovico Sforza, on the occasion of the wedding of his nephew Gian Galeazzo Sforza and Isabella of Aragon, the parents of queen Bona, and Anna Jagiellon's grandparents. It was one of the earlier and most famous Renaissance court allegorical performances and the procession described by Gruneweg might have been a distant reflection of this spectacle. The structure of both performances is based on the figures of the seven planets. It is difficult to argue convincingly that *La festa* had a direct impact on the Warsaw parade, even via Anna, but it could have influenced its creators indirectly.

The main idea of the staging was created by Duke Lodovico. The text, centred around the bride's praise, was composed by the court poet Bernardo Bellincioni. The visual and technical side was provided by Leonardo da Vinci. In the early part of the evening, there were dances and a masquerade, during which the young princess was paid homage by disguised "deputies" from various lands: Spain, Turkey (also here a figure from the Orient appeared), Poland, Hungary, Germany, and France. The tributes from the earthly kings were followed by the main part of the performance: the praises given by the lords of heaven. They were led by Jupiter, ruler of the Greco-Roman pantheon, who began to extol Isabella's virtues. He was surrounded by the other planets (and at the same time, the Olympian gods) with their attributes: Sun – Apollo, Mercury, Moon – Diana, Venus, Mars, and Saturn. The actors were arranged according to the hierarchy of the celestial spheres, and during the staging they "descended" from the title paradise to the earth to admire Izabela. The upper part of the decoration featured stars and zodiac signs in tondos. Therefore, despite all the dissimilarities, both the Milan and Warsaw spectacles had a comparable main structural axis: personified planets with attributes, in the traditional order, which were somehow accompanied by the signs of the zodiac.

Accounts of later events are also an interesting point of reference for the Warsaw march, as they provide a background for comparisons. One of them was the procession organized for the wedding of Jan Zamoyski and Griselda Báthory which took place three years after the planets parade (cf. Dubas-Urwanowicz 2011). The groom was Grand Chancellor of the Crown, Stephen Báthory's closest associate, obviously also known to the Queen. The march featured two deities associated with the planets, riding on carts, as in *trionfi*. The first was Saturn as an allegory of the golden age, preceded by symbols of time determined by days and nights. The second one was Jupiter, who appeared twice, in two carts, first as the thunderous lord of the atmospheric sky, then shown as the ruler of the pantheon and the universe, accompanied by zodiac signs. Also, Diana featured in the procession, riding on horseback, acting as the goddess of hunting rather than the Moon.

The juxtaposition with the wedding procession of 1583 shows the potential of Gruneweg's account and points to the potential uniqueness of the planets march in 1580. Unlike most of the well-known performances in the Commonwealth, it was practically fully devoted to the celestial phenomena. According to the current state of knowledge, deities identified with planets and astrological signs were popular but rather as part of spectacles, not as their main theme. As in the aforementioned example, they appeared separately, and astronomical symbolism of the figures was intertwined with other allegorical meanings taken from mythology.

Moreover, the Warsaw parade is unique in the detail of the astrological iconography. This can be seen, for example, in the description of Saturn. The author of the account of the march in 1583, published by Dubas-Urwanowicz (2011, 250), does not give too many details: "Siedział na wozie Saturnus z brodą siwą, kosę w ręce zbrojnej trzymał, która znamionowała wiek złoty za panowania króla STE-PHANA" ["Saturnus with a grey beard was sitting on a cart, holding a scythe in his armed hand, which marked the golden age during the reign of king STEPHAN" (my transl.)]. Gruneweg (2008, 603) writes in much more detail: "Saturnus hette ein grausam alt angeschitte, groe lange haer, in der rechtten hand eine sensse, in der lincken ein geschnitzt schreiende kindlein, welches er auffressen woltte, auch einen fues aufm steltzen" ["Saturn had an old, menacing face, long grey hair, a scythe in his right hand, in his left a carving of a screaming child that he intended to devour, and one leg in a stirrup"]. It is an open question to what extent the author has a phenomenal memory, aided by an ongoing diary, and to what extent he has been influenced by other images he has seen.

6 Astrological/astronomical contexts of the pageant

The Warsaw parade was also a field of transmission of mythological or astrological/astronomical motifs. As stated, they determined the structure of the procession. Its actual part had a quite clear order, based on a pattern: a richly dressed planet with attributes, in iconographically codified costume, riding a horse plus associated zodiac constellations, leading a horse. Sometimes additional characters were added, mainly for technical reasons, when a planetary house could not lead a mount. So, Mars – as a god of war – is accompanied by a pagan captive, and Sol, by a black man, representing the South, where the power of the Sun is most strongly manifested. They provide an additional oriental element. Venus is additionally assisted by her son Cupid and two nymphs, and the Moon by Day and Night, signs of the rhythm of time, which is determined by Luna.

Due to its entertaining character, the procession does not directly refer to the religious and symbolic meanings of mythology, mathematics, geometry, and cosmology. However, some symbolism is inevitable because the parade expresses a conviction about harmony and the hierarchical order of the cosmos. This was the metaphysical basis of the geocentric conception of the universe (Lewis 1964, 92–121), whose most famous mathematical model was presented by Ptolemy in the *Almagest*. This is manifested by showing the planets' sequence and the assignment of the planetary houses to them, as described in the canonical version by the same astronomer in *Tetrabiblos*.

The celestial bodies planets were arranged from top of the universe to bottom: from the highest, Saturn, to the lowest Moon, according to the geocentric model. The procession began with the planet orbiting in the most honourable place: the highest, closest to the seat of the God. Then the other inhabitants of the world above the Moon appeared one by one, less and less perfect, closer and closer to the Earth, which was tainted with changeability, evil and suffering. It was not built of ether, but of the four elements. It was incompatible with the order of the heavens, so was absent from the procession. Moreover, it was not a 'normal' planet visible in the sky, but at the centre, or rather the bottom, of the universe.

Such an order of the celestial bodies is traditional and appears, e.g., in the aforementioned *De deorum imaginibus libellus* (Albericus 1681, 301–310), one of the most famous and influential mythographic works of the time. It consists of twenty-three descriptions of deities and begins with the patrons of the planets, testifying to their importance in the reception of Greco-Roman beliefs in the Christian world. The iconography described there also appears in the Warsaw parade. The text influenced medieval astronomical representations in France and Germany, from where it may have reached Poland through pictorial representations and inspired some iconographical variants in the procession.

The planets in their proper order correspond to the various parts of the zodiac. According to Ptolemy (1940, 78–83 [I 17]), the connections between planets and their signs – the houses of the planets – influence the way they affect the sublunary world. The Sun and the Moon were assigned only one sign each (respectively diurnal Leo and nocturnal Cancer), but they were situated at the highest, closest to the zenith segment of the ecliptic. Thanks to that they most effectively generate heat and correspond to the high energy of the brightest celestial bodies. The remaining ten signs are split two by two between five planets, one diurnal and one nocturnal. The most distant and coolest Saturn was given the lowest signs, opposite Leo and Cancer: Aquarius and Capricorn. Jupiter was assigned the adjacent signs of Pisces and Sagittarius, Mars the next signs of Aries and Scorpio, Venus the next signs of Taurus and Libra, Mercury the next signs of Gemini and Virgo.

In the following chapters of Book I of the *Tetrabiblos*, Ptolemy develops a description of the relationships between the planets and their houses. These more complicated relations apparently were not taken into account by the creators of the Warsaw procession, nor by Gruneweg, who in his description of the signs did not follow the rhythm of day and night houses, or of male and female signs. As the table shows, no additional order is apparent:

Saturn	Aquarius	nocturnal	masculine
	Capricornus	diurnal	feminine
Jupiter	Pisces	nocturnal	feminine
	Saggitarius	diurnal	maskuline
Mars	Scorpius	diurnal	feminine
	Aries	nocturnal	maskuline
Sun	Leo	diurnal	maskuline
Venus	Taurus	nocturnal	feminine
	Libra	diurnal	maskuline
Mercury	Virgo	diurnal	feminine
	Gemini	nocturnal	maskuline
Moon	Cancer	nocturnal	feminine

7 Iconography of the parade

As yet, no direct source for the planets' representations in Warsaw has been identified. It was probably one of the graphic cycles of the seven planets,[5] appearing
on its own and as illustrations of books (e.g., Falimirz 1534, 50v–53r;[6] Spiczyński
1542, 196v–199r). Obviously, the appearance of the planets was influenced by literature, specifically ancient works and descriptions in mythographic treatises. Referring to the example of Albericus (1681, 301–310), clear parallels appear in the
images of Saturn, Jupiter, Mercury and Cupid, and partial ones in the case of
Mars and Diana. The Sun, identified with Apollo, differs from the type indicated
by the Gdańsk inhabitant, closer to Helios. Also Venus, shown in the text among
references to her birth from sea foam, does not match the Warsaw procession.

There were different traditions for dressing up the planets. They could wear oriental dress (e.g., Jupiter and Saturn in engravings attributed to Baccio Baldini, circa
1464), relating to Mesopotamian, Egyptian, and Arabian astronomical heritage. It
was connected also with Chaldean costume recalling Hermes Trismegistus and his
esoteric wisdom. This tradition weakened with the Renaissance, when classical cos-

---

5 For the purposes of this study, selected graphic cycles of the seven planets from fifteenth and
sixteenth century have been analysed. Most of them can be viewed in the virtual collection of the
British Museum. There are reproductions of the works, their data and bibliographic references.
For references to particular prints or cycles mentioned here, see the section 'Visual sources' in
the bibliography.
6 Pagination refers to the fifth treaty of part five.

tumes and noble nudity became popular (e.g., the series by Étienne Delaune, 1576, and by Jan Collaert II after Jan van der Straet, after 1587). Also images of planets in, or alluding to, contemporary costumes were created (e.g., a cycle by Monogrammist CG, representing the school of Lucas Cranach the Younger, 1550–1570).

In the Warsaw procession, different iconographic types intermingle. The costumes were not contemporary: Gruneweg links them to the classical manner and writes that they were "nach heidnischer weyse" (Gruneweg 2008, 602) ["in a pagan fashion"]. This most popular style seems to be more fashionable and attrac-

Figure 1: Hans Sebald Beham, *The Seven Planets: Saturn*, sixteenth century, engraving, the National Museum in Warsaw (source: https://cyfrowe.mnw.art.pl/pl/katalog/795740 [access: 29 January 2023]).

Figure 2: Virgil Solis, *The Seven Planets with Scenes from Ovid: Jupiter*, 1530–1562, engraving, the Scientific Library of the PAAS and the PAS in Cracow (source: http://pauart.pl/app/artwork?id=BGR_007196 [access: 29 January 2023]).

tive for the audience. However, the planets were not naked, due to reasons of morality and practicality, it being winter.

The deities could have no zodiac signs (e.g., the series after Johannes Wierix, 1579), stand in the company of their planetary houses (e.g., works by Hans Sebald Beham, circa 1539, and Hans Burgkmair the Elder, 1510–1560) or sit on a cart decorated with appropriate symbols (e.g., the series by Baldini, circa 1465). In the Warsaw parade, an intermediate solution was chosen: the planets did not have carts, but rode on horseback.

The description of Saturn in Gruneweg's notes does not differ from typical representations. It corresponds to Albericus' indications (1681, 301) even in the details: in his right hand, he holds a scythe (or sickle), in his left he holds a screaming child. He is a grey-haired old man with a menacing face. He has long hair, which may echo Albericus' remark about a long beard, as depicted by Hans Burgkmair (1510–1514), Hans Weiditz (1515–1536) or Heinrich Aldegrever (1533). Gruneweg also mentions a foot on a stilt or wooden limb: "einen fues aufm steltzen" (Gruneweg 2008, 603). It is absent in *De deorum imaginibus* . . . but can be seen in representations by Hans Sebald Beham (circa 1539; Figure 1) and the Monogrammist IB (1528).

Jupiter in the Warsaw procession was depicted as an earthly monarch: in royal attire, wearing a crown and with a sceptre in his hand. This is a less frequent type – more popular was a warrior, wearing a helmet (as in the images from the series by Beham circa 1539, the Monogrammist IB 1528 and Wierix 1579), or a thunderous lord of Olympus, wearing a crown but naked or half-naked (Jupiters by Weiditz 1515–1536 and Aldegrever 1533). Closer to Gruneweg's account are mixed images: a warrior wearing a crown (examples from cycles by Burgkmair 1510–1514, Delaune 1576 and in Virgil Solis' *The Seven Planets with Scenes from Ovid: Jupiter*, after Salomon Bernard, 1530–1562; Figure 2), but relatively the closest is the description of a majestic ruler on the throne, as Albericus portrays him.

Figure 3: Hans Sebald Beham, *The Seven Planets: Mars*, sixteenth century, engraving, the National Museum in Warsaw (source: https://cyfrowe.mnw.art.pl/pl/katalog/795742 [access: 29 January 2023]).

Figure 4: Virgil Solis, *The Seven Planets with Scenes from Ovid: Sol*, 1530–1562, engraving, the Scientific Library of the PAAS and the PAS in Cracow (source: http://pauart.pl/app/artwork?id=BGR_007194 [access: 29 January 2023]).

Gruneweg's description of Mars is quite typical. The deity wears full armour – his main attribute, mentioned in *De deorum imaginibus . . .* (Albericus 1681, 302–303) and shown in numerous engravings (Beham circa 1539, Baldini circa 1465, Delaune 1576, Burgkmair 1510–1514, Weiditz 1515–1536, Aldegrever 1533, Collaert after 1587, Wierix 1579, the Monogrammist IB 1528). He has also a cloak, which appears in images less frequently (Beham circa 1539, Figure 3; cp. Baldini circa 1465, Delaune 1576, Wierix 1579). The author notes that the material was decorated with stars, perhaps glittering with silver or gold thread, from which the costume was made. In the engravings, however, there are no stars on the coat. Only in Beham's work, a star shines behind the figure and its rays reach Mars and run in the same direction as the wind blowing away the hem of his coat. Gruneweg perhaps draws attention to the wealth of material because his master supplied the royal court mainly with fine materials, spices and wine (Walczak 1960, 61).

The image of the Sun in Warsaw combines references to the solar iconography of bright Helios and Apollo, and royal elements, linking the figure to Jupiter. In iconography, Sol, the ruler of the celestial bodies, wears a golden crown more often than the ruler of Olympus. However, instead of the crown, Gruneweg mentions a sceptre and the sphere of the world, corresponding with the royal apple. Sceptres often appear in representations (e.g., in the series by Beham circa 1539, Baldini circa 1465, Burgkmair 1510–1514, Weiditz 1515–1536, Aldegrever 1533, Wierix 1579, the Monogrammist IB 1528, Solis 1530–1562; Figure 4); the sphere is less frequent. Sol has it only in Weiditz's work, Delaune (1576) places it under the foot of the figure, and Burgkmair and Aldegrever depict him with an empty hand as if a sphere rested in it. Moreover, Solis (1530–1562) shows Sol with the radiant sun in his right hand.

In Gruneweg's account, the royal splendour of the golden star emphasises his brilliance: "Die Sonne riet in eittelem klarem golde [. . .]. Das angesichtte war eittel goltt unde mit langen stremen" (Gruneweg 2008, 603) ["Sun rode in burning, luminous gold. [. . .] The face was like burning gold, with long rays"]. Gruneweg must

Figure 5: Hans Sebald Beham, *The Seven Planets: Venus*, sixteenth century, engraving, the National Museum in Warsaw (source: https://cyfrowe.mnw.art.pl/pl/katalog/795745 [access: 29 January 2023]).

have been sensitive to light effects: he draws attention to them for the second time, after the starry cloak of Mars. As the author himself notes, the radial nimbus refers to the iconography (Weiditz 1515–1536, Burgkmair 1510–1514, Baldini circa 1465). Thus, he points out that the procession reproduced the images of planetary deities known to the audience and was a kind of sequence of vivid allegories.

Venus, the most beautiful of the planets – as Gruneweg states – could not be naked, according to the most popular iconographic type. However, the goddess was also depicted in a dress, longer, revealing only the lower part of her legs (Mono-grammist IB 1528), or shorter (Beham circa 1539, Figure 5; cp. Baldini circa 1465). It seems to correspond with the Warsaw parade, where probably, she was scantily clad ("fiele blos"; Gruneweg 2008, 603). She had some kind of ornaments on arms and feet, quite mysterious, because absent from the iconography, regardless of the meaning of the word used in the original (the equivalent of the contemporary German *Band*: "arme unde fusse mitt guldenen benderen unde der hals mitt ketten getziert"; Gruneweg 2008, 603). It can be understood broadly: as sashes (e.g., an element of a cloak), as more discreet ribbons (e.g., a part of footwear) or bonds (ties or shackles). It is even difficult to determine the material: most meanings of the word point to golden fabric, but the association with bonds may also point to hoops of metal – or simply jewellery, golden bracelets. The latter reading is supported by the fact that Venus wore "köstliche geschmeide" ["precious jewellery"] (Gruneweg 2008, 603). The creators of the procession foreground the richness of her costume, complementing her natural beauty – unlike in the traditional iconography, where the ornament is usually limited to an ethereal robe or hair ornaments (Weiditz 1515–1536, Beham circa 1539, Burgkmair 1510–1514, Baldini circa 1465). Numerous jewels appear only in an engraving by Monogrammist CG (1550–1570).

In the Warsaw pageant, Venus was depicted as the mistress of human hearts, the superior of Cupid who accompanies her. He was shown with his typical attributes: bow, arrows and a blindfold. The goddess holds a burning heart – a symbolic effect of her and her son's actions, a motif known from engravings (Beham circa 1539, Delaune 1576, Wierix 1515–1536, Monographist IB 1528, Monogrammist CG 1550–1570). Interestingly, Venus' nudity is less exposed in these depictions, which corresponds to her costume from the procession.

Gruneweg was much less interested in Mercury (Figure 6). He limited himself to mentioning his traditional attributes, well known from iconography: serpent sceptre, caduceus, and the wings at his helmet and sandals: he was "allenthalbe gefluegeltt" ["all winged"] (Gruneweg 2008, 603).

The Moon (Figure 7), the last 'planet', shining with silver, attracted Gruneweg's attention with her brilliance, like Sun and Mars. Gruneweg notes that the figure is dressed as a woman, which he did not record in his description of Venus. Female Moon is not obvious to the author, probably due to the grammati-

Figure 6: Hans Sebald Beham, *The Seven Planets: Mercury*, sixteenth century, engraving, the National Museum in Warsaw (source: https://cyfrowe.mnw.art.pl/pl/katalog/795746 [access: 29 January 2023]).

cal form of the word *der Mon* (now German *der Mond*), which is a masculine noun. The same is true in Gruneweg's second language, Polish: *Księżyc*. However, in Romance languages – including Latin *luna* – it is feminine, which was reflected in mythology. Thus, the Greco-Roman lunar deities are also female, including Diana and Selene. Luna, like Venus, was scantily clad ("fiele blos"; Gruneweg 2008, 603). In accordance with the iconography of the patroness of hunting (Baldini circa 1465, Burgkmair 1510–1514, Weiditz 1515–1536), she wore a short dress, held a bow in her hand, and her head was adorned with the sickle of the Moon.

Figure 7: Hans Sebald Beham, *The Seven Planets: Luna*, sixteenth century, engraving, the National Museum in Warsaw (source: https://cyfrowe.mnw.art.pl/pl/katalog/795747 [access: 29 January 2023]).

8 Conclusions

The 1580 Warsaw planets parade is an interesting example of the adaptation of the astrological/astronomical tradition for the purposes of entertainment. It was organised by Warsaw townspeople and Anna Jagiellon's circle, and combined elements of carnival city and court shows. The main theme of the performance was celestial bodies, which distinguishes it from other shows in the Commonwealth, and links it to European spectacles such as *La festa del paradiso*. The procession used cultural aspects of cosmic knowledge: the association of planets and mythological figures and their iconography. There was also a clear reference to astrological and cosmological knowledge: the order of the planets in the geocentric model of the cosmos and their relationship to the signs of the zodiac. Although the performance was primarily for play, it contains some inevitable symbolic meanings, inextricably linked to the vision of nature evoked in it, the elements of which are hierarchically ordered and connected by a complicated network of interrelations.

The event was reported in a very vivid and detailed way by Martin Gruneweg. His approach to the show was ambivalent. His description is dominated by a friendly interest in beautiful spectacle and admiration for the impressive costumes, the quality of which he could appreciate as an apprentice merchant. He seems fascinated by the lighting effects on fabrics. However, after years, from the perspective of a Dominican father, the author tries to extract moral teaching from the description of carefree game. Then, he emphasises the "paganism" of the procession and the scanty clothes of some figures. It is a kind of assessment: the world of antiquity attracts, but it remains an alien element, less perfect than Christian culture. Its attractiveness is the world's vanity. It seems more dangerous for the soul than fortune-telling astrology, which Gruneweg is apparently not concerned about. The parade, on the one hand, pleases the eyes and improves one's mood; on the other hand, it is a manifestation of the illusory splendour of this world. Not coincidentally, the actors try to show off in front of the audience: "woltte einer joe vor dem anderen gesehen warden" (Gruneweg 2008, 603) ["each one wanted to be seen before the others"]. Their fight for primacy differs from the harmonious movement of compatible planets, each of which knows its place in the cosmos.

Gruneweg ends his description of the planets with a moralistic commentary about wastefulness and seeking worldly goods instead of eternal ones. However, after the moralistic remark, he smoothly moves on to the description of Bacchus, returning to the carefree carnival atmosphere and delighting in its comical elements.

Bibliography

Visual sources

Aldegrever, Heinrich. *The Seven Planets*. Engravings (1533): https://www.britishmuseum.org/collec tion/object/P_E-4-354; https://www.britishmuseum.org/collection/object/P_E-4-353; https://www.britishmuseum.org/collection/object/P_E-4-352; https://www.britishmuseum.org/ collection/object/P_E-4-351; https://www.britishmuseum.org/collection/object/P_E-4-348; https://www.britishmuseum.org/collection/object/P_E-4-350; https://www.britishmuseum.org/ collection/object/P_E-4-349 London: The British Museum (29 January 2023).

Baldini, Baccio (attributed). *Planets*. Engravings (circa 1464): https://www.britishmuseum.org/collec tion/object/P_1845-0825-474; https://www.britishmuseum.org/collection/object/P_1845-0825- 469; https://www.britishmuseum.org/collection/object/P_1845-0825-473; https://www.britishmu seum.org/collection/object/P_1845-0825-470; https://www.britishmuseum.org/collection/ob ject/P_1845-0825-467; https://www.britishmuseum.org/collection/object/P_1845-0825-475; https://www.britishmuseum.org/collection/object/P_1845-0825-476 London: The British Mu seum (29 January 2023).

Beham, Hans Sebald. *The Seven Planets*. Engravings (circa 1539): https://www.britishmuseum.org/col lection/object/P_Gg-4C-38; https://www.britishmuseum.org/collection/object/P_Gg-4C-39; https://www.britishmuseum.org/collection/object/P_Gg-4C-40; https://www.britishmuseum. org/collection/object/P_Gg-4C-41; https://www.britishmuseum.org/collection/object/P_Gg-4C- 42; https://www.britishmuseum.org/collection/object/P_Gg-4C-43; https://www.britishmuseum. org/collection/object/P_Gg-4C-44 London: The British Museum (29 January 2023).

Burgkmair, Hans the Elder. *The Seven Planets*. Woodcuts (1510–1514): https://www.britishmuseum. org/collection/object/P_1845-0809-641-647 London: The British Museum, (29 January 2023).

Collaert, Jan II [after Jan van der Straet]. *Septem planetae*. Engravings (after 1587): https://www.british museum.org/collection/object/P_1873-0809-459; https://www.britishmuseum.org/collection/ob ject/P_1957-0413-30; https://www.britishmuseum.org/collection/object/P_1957-0413-33; https://www.britishmuseum.org/collection/object/P_1957-0413-31; https://www.britishmuseum. org/collection/object/P_1873-0809-458; https://www.britishmuseum.org/collection/object/P_ 1957-0413-32; https://www.britishmuseum.org/collection/object/P_1957-0413-34 London: The British Museum (29 January 2023).

Delaune, Étienne. *Planets*. Engravings (1576): https://www.britishmuseum.org/collection/object/P_ Gg-4D-133; https://www.britishmuseum.org/collection/object/P_Gg-4D-133-A; https://www.brit ishmuseum.org/collection/object/P_Gg-4D-132; https://www.britishmuseum.org/collection/ob ject/P_Gg-4D-132-A; https://www.britishmuseum.org/collection/object/P_Gg-4D-134; https://www.britishmuseum.org/collection/object/P_Gg-4D-134-A London: The British Museum (29 January 2023).

Monogrammist CG. *The Seven Planets*. Woodcuts (1550–1570): https://www.britishmuseum.org/collec tion/object/P_1904-0519-4; https://www.britishmuseum.org/collection/object/P_1904-0519-5; https://www.britishmuseum.org/collection/object/P_1904-0519-6; https://www.britishmuseum. org/collection/object/P_1904-0519-7; https://www.britishmuseum.org/collection/object/P_1904- 0519-8; https://www.britishmuseum.org/collection/object/P_1904-0519-9; https://www.british museum.org/collection/object/P_1904-0519-10 London: The British Museum (29 January 2023).

Monogrammist IB, *The Seven Planets*. Engravings (1528): https://www.britishmuseum.org/collection/ object/P_1874-0808-174; https://www.britishmuseum.org/collection/object/P_1845-0809-1252;

https://www.britishmuseum.org/collection/object/P_1837-0616-173; https://www.britishmu
seum.org/collection/object/P_1837-0616-174; https://www.britishmuseum.org/collection/object/
P_1837-0616-175; https://www.britishmuseum.org/collection/object/P_1837-0616-176;
https://www.britishmuseum.org/collection/object/P_1837-0616-177 London: The British Museum
(29 January 2023).
Solis, Virgil [after Salomon Bernard]. *The Seven Planets with Scenes from Ovid.* Engravings (1530–1562):
http://pauart.pl/app/artwork?id=BGR_007196; http://pauart.pl/app/artwork?id=BGR_007195;
http://pauart.pl/app/artwork?id=BGR_007194; http://pauart.pl/app/artwork?id=BGR_007193;
http://pauart.pl/app/artwork?id=BGR_007192; http://pauart.pl/app/artwork?id=BGR_007191.
Cracow: The Scientific Library of the PAAS and the PAS (29 January 2023).
Wierix, Johannes (after). *The Seven Planets.* Engravings (1579): https://www.britishmuseum.org/collec
tion/object/P_Gg-4Q-1; https://www.britishmuseum.org/collection/object/P_Gg-4Q-2;
https://www.britishmuseum.org/collection/object/P_Gg-4Q-3; https://www.britishmuseum.org/
collection/object/P_Gg-4Q-4; https://www.britishmuseum.org/collection/object/P_Gg-4Q-5;
https://www.britishmuseum.org/collection/object/P_Gg-4Q-7; https://www.britishmuseum.org/
collection/object/P_1973-U-518 London, the British Museum (29 January 2023).
Weiditz, Hans. *The Seven Planets.* Woodcuts (1515–1536):https://www.britishmuseum.org/collection/
object/P_1927-0614-304-310 London: The British Museum (29 January 2023).

Textual sources

Albericus. "De deorum imaginibus libellus." *Mythographorum Latinorum.* Vol. 2. Ed. Thomas Muncker.
Amsterdam: Joannis à Someren, 1681. 301–330.
Besala, Jerzy. *Stefan Batory.* Warsaw: PIW, 1992.
Bogucka, Maria. *Anna Jagiellonka.* Wrocław: Ossolineum, 2009.
Bogucka, Maria. *Staropolskie obyczaje w XVI–XVIII wieku.* Warsaw: PIW: 1994.
Bues, Almut, and Eliza Borg. "Warszawa z lat 1579–1582 w zapiskach gdańszczanina Martina
Grunewega." *Rocznik Warszawski* 35 (2007): 151–178.
Crăciun, Maria. "Conversion in the Confessional Age." *Martin Gruneweg (1562–after 1615): A European
Way of Life.* Ed. Almut Bues. Wiesbaden: Harrassowitz Verlag, 2009. 241–262.
Dobrzycki, Jerzy, Mieczysław Markowski, and Tadeusz Przypkowski. *Historia astronomii w Polsce.* Vol. 1.
Ed. Eugeniusz Rybka. Wrocław: Ossolineum, 1975.
Dubas-Urwanowicz, Ewa. "Wesele Jana Zamoyskiego z Gryzeldą Batorówną." *Białostockie Teki
Historyczne* 9 (2011): 237–251.
Dudzik, Wojciech. *Karnawały w kulturze.* Warsaw: Sic!, 2005.
Falimirz, Stefan. *O ziołach i o mocy ich [. . .].* [Cracow:] Florian Ungler, 1534.
Garai, Luca. *La festa del paradiso di Leonardo da Vinci: Quattro ipotesi per la construzione di una
macchina teatrale per Ludovico il Moro.* Milan: La vita felice, 2014.
Garin, Eugenio. *Lo zodiaco della vita. La polemica sull'astrologia dal Trecento al Cinquecento.* Bari: Editori
Laterza, 1982.
Gruneweg, Martin. *Die Aufzeichnungen des Dominikaners Martin Gruneweg (1562–ca. 1618) über seine
Familie in Danzig, seine Handelsreisen in Osteuropa und sein Klosterleben in Polen.* Vol. 2. Ed. Almut
Bues. Wiesbaden: Harrassowitz Verlag, 2008.
Hapanowicz, Piotr. "Kraków w zapiskach dominikanina Martina Grunewega (1562–ca. 1618)." *Rocznik
Krakowski* 80 (2014): 39–56.

Komza, Małgorzata. *Żywe obrazy. Między sceną, obrazem i książką*. Wrocław: Wydawnictwo UWr, 1995.

Lewis, C.S. *The Discarded Image: An Introduction to Medieval and Renaissance Literature*. Cambridge: Cambridge University Press, 1964.

Osiecka-Samsonowicz, Hanna. "Gwiazdy, planety i znaki zodiaku w scenografiach włoskich spektakli teatralnych w końcu XVI i na początku XVII wieku." *Poezja i astronomia*. Eds. Bogdan Burdziej and Grażyna Halkiewicz-Sojak. Toruń: Wydawnictwo UMK, 2006. 207–214.

Ptolemaeus, Claudius. *Tetrabiblos*. Ed. and transl. F.E. Robbins. Cambridge, MA: Harvard University Press, 1940.

Spiczyński, Hieronim. *O ziołach tutecznych i zamorskich [. . .]*. Cracow: Florian Ungler, 1542.

Walczak, Ryszard. "Pamiętniki Marcina Grunewega." *Studia Źródłoznawcze* 5 (1960): 57–77.

Wrede, Marek. *Itinerarium króla Stefana Batorego: 1576–1586*. Warsaw: DiG, 2010.

Zadrożyńska, Anna. *Świętowania polskie*. Warsaw: Twój Styl, 2002.

Gábor Kutrovátz

Anatomical Descriptions in Star Catalogues: Ptolemy, Brahe, Halley, and Hevelius

Abstract: This chapter focuses on the textual descriptions of stellar objects used in the influential star catalogues of Ptolemy, Brahe, Halley, and Hevelius. The primary purpose of this work is to provide a quantitative survey of these descriptions, and especially anatomical descriptions, i.e., ones that describe their objects by spatially relating them to body-parts of constellation figures. After identifying the functional components of descriptions, some basic statistical results are presented concerning the absolute and relative occurrences of these components and of the terms that comprise them. These results highlight both the differences and the similarities between the use of descriptions in the studied catalogues, and they allow us to form hypotheses about the individual preferences of the authors, the changes resulted by historical and cultural distance, and the general function of constellation lore within astronomy. Identifying stars as structural or anatomical elements of constellations served a cognitive purpose: it provided the means to memorize and recognize stellar objects, hence providing them with identity.

1 Introduction

The astronomical tradition from antiquity to the early modern era recorded visible stars in catalogues. For each star, these catalogues provide both a textual description and numerical data. The data – coordinates of celestial position and degree of brightness – have been studied with great care by scientists and historians of science.[1] However, only a minor interest has been devoted to the descriptions employed to identify stellar objects.

This chapter focuses on these descriptions. It is an extension of my earlier analysis of the language of descriptions in Ptolemy's catalogue (Kutrovátz 2022),[2]

---

[1] For a concise history and evaluation of such approaches in case of Ptolemy's catalogue, cf. Graßhoff (1990). For analyses of two of the Latin catalogues, cf., e.g., Verbunt and van Gent (2010a, 2010b).

[2] Therefore, this chapter contains some passages that are common with that paper, where the results and conclusions are the same, but tries to keep them to a minimum.

here compared to three important catalogues published in the seventeenth century. The primary purpose of this work is to provide a quantitative survey of these descriptions, and especially anatomical descriptions, i.e., ones that describe their objects by spatially relating them to body-parts of constellation figures. After identifying the functional components of descriptions, I present, in a table format, some basic statistical results concerning the absolute and relative occurrences of these components, and of the terms that comprise them. These results highlight both the differences and the similarities between the use of descriptions in the studied catalogues, and allow us to form hypotheses about the individual preferences of the authors, the changes resulted by historical and cultural distance, and the general function of constellation lore within astronomy.

2 The catalogues

The oldest extant star catalogue[3] is found in Books VII and VIII of Ptolemy's groundbreaking astronomical work, the *Syntaxis mathematica* or *Almagestum.*[4] Ptolemy lists 1,028 entries representing 1,022 individual stars,[5] grouped into 48 constellations. The constellations are not definite areas of the sky, like in contemporary astronomy, but rather imaginary figures representing and conceptualizing star patterns. The identity of individual stars is usually defined by their position within these constellation figures, i.e., on (or near) which part of their figure they are found. This information is given by the textual descriptions, two to 16 words of length and around six on average, that accompany the numerical data of coordinates and 'magnitude' (degree of brightness). For example, the first entry defines its object as "[t]he star on the end of the tail" of Ursa Minor, the Little Bear (Toomer 1984, 341).

Most constellation figures are associated with Greek mythology. There are human figures (even if often deities) like Andromeda, Perseus, Orion, or Virgo, then there are animal figures like the Bull, the Lion, or the Crab, and there are some more irregular "monsters" like the Pegasus, the Centaur, or the Archer (also a centaur). In addition to these animate (live) figures, there are nine further figures representing inanimate objects, like a Crown, a river (Eridanus), or a Lyre.

---

3 For a detailed summary of the history of catalogues and constellation lore, cf. Ridpath (2018). Kanas (2007) is useful for relating this lore to its manifestation in historical star maps.

4 The analysis below is based on the standard edition of Heiberg (1903), 38–169.

5 Three stars are listed twice as belonging to two constellations simultaneously, and three items are treated as nebulous objects rather than stars.

All of these constellations are among the ones that are in official use today, except for the ship Argo that was divided into three smaller constellations in the eighteenth century.

While almost all star catalogues of the medieval and renaissance periods were close copies of Ptolemy's list, perhaps slightly updating or modifying details, the seventeenth century saw the publication of novel star lists. The first one we study is that of Tycho Brahe, mostly comprised in the last decades of the sixteenth century, but eventually published entirely, and with amendments, by Kepler (1627). This had approximately the same quantity of stars as Ptolemy's list, but the set of listed stars is not the same: Tycho observed more stars in the northern constellations and less in the southern ones. Also, he omitted some southern constellations he could not see,[6] but added two more to the northern list.

Halley's catalogue (1679) was specifically devoted to constellations in the southern hemisphere, observed from the island of Saint Helena. It was meant to be a supplement to Tycho's catalogue. While it partly overlapped with the regions surveyed by the previous two, its primary focus was the part of the sky surrounding the southern celestial pole, invisible from the geographical latitudes hosting the classical astronomical tradition. It contained 14 novel constellations, mostly introduced at the beginning of the century, but altogether far less stars than its predecessors.

Hevelius (1690) re-surveyed the northern and equatorial regions of the sky (as seen from Gdansk in Northern Europe), listing half as many more stars as Ptolemy or Tycho, and introducing eleven new constellations. Its main point of reference was Tycho's catalogue, but it often deviated from it. It was supplemented with a star atlas, published posthumously together with the catalogue.

The survey presented in this chapter is not based on the original editions, but on Baily (1843),[7] which publishes them in one volume. While I found this publication to be faithful to the originals in terms of textual descriptions, it also includes a lot of useful editorial notes concerning individual stars, as well as the so-called Baily numbering, which is handy for references to individual items.

---

6 Partly due to the difference in geographical latitude (Ptolemy lived in Alexandria, while Tycho in Denmark), and partly due to position shifts caused by the precession of the Equinoxes.
7 Except for Ptolemy, as noted above.

3 Anatomical descriptions and references

Taking Ptolemy's text as the paradigm, descriptions fall into three kinds according to their relations to other descriptions. Independent descriptions define their objects in a self-sufficient way (see [P44] below),[8] while linked descriptions define groups of objects via interconnected textual entries. Members of linked descriptions are either premier descriptions (see [P15]) or subsequent descriptions (see [P16]). The examples below are pure cases, i.e., without any further complication:

[P44] "ὁ ἐπὶ τῆς γλώσσης" ["The star on the tongue"][9]

[P15] "τῶν ἐν τῷ τραχήλῳ β ὁ προηγούμενος" ["The more advanced of the two stars in the neck"]

[P16] "ὁ ἑπόμενος αὐτῶν" ["The one to the rear"]

As discussed in section 4.6, boundaries between these kinds would become less clear in the Latin catalogues of the early modern era.

These examples testify that textual descriptions tend to define their objects by relating them to body-parts of the constellation figures. In this chapter these are called anatomical descriptions. Not all descriptions are anatomical, however. On the one hand, we have inanimate descriptions where stars are identified with objects rather than body-parts. This is the case not only with stars in inanimate constellations without anatomical parts (e.g. [He1014], "In aplustri" ["On the stern" (of the ship Argo)]), but also with stars in animate constellations that represent artefacts held or worn by the figures (such as weapons, e.g. [T870], "Quæ in manubrio ensis" ["Which <is> on the sword's hilt" (in Orion)]). Also, star positions are occasionally defined with respect to previously identified stellar objects, rather than via direct correspondence to parts of the figures; these are relational descriptions (e.g. [T347], "Illa quæ supra hanc" ["That which <is> above this"], referring to the previous entry). Finally, stars are sometimes described as constituting abstract geometrical figures instead of fictional entities; these are geometrical descriptions (e.g. [He640], "Apex rhombi occidentalis" ["The western apex of the rhombus"]). Moreover, these methods of identification are not mutually exclusive: a star may by defined in multiple ways simultaneously, e.g., by providing both an anatomical and a relational reference (see below).

---

8 Throughout the chapter, when referring to individual descriptions, the numbers in square brackets provide the Baily-ID's of the objects, which is simply the ordinal number of the entry in the catalogue as introduced in Baily (1843), prefixed here by 'P' for Ptolemy, 'T' for Tycho, 'Ha' for Halley, and 'He' for Hevelius.

9 English translations given here (and only here) are cited from Toomer (1984), 342–344. All other translations are mine.

The general structure of anatomical descriptions, based on the analysis of Ptolemy's text in Kutrovátz (2022), is as follows: {Description} = {Subject} + {Reference} + {Additional}, where {Subject} is the subject phrase denoting the stellar object, {Reference} is the referential phrase providing an external identification, and {Additional} consists of additional components. This formula expresses a structure at the level of semantical functions, rather than the level of syntax, even if many components and relations below tend to correlate with syntactical units and structures.

For us, the most important part is the referential phrase. In general, it can have four components: {Reference} = {Preposition} + {Term} + {Specification} + {Qualification}, where {Preposition} and {Term} are the core components, while {Specification} and {Qualification} are optional. For a full example, observe [He253], "In extremitate dextri pedis" ["On the end of the right leg"]. Here, {Preposition} is 'in', expressing the spatial relation between the star and the associated body-part. {Term} is 'pes', naming the body-part. The latter is always present, since by definition, an anatomical description is one that has an anatomical term in its referential phrase, while {Preposition} tends to become less ubiquitous in the Latin catalogues, as seen below. {Specification} specifies which one of the entities denoted by the referential term is referred to, if more than one exists (e.g., hands, legs); here it is 'dexter' ["right"]. {Qualification} qualifies which part of the entity denoted by the referential term is referred to, if it is given (quite rarely, as we shall see), in this case it is 'extremitas' ["end"].

Descriptions with only one {Term} contain single references, while those with two (or more) instances of {Term} contain double (or multiple) references. While the majority of descriptions provide a single reference, there are various reasons for using double references. For example, the description may define its object with respect to two separate body-parts simultaneously, e.g. [He256], "Sub sinistro brachio, in latere" ["Below the left arm, on the side"]. Or else it may provide both a larger body-part and a more specific area, e.g. [T878], "Quæ in surâ sinistri pedis" ["Which <is> on the calf of the left leg"].[10]

Next we need to make a distinction between explicit and implicit references. The latter are used in subsequent descriptions. For example, in the case of [P15]

---

[10] While an anatomical term is taken to be expressed by a single word, and this results in intuitively clear instances most of the time, there are genitive structures that complicate matters. For example, 'ancon alae' (e.g., [T200], [Ha254]) is an expression where 'ancon' is meant to be the bend of the wing, while when used separately it means elbow (e.g., [T336]) – and the same ambiguity applies to Ptolemy's 'ἀγκών' (e.g., [P164] vs. [P186]). While this composite expression refers to one specific body region, it contains two terms with anatomical meaning, and therefore is treated in this study as a double anatomical reference.

and [P16] cited above, it is clear that the referential phrase of [P15] also applies to the object of [P16], so that the latter is also on the neck, even if it is not explicitly stated in the corresponding description, but rather implied by the context. Since the purpose of linked descriptions is to avoid repetition in cases where multiple stars are associated with the same part of the figure, it should be assumed that all components of the premier referential phrase have an implicit occurrence in the subsequent description, including the {Term}, the {Preposition}, and – if they are given – the {Specification} and the {Qualification} as well.

On top of the information contained in the referential phrase, further details are occasionally provided in the additional parts of the description: {Additional} = {Property} + {Name} + {Relational} + {Objectual} + {Geometrical}. The relevant parts are italicized in the following examples. {Property} is a distinguished property of the denoted star, such as in [Ha252], "In capite *lucida*" ["The *bright* <star> on the head"], with "*lucida*" constituting {Property}. {Name} is the name of the star, e.g. [Ha86], "In ore. *Fomalhaut*" ["In the mouth. *Fomalhaut*"]. The remaining three components are additional references (in addition to the anatomical one) of non-anatomical kinds: relational (e.g. [He1360], "In sinistro genu *sub Palilicio*" ["Below the left knee, *under Palilicium*" (i.e., Aldebaran)]), objectual (e.g. [He483], "In sinistro brachio, *ad sceptrum*" ["In the left arm, *by the sceptre*"]), and geometrical (e.g. [He511], "In *quadrato* pectoris præcedens borealis" ["The leading [= western] <and> northern on the *rectangle* of the chest"]).

The subject phrase is plain for independent descriptions. In Ptolemy's catalogue, this kind of {Subject} is simply the masculine definite article in the singular nominative case: 'ὁ'. The article belongs to the term 'ἀστήρ' ["star"] which is, however, absent from the text, making the syntax elliptical, see, e.g., [P44] above.[11] Latin catalogues are more flexible since several forms coexist, with the common characteristic that the term for 'star' remains absent as well. Some examples from Tycho's catalogue, with the subject phrase italicized: [T78] "In ore" ["In the mouth"] – here the subject phrase actually disappears; [T80] "*Quæ* ad genam" ["*Which* <is> by the face"] – note that since Latin does not have articles, here the gender (feminine) of the relative pronoun indicates that it corresponds to 'stella'; [T77] "*Quæ est* in lingua" ["*Which is* on the tongue"] – here the copula is present (a rare case); [T117] "*Illa quæ* in humeris" ["*That which* <is> on the shoulders"] – here a demonstrative pronoun occurs. Similar solutions are used by Halley and Hevelius too – with a tendency to omit {Subject} altogether –, implying that language use in their descriptions was less formal or technical than in Ptolemy's text.

---

11 Such elliptical formulae were not only characteristic of common vernacular, but also of technical, e.g., mathematical, texts. For the latter, cf. Netz (1999), 127–167.

For premier descriptions, {Subject} contains the part {Quantity} referring to the quantity of the group members, plus at least one selective term, {Selective}, telling which one of the members is referred to, usually by stating its position within the group. E.g. [P15], "τῶν ἐν τῷ τραχήλῳ β ὁ προηγούμενος" ["*Of the two* <stars> on the neck, *the leading*" (i.e., western)], with elements of the subject phrase italicied: "τῶν [. . .] β" ["of the two (. . .)"] is {Quantity}, and "ὁ προηγούμενος" ["the leading"] is {Selective}. In the corresponding subsequent description, [P16], "ὁ ἑπόμενος αὐτῶν" ["The following [= eastern] of them"], "αὐτῶν" ["of them"] refers back to the {Quantity} of the premier description (and it always does so in this form), while {Selective} ("ὁ ἑπόμενος" ["the following"]) is a counterpart of the {Selective} in the premier description. In the Latin catalogues, similarly to individual descriptions, instances of {Subject} are often less formal and procedural than in Ptolemy's Greek, as seen below.

4 Quantitative analysis

4.1 General features

Equipped with all the definitions and distinctions introduced in the previous section, we are now ready to perform a quantitative analysis on the catalogues. Before we begin, it is essential to note that descriptions of the same stars in different catalogues often depend on each other. For many stars in the Latin catalogues, identification is determined by the tradition stemming from Ptolemy's list.[12] But that does not mean that individual variances are excluded. Even among very prominent stars, the star Antares is the heart of the Scorpion for Tycho ([T695]), Halley ([Ha9]), and Hevelius ([He1235]), but it is the middle of the body for Ptolemy ([P553]). The star Regulus is the heart of the Lion for Ptolemy ([P469]) and Tycho ([T598]), but for Hevelius it is only named without any anatomical or other kind of descriptive iden-

---

12 That is not to say that Ptolemy was the one to actually found this tradition. On the one hand he had Hipparchus' catalogue at his disposal, but since the latter is not extant, we, together with early modern astronomers, have no direct way to consult it. On the other hand, anatomical identification of stars was a central feature of the descriptive tradition represented by Aratus, Eratosthenes, and Hyginus, cf. Condos (1997) and Hard (2015). However, these texts do not identify stars individually as Ptolemy does, but rather list the number of stars representing specific body-parts for each constellation, and thus it is not possible to find 'descriptions' the way we do in this chapter, so we cannot provide a straightforward comparison between these texts and the catalogues we study. A number of anatomical references to individual stars can be found in Babylonian texts (Watson and Horowitz 2011, 187–205), but that period falls outside the scope of this study.

tification ([He852]), and it is missing from Halley's catalogue of the southern constellations.

We expect variations for the following reasons:

1. As we saw, each of these catalogues introduces some constellations unknown to the previous ones, while other constellations are omitted.
2. The number of stars in any specific constellation varies, meaning that the set of stars they work with differs.
3. Even when they discuss the same star, individual choices in the descriptions can occur, as seen above.
4. Anatomical identifications depend on the way the authors envisioned constellation figures, which in turn depends on which star charts they consulted.

However, it would be a daunting task to try and identify all individual entries across the catalogues, owing to the inaccuracy of coordinates (especially for the fainter stars that comprise the majority), and this chapter does not incorporate such considerations.[13] It is sufficient for our purposes to assume that, while there is an obvious convergence between the catalogues imposed by the tradition, this tradition is alive and it allows for considerable variations and differences that leave plenty of room for a comparative quantitative analysis.

Some general features of the catalogues are summarized in Table 1. Since this chapter focuses on anatomical descriptions (AD), i.e., descriptions with at least one anatomical term in their referential phrase, it is useful to note not only the size of the overall stellar population,[14] but also the number of entries belonging to animate constellation figures ('AC entry'), since those are the ones where ADs are primarily expected.[15] As we see, the proportion of AD to the overall stellar

---

13 Of course, modern editions usually attempt to identify the entries by assigning designations of standard catalogues in use today. However, the number of remaining uncertainties is always far from negligible, cf., e.g., the numerous notes in Baily (1843). For a heroic attempt to identify stars across many historical catalogues, cf. the useful book by Helmut Werner and Felix Schmeidler (1986).

14 The number of relevant entries in Hevelius is open to interpretation: there are 23 entries where he borrows the descriptions from Tycho's catalogue, but is unable to find the corresponding star. While these are "empty" catalogue entries not representing objects with assigned numerical data, nevertheless they contain descriptions – even if only repetitions of Tycho's descriptions – and, therefore, are included in this survey.

15 However, some descriptions are refined by giving the stellar position with respect to a nearby part of an adjacent constellation figure as well, although the number of these cases is relatively small (17 in Ptolemy, 26 in Tycho, 9 in Halley, 35 in Hevelius, all within the corpus of anatomical descriptions). For this reason, inanimate constellations can occasionally contain stars with anatomical descriptions.

population varies between 55% (Halley) and 76% (Hevelius), indicating differences between the authors in terms of favour for anatomical identifications. However, the possibilities for AD depend on the set of constellation figures, and if we narrow down our scope to animate constellation figures then we find more coherence, with the proportion of AD to this reduced population ranging between 73% (Tycho) and 85% (Halley). This means that all these astronomers had a similar tendency to apply ADs as the default means of identification, but without persistence to apply the method as a rigid rule.

Table 1: General features of the catalogues. Entry: № of objects listed. Constellation: № of constellations. Anim. const's (AC): № of live/animate constellation figures. AC entry: № of stellar objects contained by all AC. Anat. desc. (AD): № of anatomical descriptions. AD to entries: proportion of AD to all descriptions or entries. AD to AC entries: proportion of AD to AC entries. Anat. ref. (AR): № of anatomical references. Multiple AR: № of multiple anatomical references. Multiple AR to AD: proportion of multiple AR to AD.

GENERAL	Ptolemy	Tycho	Halley	Hevelius
Entry	1028	1005	341	1564
Constellations	48	46	23	56
Anim. const's (AC)	39	37	17	44
AC entry	895	904	222	1417
Anat. desc. (AD)	721	658	189	1182
AD to entries	70.14%	65.47%	55.43%	75.58%
AD to AC entries	80.56%	72.79%	85.14%	83.42%
Anat. ref. (AR)	785	702	202	1336
Multiple AR	64	41	13	147
Multiple AR to AD	8.88%	6.23%	6.87%	12.44%

We also indicate the number of anatomical references (AR).[16] The quantity of multiple ARs seems to indicate an attitude toward descriptive precision: multiple references make descriptions both more cumbersome and more informative. While the proportion of multiple AR to AD is slightly lower in Tycho and Halley than in Ptolemy, suggesting a decreasing preference for descriptive identifications, the difference is too small to be significant. However, Hevelius is definitely more meticulous than the others, not only because the proportion of double ARs is notably higher, but also because he is the only one to employ triple ARs (albeit

---

16 In some descriptions, Hevelius contrasts his anatomical identification with Tycho's different anatomical identification of the same star. Since Tycho's cited anatomical references are not accepted by Hevelius, the latter references are ignored in this survey.

only in 5 cases altogether). While this may be attributed to the fact that he has the highest number of entries and, therefore, he needs to deal with numerous faint stars[17] that are difficult to describe (as opposed to Halley with his relatively small population), the overall impression based on several further results below is that Hevelius' pedantry in descriptions is a personal preference. His descriptive precision is also shown by the many cases where he separates references within the same description with the terms 'sive,' 'seu' and 'vel' ['or'], suggesting that he offers alternative references for better identification.

4.2 Terms

When identifying anatomical terms, a number of methodological decisions had to be made. For example, on top of terms signifying actual body-parts, I chose to include a few terms referring to pieces of garment where the general positions are obvious enough to be seen as anatomical specifications, e.g. 'tiara' (*tiara*) instead of 'head' or 'forehead,' or 'belt' (*cingulum*) instead of waist. On the other hand, terms denoting non-regular parts of a body (such as the cut-offs of half-figures like Taurus or Pegasus) were ignored, similarly to 'curves' (*flexura*), like in Draco or Serpens, without stable anatomical identity, in contrast to 'segments' (*spondylus*) of Scorpio. Moreover, reference instances are considered only when terms are applied in their anatomical sense: for example, 'side' (*latus*) counts when referring to parts of organic bodies but does not count when referring to geometrical objects.

Table 2 summarizes the results. While the number of different anatomical terms (i.e., ignoring in how many instances they are used) is basically the same for Ptolemy, Tycho, and Hevelius (despite the latter dealing with substantially more entries), Halley employs less than half as many terms. The reason is primarily not the significantly smaller size of his sample,[18] but mainly two factors: first, that the new constellations he surveys are mostly animals (such as birds), instead of human figures with plenty of anatomical details, and second, that these new

---

[17] While this paper does not discuss the numerical data contained in the catalogues, let us note that Hevelius is the first to attribute a brightness of 7 magnitudes to some stars (11 cases, plus 9 cases of 6.5 magnitudes), while previous catalogues had 6 as the value for the faintest objects.

[18] For a graph displaying the aggregated sum of already introduced terms against the number of descriptions in Ptolemy's catalogue, cf. Kutrovátz (2022), 99, Figure 2. This shows that the number of different terms is far from being directly proportional to the number of descriptions. By the point where Ptolemy's catalogue reaches the number of descriptions equal to Halley's overall population, he already introduced 49 different terms, in contrast with Halley's 34.

Table 2: Anatomical terms in the catalogues. Instances: № of overall occurrences of anatomical terms (equal to the № or AR in Table 1). Explicit: № of overall occurrences of explicitly stated anatomical terms. Explicit prop.: proportion of explicit occurrences to all instances. Kinds: № of different anatomical terms used. Instance/Term: № of times one term is used on average. 10 most frequent: the most frequently used anatomical terms, with the № of occurrences in parentheses.

TERM	Ptolemy	Tycho	Halley	Hevelius
Instances	785	702	202	1336
Explicit	553	557	147	1313
Explicit prop.	70.45%	79.34%	72.78%	98.28%
Kinds	73	73	34	76
Instance/Term	10.75	9.62	5.94	17.58
10 most frequent	κεφαλή (53)	pes (69)	cauda (46)	pes (156)
	πούς (48)	caput (67)	ala (24)	cauda (128)
	οὐρά (47)	genu (43)	caput (17)	caput (69)
	γόνυ (42)	cauda (41)	collum (14)	collum (63)
	ὦμος (41)	manus (38)	pes (12)	ala (58)
	τράχηλος (28)	humerus (32)	femur (8)	manus (49)
	κέρας (23)	collum (31)	cor (7)	humerus (48)
	μηρός (23)	ala (30)	spondylus (7)	genu (47)
	πτέρυξ (21)	dorsum (27)	dorsum (6)	femur (44)
	χηλή (21)	brachium (26)	frons (6)	venter (41)

constellations lack the mythological context that would highlight specific body-parts. The former reason is also evidenced by the overwhelming dominance of tails over other body-parts.

Also note that there are significant differences in the proportion of explicit instances to the overall cases (i.e., to the number of AR in general). While Ptolemy, Tycho, and Halley display a tendency to be explicit (in 70–80% of the cases) but apply implicit references quite regularly, Hevelius keeps implicit references to a minimum. Implicit references are tied to subsequent descriptions, and this result may seem to indicate that Hevelius avoids this kind of description, but as we shall see in section 4.6, this is not the case. Also, this difference can partly be explained by the following consideration: Previous catalogues arrange their stars according to proximity, following the outline of the constellation figure along an arbitrary path (often from head to foot/tail) and therefore juxtaposing entries that represent the same body-parts. Hevelius, on the other hand, mixes this figural ordering with the governing practice of listing stars of an asterism in descending order of brightness. As a result, nearby stars often become separated in the list, making it impossible for the elements of the referential phrase to remain implicit and implied by the directly previous entries. However, even when members of linked descriptions are next to each other in the catalogue, Hevelius prefers to explicitly repeat every detail, which is another sign of his descriptive pedantry.

Figure 1 illustrates the proportion of references to body regions in the four catalogues, combined from all animate constellation figures, presented on a human figure with some animal parts. Note that

1. several localizations are questionable, both because attributing animals' anatomical terms to human body-parts are sometimes problematic, and because of the ambiguity regarding the precise meanings of terms (especially in the Greek).[19]
2. Multiple different terms can refer to the same region (e.g. synonyms).
3. Some terms were ignored as not fitting any of the depicted regions.

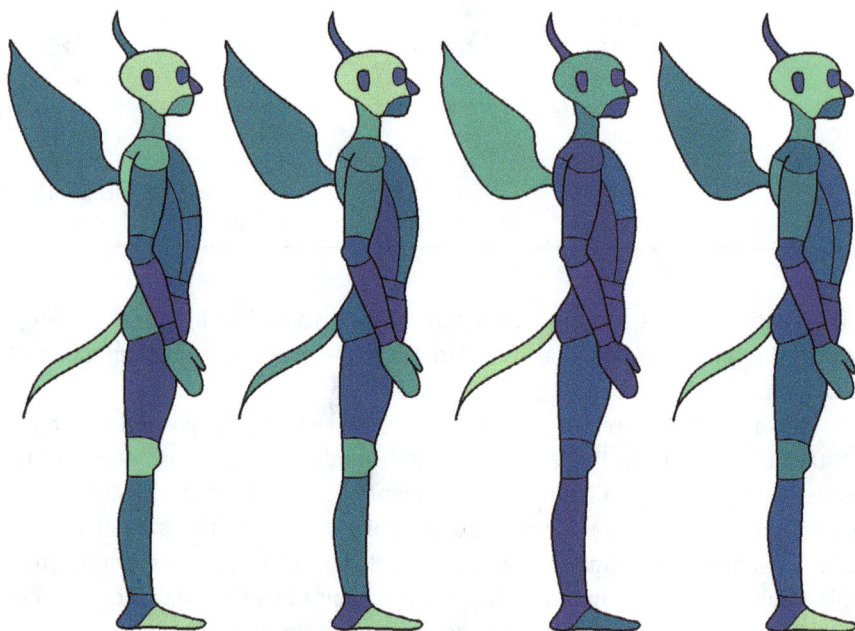

Figure 1: The proportion of references to body regions in the catalogues. From left to right: Ptolemy, Tycho, Halley, Hevelius. The colour scale ranges between pure blue (0 references) and pure green (maximum number of references). Note that the scale is slightly more sensitive toward the lower end of the spectrum.

---

[19] Ancient Greek anatomical vocabulary, found to be largely flexible and variable, is surveyed by Lloyd (1983), 149–167. His conclusions are recited and summarized by Netz (1999), 121–122.

4.3 Prepositions

While Ptolemy almost always attaches a preposition to an anatomical term, the Latin catalogues are more flexible. One way to avoid a preposition is to simply name the body-part (in the nominative case) as the identity of the star, e.g. [T420], "Caput" ["Head"]. Another way is to use a genitive structure, e.g. [T424], "Lucida colli" ["The bright <star> of the neck"]; or [Ha38], "Sequens capitis" ["The following [= eastern] <star> of the head"]. We can see in Table 3 that these irregular solutions become rather frequent, reducing the quantity of actual prepositions relative to anatomical references. Also, quite unsurprisingly, we can observe that the proportion of explicit prepositions is very similar to (but slightly lower than) the proportion of explicit terms, as terms and prepositions usually stick together.

Prepositions indicate the spatial relation between the star and the mentioned body-part. In the majority of cases stars fall on the named body-part, as expressed by prepositions that serve as exact markers: "ἐν" ["in"] and "ἐπὶ" ["on"] in Greek, and "in" in Latin. However, some stars are only nearby the named anatomical region, as expressed by prepositions – e.g. 'above,' 'below,' 'between' etc. – that are called approximate markers here. The latter cases usually involve fainter stars, when body-parts are already defined by more prominent objects. Note that nominative or genitive references are also exact markers, identifying the star with the body-part. Table 3 presents the proportion of exact markers (i.e., the actual prepositions 'in' and 'on,' plus nominative and genitive markers) to the number of anatomical references. There is an interesting but understandable correlation between the size of the stellar population and the proportion of exact markers: the larger the population, the smaller this proportion, since the more stars we have, the more frequently they fall nearby (and not exactly on) the already defined figures.

The relatively wide range of preposition kinds (i.e., different prepositions) is somewhat unexpected. Not all of them are prepositions in the grammatical sense – e.g., Tycho uses expressions such as "præcedit" ["precedes"] ([T241]) or "contingit" ["touches"] ([T357]), which have the same descriptive function here as actual prepositions –, but most of them are. This suggests that the authors intended to provide accurate relations, especially Ptolemy and Tycho. Hevelius, however, seems less meticulous in this matter.

A possible objection to the above results could be that it is worth making a distinction between constellation stars and unformed stars (Greek: 'ἀμόρφωτοι', Latin: 'informata'). For many constellations, Ptolemy first lists those stars that actually constitute the constellation figure, and then separates those that are unformed in the sense that they are near, but not actually on, the figure. Out of the 48 constellations, 22 have unformed companions, totalling 108 in number (10.5% of all the objects). One could expect that exact markers are present in constella-

Table 3: Prepositions in the catalogues. Null – nominative: № of instances with a nominative case instead of a preposition. Null – genitive: № of instances with a genitive case instead of a preposition. Instances: № of actual prepositions. Explicit: № of explicitly stated prepositions. Explicit prop.: proportion of explicit prepositions to all actual prepositions. Exact markers: № of instances suggesting a spatial coincidence of body-parts and stars. Exact m. to AR: proportion of exact markers to all anatomical references. Kinds: № of different prepositions used. 6 most frequent: the most frequently used prepositions, with the № of occurrences in parentheses.

PREPOSITION	Ptolemy	Tycho	Halley	Hevelius
Null – nominative	3	22	12	99
Null – genitive	7	76	31	149
Instances	774	604	159	1078
Explicit	531	473	117	1052
Explicit prop.	68.60%	78.31%	73.58%	97.59%
Exact markers	619	564	172	964
Exact m. to AR	78.85%	80.77%	85.15%	72.16%
Kinds	25	24	12	14
6 most frequent	ἐν (327)	in (466)	in (129)	in (716)
	ἐπὶ (282)	ad (28)	ante (5)	ad (135)
	ὑπὸ (51)	inter (26)	sub (5)	sub (89)
	ὑπὲρ (23)	super (20)	ad (3)	super (60)
	μεταξύ (12)	sub (15)	inter (3)	inter (45)
	κατά (11)	infra (13)	super (3)	infra (14)

tion figure descriptions, while approximate markers in unformed star descriptions. However, we ignore this distinction for two reasons. First, Latin catalogues do not follow Ptolemy in this practice. They occasionally use the term "informis" in individual descriptions, but no systematic distinction is made, and the difference seems to lose its significance as the relative number of cases drops.[20] Second, even in Ptolemy, while he does use anatomical descriptions for unformed stars frequently,[21] discounting these cases from the corpus of descriptions yields 86.7% for the proportion of exact markers among the restricted population of constellation stars, which is not radically higher than the proportion of 78.9% among the overall list of anatomical references.

---

20 71 in Tycho, 10 in Halley, and 29 in Hevelius, all within a sample restricted to anatomical descriptions.

21 While the proportion of AD is altogether 70.1%, the proportion of AD in the limited corpus of unformed entries is still 63.0%.

4.4 Specifications and qualifications

Table 4 shows the results concerning specifications, i.e., which one of the possible multiple body-parts denoted by the anatomical term is specifically referred to. As we see, Ptolemy, Tycho, and Hevelius use specifications in around half of all anatomical references, while for Halley, this proportion is less than a third. This difference is probably not primarily due to individual variations in descriptive precision, but due to the set of constellation figures and named body-parts the authors deal with. Classical constellations are often human figures, viewed from the front, and bilateral organs such as arms, hands, legs, and feet, play a prominent role in these figures. It is not surprising that for all the catalogues, the specifications 'left' and 'right' are predominant. The southern constellations surveyed by Halley, on the other hand, are mostly animals viewed typically from the side, with less anatomical detail, leaving a narrower room for specification. Admittedly, limbs of quadrupedal animals require more specification so that the cases of double specifications often correspond to these cases (like 'right front leg'), but, e.g., tails (very frequent in Halley's case, and for animals in general) are unique, and the same goes for necks that are often long enough to host multiple stars.

Table 4: Specifications in the catalogues. Single spec.: № of anatomical descriptions with one specification. Double spec.: № of anatomical descriptions with two specifications. Sum total: overall № of specifications. Explicit: № of explicitly stated specifications. Explicit prop.: proportion of explicit to all specifications. Freq. per AR: proportion of the overall № of specifications to the № of anatomical references. Kinds: № of different specifications used. Most frequent: the most frequently used specifications, with the № of occurrences in parentheses.

SPECIFICATION	Ptolemy	Tycho	Halley	Hevelius
Single spec.	337	277	49	515
Double spec.	29	24	5	64
Sum total	395	325	59	643
Explicit	282	246	45	620
Explicit prop.	71.39%	75.69%	76.27%	96.42%
Freq. per AR	50.32%	46.30%	29.21%	48.13%
Kinds	17	16	14	18
Most frequent	ἀριστερὸς (139)	dextra (113)	dextra (26)	sinistra (237)
	δεξιὸς (139)	sinistra (110)	sinistra (12)	dextra (220)
	βόρειος (32)	superior (19)	laeva (5)	australis (44)
	νότιος (25)	borealis (16)	prior (4)	posterior (41)
	ὀπίσθιος (21)	praecedens (14)		borealis (38)
	ἐμπρόσθινος (15)	australis (13)		anterior (19)

For the same reasons, qualifications (i.e., which part of the named organ the star falls on) are significantly more frequent in Halley than the others, see Table 5. In case of larger body-parts with multiple references, one needs to qualify which portion represents the star in question. These cases are usually handled by using linked descriptions (see below), but qualifications can also serve this purpose. Also, even when there is only one star coinciding with the named large body-part, and linked descriptions are not feasible, it is informative to qualify whether the star is, e.g., on the tip or the root of the long tail. Moreover, these latter cases are more typical when relatively few stars are listed (like in the case of Halley) than when we have many small stars and, therefore, individual body-parts are more likely to be represented by several of them (e.g., Hevelius), prompting linked descriptions with selective terms instead of qualifications.

Table 5: Qualifications in the catalogues. Instances.: № of anatomical references with qualification. Explicit: № of explicitly stated qualifications. Explicit prop.: proportion of explicit to all qualifications. Freq. per AR: proportion of the overall № of qualifications to the № of anatomical references. Kinds: № of different qualifications used. Most frequent: the most frequently used qualifications, with the № of occurrences in parentheses.

QUALIFICATION	Ptolemy	Tycho	Halley	Hevelius
Instances	92	49	36	111
Explicit	75	44	26	111
Explicit prop.	81.52%	89.80%	72.22%	100%
Freq. per AR	11.72%	6.98%	17.82%	8.31%
Kinds	11	16	6	16
Most frequent	ἄκρα (51)	medium (10)	medium (11)	eductio (29)
	ἔκφυσις (21)	extremitas (6)	eductio (10)	medium (25)
	μέσος (13)	extremus (6)	extremus (10)	extremitas (17)
		eductio (4)	summus (2)	cuspis (11)

For specifications, it would be tempting to address the question, in which instances they are meaningful at all, and restrict the above results to those body-parts where specifications are required. In Ptolemy's catalogue, out of the 73 anatomical term-kinds, only around 30 are such that can have multiple referents within the figure. But, on the one hand, not all instances are clear: e.g., for animals with a side view, it is not obvious whether distinguishing between 'left' and 'right' limbs is always feasible. On the other hand, specifications are often missing even for those body-parts that have multiple specimens. In Ptolemy, the percentage of specifications within the corpus of those references that have several possible referents is still only around 60%, showing that this tradition is somewhat negligent with regards to specifications. Surely, the discursive context often provides clues: when following the outline of a constellation figure, if a 'right knee' is mentioned in one description, then it is sufficient to refer to the 'shin' in the next one without the specification, but my impression is that there remains a considerable body of instances where not even the context is informative.

All in all, we can conclude that specifications, when theoretically required, are more often used than not. Qualifications on the other hand are quite rare, indicating a tendency to identify stars with body-parts, as opposed to placing stars on specific locations of body-parts.

4.5 Additional components

Descriptive components outside the referential and subject phrases are altogether rare. Properties (i.e., of the star itself) are shown in Table 6. We need to add some notes here. First, only Ptolemy employs descriptions with two properties mentioned, in 5 cases, while the others are always content with naming one property, if at all. Second, variants of the same term can appear (e.g., "lucida," "lucidior," "lucidissima" ["bright," "brighter," "brightest," respectively]), and while these are counted separately in the number of term kinds, they are lumped together in the results showing frequencies (see the instances where only the roots are given). Third, the meaning of implicit instances requires some explanation. In the cases so far, implicit instances were made possible by assuming that the referential phrase of the premier description simultaneously applies to the object(s) of the subsequent description(s) as well. However, properties belong to individual stars, and are not part of the anatomical reference. Nevertheless, there are cases where properties are attributed to groups of objects via linked descriptions, e.g. [T133], "Præcedens trium obscurarum in pede sinistro" ["The leading of the three obscure <stars> in the left leg"], where 'obscure' obviously applies to the two subsequent descriptions as well.

While Tycho has both the highest number of instances and the widest range of expression (implying a personal preference), but all in all, the number of instances is always small (less than 10% of descriptions), and this is not surprising. One the one hand, while providing information about the brightness or faintness of the object is informative (see the relatively high frequency of the corresponding terms), it is nevertheless redundant since that information is already contained in the data section for each star (by giving its magnitude). On the other hand, stars appear as light points, and they rarely have any notable property, except for some scarce cases when they seem nebulous, double, or slightly reddish.

Table 6: Properties of stars in the catalogues. Instances.: № of descriptions stating properties. Explicit: № of explicitly stated properties. Freq. per AD: proportion of the № of AD with properties to the overall № of AD. Kinds: № of different terms for properties. Most frequent: the most frequently attributed properties, with the № of occurrences in parentheses.

PROPERTY	Ptolemy	Tycho	Halley	Hevelius
Instances	46	65	4	55
Explicit	37	55	4	52
Freq. per AD	6.38%	9.88%	2.12%	4.65%
Kinds	7	15	3	9
Most frequent (root)	λαμπρὸς (19)	lucid . . . (29)	nebula (2)	parv . . . (20)
	ὑπόκιρρος (6)	parv . . . (12)	clara (1)	lucida (12)
	νεφελοειδὴς (6)	obscur . . . (9)	lucida (1)	nebulosa (11)

For other additional components, see Table 7. Providing names for stars is scarce. Let us mention, first, that Tycho provides five descriptions offering two alternative names each, so the total number of names is 31 in his case.[22] Second, Hevelius tends to omit descriptions entirely when they can be substituted with star names, and that is why his anatomical descriptions contain significantly less names than the whole catalogue.

Table 7: Further additional components in descriptions. Name (AD): № of AD containing individual star names. Name (all): № of all descriptions containing individual star names. Relational: № of AD containing an additional relational reference. Geometrical: № of AD containing an additional geometrical reference. Objectual: № of AD containing an additional inanimate reference.

ADDITIONAL	Ptolemy	Tycho	Halley	Hevelius
Name (AD)	11	22	2	7
Name (all)	15	26	4	24
Relational	85	46	6	34
Geometrical	56	28	0	15
Objectual	9	6	5	36

Cases where additional non-anatomical references are provided, on top of the anatomical ones within the same description, are also relatively rare (not more than a few per cent for each type). The authors are mostly content with offering

---

22 These include "Nova anni 1572" and "Nova anni 1600" as well, which are not strictly names but serve a similar purpose.

anatomical references, and complementing these with further clues for identification is rather uncommon.

4.6 Selective terms

Finally, we examine selective terms that are used to differentiate between the members of groups covered by linked descriptions. This is complicated by the fact that identifying these cases is more obvious in Ptolemy than the others. First, Ptolemy almost always uses the {Quantity} phrase (in premier descriptions) to indicate how many members the group has. Second, subsequent descriptions in his catalogue are always juxtaposed with the premier ones. Third, descriptive phrases in subsequent descriptions nearly always remain implicit, guiding the identification of groups. For the Latin authors however, indication of the quantity of group members is quite rare, implicit references are treated more casually, and group members are often separated by other objects in the list (especially for Hevelius). I found a significant number of instances where a description seems to have a form of a linked description, but other members of the group cannot be found (or have to be assumed based on vague circumstantial clues). As the structure of descriptions becomes less formulaic, the precise function of some terms is more flexible, and the boundary between instances of {Selective} on the one hand, and instances of {Property}, {Qualification}, and {Relational} on the other, becomes unclear (see Tycho's large number of selective term-kinds in Table 8).

Table 8: Selective terms in the catalogues. 1 selective term: № of AD containing one selective term. 2 selective terms: № of AD containing two selective terms. Instances SUM: overall № of selective terms in AD. Freq. per AD: Proportion of AD with selective term(s) to AD in general. Kinds: № of different selective terms. Most frequent: the most frequently used selective terms, with the № of occurrences in parentheses.

SELECTIVE	Ptolemy	Tycho	Halley	Hevelius
1 selective term	337	295	96	514
2 selective terms	11	54	0	142
Instances SUM	359	403	96	798
Freq. per AD	49.79%	55.89%	50.79%	67.51%
Kinds	11	42	20	29
5 most frequent (root)	ἑπόμενος (87)	praecedens (65)	borea . . . (20)	praecedens (156)
	προηγούμενος (80)	sequens (64)	austr . . . (19)	sequens (133)
	νοτιώτερος (74)	austr . . . (62)	praecedens (16)	borea . . . (78)
	βορειότερος (71)	borea . . . (56)	sequens (15)	austr . . . (72)
	μέσος (40)	media (31)	media (8)	superior (65)

Despite these uncertainties, the results shown in Table 8 illustrate variations in discursive preferences. Again, Hevelius is the most precise as he applies selective terms the most often (two thirds of descriptions, as opposed to half in case of the others). We may assume that it is because he deals with the highest number of stars, and therefore it is more frequent to have multiple stars falling on the same body-part. But the fact that Halley's proportion is so close to those of Ptolemy and Tycho, despite listing third as many stars, seems to contradict this explanation. The use of double selective terms also indicates intent on descriptive care, and here Hevelius is clearly superior to the others, especially Halley who never uses two selective terms in one description. However, a direct correspondence between the number of these instances and the overall number of stellar objects is made implausible by the huge difference between Ptolemy (eleven objects) and Tycho (54 objects).

Altogether, selective terms are used frequently, providing descriptive information about the objects, and especially about their relative locations.

5 Conclusions

While Ptolemy's language use is so formal and, to some degree, ritual that it can be considered technical jargon, the Latin catalogues are more flexible and narrative in the way descriptions are articulated. Ptolemy's descriptions conform to a generic formula that we took as our starting point, but this formula falls apart in the later catalogues, especially due to the variability (and even complete disappearance) of the subject phrase. I do not see any grounds for blaming this divergence on the difference between Greek and Latin, but rather, it seems to depend on the different eras and scientific cultures. However, this survey seems insufficient to support any detailed explanation for this contrast.

Apart from the formality of language use, there are further differences between individual catalogues in terms of the tools they prefer to employ to describe objects. The range of {Term} and the frequency of {Specification} are influenced by the set of constellation figures, while the percentage of exact markers and {Qualification} partly depend on the overall number of stars. But the correlations are never strict, and more often than not, variations seem to have more to do with individual preferences than with the properties of the surveyed stellar population.

Despite the differences, textual descriptions in general, and anatomical descriptions in particular, remained an integral part of star catalogues up to the eighteenth century. These descriptions indicate a genuine relation between body-parts and corresponding stars, which is evidenced by a number of common discursive features. First, anatomical descriptions predominantly contain one anatomical reference each, i.e., they define their object by identifying it with one body-part only, without specifying its position by relating it to another body-part or providing a smaller part of the larger one mentioned. Then, the great majority of {Preposition} are exact markers, as opposed to approximate ones, showing that body-parts and star positions tend to coincide when possible. Furthermore, body-parts with structural significance (such as head, knee, foot, shoulder, or hand) are overrepresented as opposed to otherwise larger areas of the body (belly, back, chest). Also, details about the position within the named body-part, i.e., {Qualification}, are rarely given. Moreover, textual descriptions seldom provide further clues (i.e., other than anatomical, such as {Property} or {Name}) to single out their object, and additional identifications are also infrequent. On the other hand, the relative abundance of {Specification} indicates that the correspondence is meant to be unambiguous: when applicable, there is a tendency to specify which one of the multiple body-parts is referenced in the description. In sum, language use testifies to constellation figures as the conceptual framework to identify stars.

Modern readers are often inclined to view constellations as accidental cultural garnish, products of excessive imagination, or remnants of lingering superstition distinct from astronomy proper. But the results presented in this chapter indicate that constellations were functional within classical astronomy. Identifying stars as structural or anatomical elements of constellations served a cognitive purpose: it provided the means to memorize and recognize stellar objects, and that is how stars were endowed with identity.

Bibliography

Baily, Francis. *The Catalogues of Ptolemy, Ulugh Beigh, Tycho Brahe, Halley, Hevelius* [. . .]. London: Memoirs of the Royal Astronomical Society, 1843.

Condos, Theony. *Star Myths of the Greeks and Romans. A Sourcebook*. Grand Rapids: Phanes Press, 1997.

Graßhoff, Gerd. *The History of Ptolemy's Star Catalogue*. New York: Springer, 1990.

Halley, Edmond. *Catalogus stellarum australium, sive supplementum catalogi Tychonici* [. . .]. London: Thomas James, 1679.

Hard, Robin. *Eratosthenes and Hyginus Constellation Myths, with Aratus's 'Phaenomena'*. Oxford: Oxford University Press, 2015.

Heiberg, Johan Ludvig. *Claudii Ptolemaei opera quae exstant omnia. Volumen I: Sytaxis mathematica. Pars II: Libros VII–XIII continens.* Leipzig: D.B. Teubner, 1903.

Hevelius, Johannes. *Prodromus astronomiæ cum catalogo fixarum, & firmamentum Sobiescianum.* Gdansk: Johannes-Zacharia Stolli, 1690.

Kanas, Nick. *Star Maps. History, Artistry, and Cartography*. Dordrecht: Springer, 2007.

Kepler, Johannes. *Tabula Rudolphina, quibus astronomica scientia, temporum longiquitate collapsa Restauratio continetur*. Frankfurt: Jonas Saurius, 1627.

Kutrovátz, Gábor. "Anatomical Identifications of Stars: Textual Descriptions in Ptolemy's Star Catalogue." *Studies in History and Philosophy of Science* 91 (2022): 94–102.

Lloyd, Geoffrey E. R. *Science, Folklore, and Ideology. Studies in the Life Sciences in Ancient Greece*. Cambridge: Cambridge University Press, 1983.

Netz, Reviel. *The Shaping of Deduction in Greek Mathematics*. Cambridge: Cambridge University Press, 1999.

Ridpath, Ian. *Star Tales: Revised and Expanded Edition*. Cambridge: Lutterworth Press, 2018.

Toomer, Gerald James. *Ptolemy's Almagest*. London: Duckworth, and New York: Springer, 1984.

Verbunt, F., and R. H. van Gent. "Three Editions of the Star Catalogue of Tycho Brahe: Machine-Readable Versions and Comparison with the Modern Hipparcos Catalogue." *Astronomy and Astrophysics* 516, A28 (2010[a]): 1–24.

Verbunt, Frank, and Robert H. van Gent. "The Star Catalogue of Hevelius. Machine-readable Version and Comparison with the Modern Hipparcos Catalogue." *Astronomy and Astrophysics* 516, A29 (2010[b]): 1–22.

Watson, Rita, and Wayne Horowitz. *Writing Science before the Greeks*. Leiden and Boston: Brill, 2011.

Werner, Helmut, and Felix Schmeidler. *Synopsis der Nomenklatur der Fixsterne. Synopsis of the Nomenclature of the Fixed Stars*. Stuttgart: Wissenschaftliche Verlagsgesellschaft, 1986.

III **The Long Eighteenth Century**

Hania Siebenpfeiffer

Imagining the Extra-Terrestrial 'Other' in Early Modern Literature

Abstract: The controversy over the plurality of the worlds particularly shaped seventeenth- and eighteenth-century cosmologies. Nevertheless, the conjecture of extra-terrestrial species was no genuine invention of the early modern period. Already in the early fifteenth century, Nicolaus Cusanus expressed the idea that the finite infinity of the universe inevitably led to the multiplication of inhabited worlds, thus establishing three pioneering principles of alien life forms: (i) everything that can exist, does in fact exist; (ii) everything created fulfils a larger purpose in legitimizing inference from one existence to another; (iii) the order of the cosmos is not alien but can be grasped by human reasoning. The chapter links Cusanus' late medieval cosmological speculations to the early modern debate about the extra-terrestrial other, namely to Bernard le Bovier de Fontenelle's *Entretiens sur la pluralité des mondes* (1686) and Christiaan Huygens' *Kosmotheóros* (1698). They were the first to realize that establishing the verisimilitude of multiple inhabited planets and the probability of extra-terrestrial anthropology requires an all-encompassing natural philosophy. Its basic principles were defined as the material equivalence of all celestial bodies and the universal causal relationship between environment and species. Huygens' extra-terrestrial anthropology especially was deeply influenced by preceding literary imaginations of the extra-terrestrial 'other' in Johannes Kepler's *Somnium* (1634), Francis Godwin's *The Man in the Moone* (1638) or Hector Savinien Cyrano's *L'autre monde ou les États et Empires de la Lune et du soleil* (1657/1662). At the same time, Fontenelle's and Huygens' cosmic anthropology helped to shape the enlightenment concept of the cosmic extra-terrestrial as the future of humankind fictionalized i.e. in Eberhard Christian Kindermann's *Die Geschwinde Reise mit dem Luft=Schifff nach der obern Welt* (1744).

Theses Quadragesimales in Scholis Oxonii publicis:
Coeli sint Fluidi. Terra Moveatur.
Terra non sit centrum universi. Luna sit Habitabilis.
Radii Luminosi sint Corporei. Sol sit Flamma.
(*Oxford College Dissertation by John Wallis*, 1684)

Even though the controversy over the plurality of the worlds particularly shaped seventeenth- and eighteenth-century cosmologies, the conjecture of extra-terrestrial species was no genuine invention of the early modern period. Already in *De docta ignorantia* (1438/1439), Nicholas of Cusa by a complex metaphysical conjecture completed by mathematical calculation argues that the universe by necessity is neither finite nor infinite and that it is entirely populated by celestial species (*stellarum habitatores*). Contrasting two different concepts of infinity – the absolute infinity of God alone (*negative infinitum*) and the relative or derived infinity of space (*privative infinitum*)[1] – he points out that the universe as the largest space given is without limit, but at the same time, it cannot possess the absolute infinity of God. Hence, the universe should be envisioned as an infinite robbed of the absolute infinity, in other words: a finite-infinite. For Cusanus, the finite infinity of cosmic space is at the same time the prerequisite for the idea that the universe consisted of more than one cosmos and that it is inhabited by more than one species. In a compelling reflection on the relation of part and whole he concludes that even if to every part in cosmic space there were a respectively bigger and smaller one, without the ascending or descending relation ever reaching an end, the universe necessarily would be potentially infinite. Since, however, it is at the same time restricted by the limited capacity of its materiality – an argument that Cusanus took directly from Aristotle – its infinity, although potentially given, would not be an absolute infinity as in God, but a limited one. As a finite-infinite universe, it no longer follows the hierarchical order of smallest and largest or centre and periphery.[2]

Abandoning the Aristotelian geometric order that defined the universe as a continuous extension from the smallest to the largest, the deepest to the highest, and the centre to the outer circumference, the most important paradigm of Cusanus' universe instead becomes the existence of God as cosmic centre and circumference simultaneously. Considering that the very concept of limited creative power is incompatible with the idea of God itself, he then consequently derives the assumption that God could (and would) not have created the universe as a

---

1 "Universum vero cum omnia complectatur, quae Deus non sunt, non potest esse negative infinitum, licet sit sine termino et ita privative infinitum; et hac consideratione nec finitum nec infinitum est" (Cusa 1932, vol. II, cap. 1, §97).

2 "Non habet igitur mundus circumferentiam. Nam si centrum haberet, haberet et circumferentiam, et sic intra se haberet suum initium et finem, et esset ad aliquid aliud ipse mundus terminatus, et extra mundum esset aliud et locus; quae omnia veritate carent. [. . .] Et cum non sit mundus infinitus, tamen non potest concipi finitus, cum terminis careat, intra quos claudatur. Terra igitur, quae centrum esse nequit, motu omni carere non potest" (Cusa 1932, vol. II, cap. 11, §156). If there is a smallest, which formed its centre, then also a largest exists, but then the universe is evidently no longer unlimited. Therefore, the infinite universe, deprived of its absolute infinity, except in God, would have neither centre nor periphery, but would be in all its parts, the Earth included, in incessant, decentring motion, which is God. Cf. Omodeo (2014).

space of finite infinity without filling its vastness with other (intelligent) beings, thus preventing that the universe with its innumerable stars appears a waste of space and matter: "[. . .] God is the centre and circumference of all stellar regions and [. . .] natures of different nobility proceed from him and inhabit each region (lest so many places in the heavens and on the stars be empty and lest only the Earth – presumably among the lesser things – be inhabited) [. . .]" (Hopkins 1985, 96).[3]

In *De docta ignorantia*, the finite infinity of the universe and the dual property of God as the centre and circumference of cosmic space inevitably leads us to consider the multiplication of worlds, including their inhabitants (*stellarum habitatores*). To them, Cusanus devotes an entire chapter entitled *De condicionibus terrae*,[4] in which he states that although the celestial inhabitants are unknown to us (as we are to them), and neither their existence can be proven nor their appearance foreseen, it is imperative still that they exist. This concept of *plenitude*[5] as a necessary, ubiquitous fullness brought into play by Cusanus is the first of three presuppositions that structured the debate about the inhabitation of further planets in the early modern period. The second presupposition justifying the necessity of thinking about extra-terrestrial life forms is that of absolute meaningfulness, according to which nothing in God's creation was senseless or purposeless. God never would have created more than one sun visible to men if it did not serve to give light and warmth to the inhabitants of the planets surrounding it the same way the terrestrial sun does for us. For the first time, the autoptic evidence of the fixed stars becomes conjectural proof of the habitability of the heavenly bodies orbiting them. Both principles – the fullness and the meaningfulness of all creation likewise – draw their legitimacy from the third principle, God's rationality, stating that everything in God's creation obeys a rational order. Since human reason behaves congruent to this order, there is only a gradual, but no principal difference between the order of the universe and the order of human thinking. Human understanding thus can grasp the divine order of things at least remotely

---

3 "Nam etsi Deus sit centrum et circumferentia omnium regionum stellarum et ab ipso diversae nobilitatis naturae procedant in qualibet regione habitantes, ne tot loca caelorum et stellarum sint vacua [. . .]" (Cusa 1932, vol. II, cap. 12, §168).

4 Cusa (1932), vol. II, cap. 12, §§162–174.

5 In contrast to Plato, Cusanus understands plenitude as an indispensable necessity according to which the infinite creative power of God proves itself precisely in the fact that it transforms potentiality into actuality while at the same time leaving no potentiality unused. Cf. the discussions in Pasqua (2017).

even when it is withdrawn from its immediate sensual experience and hence is capable of an 'extra-terrestrial thinking' without any empiric proof.

Although Cusanus' conjecture obviously is still deeply rooted in Aristotelian Christian reasoning, it establishes three pioneering principles of alien life forms: *Firstly*, that everything that can exist does exist somewhere in the cosmos. *Secondly*, that everything created fulfils a larger purpose in legitimizing inference from one existence to another. And *thirdly*, that the order of the universe and that of thought are congruent; hence the cosmos is not alien to human reasoning, but can be grasped by human cognition in its essential features. Almost a century would pass before the 1543 publication of Nicolaus Copernicus' *De revolutionibus orbium coelestium* confirmed Cusanus' still metaphysical contemplation about the rotation of celestial bodies and the plenitude of stars and bolstered a crucial methodological aspect of early modern extra-terrestrial anthropologies: the assumption that the cosmos in its entirety is determined by universal laws and therefore can be understood by means of mathematical calculation and empirical analogies. The transformation of the Christian-Aristotelian cosmos still pivotal in *De docta ignorantia* into the modern infinite, decentred and pluralized multiverse that took place between the sixteenth and eighteenth centuries centred around the question of the *stellarum habitatores*, while focusing on three main aspects: the ability to prove the existence of other inhabited celestial bodies, the appearance and moral condition of their residents and the genealogical, physiognomic and moral relationship of extra-terrestrial species to humanity.

Thus, a line can be drawn from Cusanus to Christiaan Huygens' *Kosmotheóros* (Huygens 1698a), a cosmological treatise that after its initial publication in 1698 shaped the philosophical debates about existence and anthropology of aliens like no other work at that time. Intentionally published only after Huygens' death, it was almost immediately translated into English, French, Dutch, and German (Huygens 1698b; 1699; 1702; 1703). In its essence, the *Kosmotheóros* offers a continuation of Galileo's discoveries in the *Sidereus Nuncius* of 1610, the subsequent observations by Christoph Scheiner, Johann Fabricius and Thomas Harriot, who between December 1610 and March 1611 independently had discovered the sunspots,[6] as well as by Jan Heweliusz (Hevelius), the famous lunar cartographer from Gdańsk. Huygens, too, was an acclaimed inventor of telescopes and a gifted astronomical observer (cf. van Helden 1980): between 1655 and 1680 he discovered the Saturn moon Titan, identified the supposed ears of Saturn as its rings (Huygens 1659) and the Orion Nebula as a star cluster, recognized the rotational movement of Mars, and precisely calculated its rotational period as 24 hours.

---

6 Cf. Reeves and van Helden (2010). On the literary impact, cf. Siebenpfeiffer (2015).

Once he discovers Johannes Kepler's conjecture about the inhabitants of the moon, the *Somnium sive astronomia lunaris*, Huygens starts exchanging private notes on the possibility of other inhabited planets.[7] Even though he partly rejects Kepler's assumption that the moon is inhabited by two species, the "privolvans" and the "subvolvans," claiming that she has mountains but "neither a sea, nor rivers, nor clouds, nor airs and waters,"[8] he nevertheless adopts Kepler's poetic invention presenting bold astronomical conjectures in a compelling theatrical narrative. In accordance with the well-chosen title,[9] the cosmic system in the *Kosmotheóros* thus unfolds like a play on stage. Having introduced the methodological framework of his treatise, each planet is venerated with an individual entrance followed by a dense description of its material composition and visual appearance before the narrator turns to the characteristics of the species inhabiting a particular celestial body. Referring to the same basic qualities such as the existence of water and heat as fundamental preconditions for biological life, the existence of various animals and plants as potential alimentary sources, the actuality of sensorial perceptions such as sight, hearing, taste and smell along with a humanlike physiognomy including hands and feet, Huygens lastly even ascribes to them a rational soul enabling them to live in societies not very different to his own, practising science, navigation, arts and music.

The significance of Huygens' conjecture about the extra-terrestrials does not primarily derive from the fact that he populates the entire universe, for Bernard le Bovier de Fontenelle had already lavishly done so in his *Entretiens sur la pluralité des mondes* in 1686. The true importance of the *Kosmotheóros* originates from Huygens' methodological predefinition, *how* to anticipate alien life, as he was first to realize that establishing the verisimilitude of inhabited planets as well as the probability of extra-terrestrial anthropology requires regularities universally valid, and based on natural philosophy. In consequence, Huygens defines two empirical principles to prove his pluralized universe: the material equivalence of all celestial bodies and the universal causal relationship between environment and species. His conjectural reasoning roughly runs as follows: As the physiognomy and intelligence of a species are related to its natural environment, the observable condition of a

---

7 Cf. Huygens' correspondence with his brother Constantijn Huygens (Huygens 1967–1970, vols. 1–10).

8 Huygens: *The celestial worlds discover'd*, p. 130. Cf. the online edition of the first English translation of 1698 https://webspace.science.uu.nl/~gent0113/huygens/huygens_ct_en.htm (20.05.2022; 17:32).

9 *Kosmostheóros* can be translated as "beholder of the cosmos": *cosmos* meaning 'world' or 'universe' (in particular, when viewed as an ordered and harmonious system) and *theóros* 'spectator,' both in a theoretical and in a theatrical way.

celestial body convincingly allows conclusions about whether it is inhabited and by which species. If a celestial body, for instance, resembles Earth (as in fact all celestial bodies in *Kosmotheóros* do) because it receives heat from a sun, its surface is covered by land masses, shadows on its surface point to water and atmospheric opacity indicates air and moisture, the planet certainly will be inhabited by a species highly resembling human mankind possessing reason, understanding, morality and the ability to practise astronomy:

> If therefore the Principle we before laid down be true, that the other Planets are not inferior in dignity to ours, what follows but that they have Creatures not to stare and wonder at the Works of Nature only, but who employ their Reason in the examination and Knowledge of them, and have made as great advances therein as we have? They do not only view the Stars, but they improve the Science of Astronomy [. . .]. (Huygens 1698b, 61)

Huygens' main argument for such avowedly far-reaching conclusions about the nature of the universe that cannot be observed but only conjectured was the lack of alternatives arising from his very cosmological thinking. If the principles of material equivalence and causality between environment and species are rational as well as universal, Earth can no longer claim to be the only planet inhabited. Thus, humankind loses its unique position in God's creation with almost the same argument Cusanus had given 250 years earlier:

> For everything in them [the planets; H.S.] is so exactly adapted to some design, every part of them so fitted to its proper life, that they manifest an Infinite Wisdom, and exquisite Knowlege in the Laws of Nature and Geometry, as, to omit those Wonders in Generation, we shall by and by show; and make it an absurdity even to think of their being thus haply jumbled together by a chance Motion of I don't know what little Particles. Now should we allow the Planets nothing but vast Deserts, lifeless and inanimate Stocks and Stones, and deprive them of all those Creatures that more plainly speak their Divine Architect, we should sink them below the Earth in Beauty and Dignity; a thing that no Reason will permit, as I said before. (Huygens 1698b, 20–21)

A significant shift becomes visible between, on the one side, Cusanus' assertion that the celestial inhabitants (*stellarum habitatores*), even though they exist, are incommensurable with the human being and therefore not to be known,[10] and Huygens' emphasis on the similarity of all rational beings throughout the whole universe on the other. At the end of the seventeenth century, Cusanus' comparatively vague assumption of a general existence of extra-terrestrial beings has given way to a self-conscious conjecture not only about their existence, but their corporal appearance,

---

10 "Improportionabiles igitur sunt illi aliarum stellarum habitatores, qualescumque illi fuerint, ad istius mundi incolas [. . .]" (Cusa 1932, vol. II, cap. 12, §170).

mental abilities and cultural achievements that boldly relies on the human species as a blueprint for all species throughout space. The *Kosmostheóros* hence not only introduces the material and, in a way, proto-evolutionary principles of material equivalence and environmental causality into the enlightenment discussion of the *plurality of worlds*, but at the end inaugurates a powerful anthropological matrix that governs the imaginations of the alien yet to come: a matrix in which the erudite, white, male Central European serves as an unquestioned role model for an extra-terrestrial anthropology that conceives alien life-forms simultaneously as strange *and* familiar. By doing so, Huygens reveals himself as someone who, even though without using the argument of Christian superiority, does not substantially differ from early (Christian) Enlightenment anthropology. And even if he explicitly criticizes at one point the Christian ideology of human perfection that defines man as the highest of all real and potential creatures, it becomes clear that Huygens's criticism is directed against the mechanical materialism of La Mettrie, whose 'clockwork hypothesis' of the soulless animal he had sharply rejected shortly before. Hence, he disregarded the possibility of actually imagining radically different forms of life in the still undefined and open cosmic space in full purpose of him arguing as a scholar and not a poet. Creating a cosmos of various lifeforms is not in the sense of his writing, that is not concerned with radical difference between terrestrial and extra-terrestrial species, but with the greatest possible similarity in combination with the greatest possible diversity, which allows gradual, but no fundamental deviations from the law of similarity. Therefore, the inhabitants of all planets, but especially those of the outer larger celestial bodies, all possess the same physique, the same limbs, the same abilities, the same affects and the same capacities as the European paradigm of human mankind: They possess reason, have hands and feet as well as five senses, live in social societies and obey their laws, inhabit palaces, houses and huts, feed on fruits and animals, engage in barter trade and merchandise economy, have a written language, master geometry besides the already mentioned astronomy and finally they are musical.

As one can see in Huygens' cosmological master narrative, from now on, inhabitants of other planets were supposed to be familiar because they were just like us humans, merely living on other planets. Still, "just like us humans," they incarnate a small degree of strangeness, because the law of *similitudo*, which had already shaped the colonial debates about the indigenous people of the earthly new worlds in the sixteenth century, now assigns the same position to the aliens. Positioned between identity and alterity, early modern exoplanetary anthropology creates a conjectural space of ambiguity between 'us' and 'them,' between the terrestrial and the extra-terrestrial, the familiar and the unfamiliar. But whereas the *astronomia nova* of Copernicus, Kepler and Galilei provided the cosmological framework required, it was the new genre of literary space travels that became the true master medium of the early modern cosmic alien.[11]

1 Literary aliens

> SECOND HERALD: Certain and sure news –
> FIRST HERALD: Of a new world –
> SECOND HERALD: And new Creatures in that world.
> FIRST HERALD: In the Orb of the moon –
> SECOND HERALD: Which is now found an Earth to be inhabited!
> FIRST HERALD: With navigable seas and rivers!
> SECOND HERALD: Variety of nations, polities, laws!
> FIRST HERALD: With havens in't, castles and port towns!
> SECOND HERALD: Inland cities, boroughs, hamlets, fairs and markets!
> FIRST HERALD: Hundreds and wapentakes! Forests, parks, cony-ground, meadow pasture, what not?
> SECOND HERALD But differing from ours.
> (Jonson 2012, 435–436, ll. 102–113)

In the seventeenth and eighteenth centuries, not all literary space travels focused upon an experimental exploration of the anthropological dis/similarities between terrestrial and extra-terrestrial. In fact, some texts still favored the Menippean narrative of satirical exaggeration and parodical imitation deriving from Lucian's *Verae Historiae* and his *Icaromenippus*,[12] while others abstained completely from exploring the narrative potential of *similitudo*. Instead of making the most of the tension between "us and them," their cosmic worlds are populated with more or less accurate, though rather boring revenants of male central Europeans.[13] Finally, likewise ineffectual are those French novels that proceeded from a universally spiritualized, often panspermic cosmos consisting of animated matter.[14] In contrast to these texts, Francis Godwin's *The Man in the Moone* (1638), Hector Savinien Cyrano's *L'autre monde ou les États et Empires de la Lune et du soleil* (1657 and 1662) and Eberhard Christian Kindermann's *Die Geschwinde Reise mit dem Luft=Schifff nach der obern Welt* (1744) invented a new cosmic narrative grounded

---

11 For a detailed epistemological and poetological analysis of the genre of early modern literary space travel, cf. Siebenpfeiffer (2025, forthcoming).

12 Cf., for example, Cavendish (1666), or McDermot (1728). William Hogarth's engraving *Some of the Principal Inhabitants of the Moon* (1724) is a striking example for the visual tradition of mock cosmic travels.

13 Cf., for instance, Venator (1660), Hertel (1758), Villeneuve (1761), or Defoe (1741).

14 Cf., for example, de la Roche (1754).

in early modern natural philosophy and based on empirical observation. Even though the scientific backgrounds of the cosmological knowledge presented differ in many aspects, all three novels use the imaginative potential of fiction to create a reliable cosmic world, verisimilar to Earth in many aspects, firstly the inhabitants. Their cosmic aliens are best considered as species that bear resemblances and differences at the same time: their aliens still appear strangely familiar today because they are familiar strangers. Even more striking is that all three novels supplement their extra-terrestrial anthropology with a temporal dimension: in Godwin's *The Man in the Moone* it is the past; in Cyrano's *L'autre monde ou les États et Empires de la Lune et du soleil* the present; and in Kindermann's *Die Geschwinde Reise* the future. The hitherto unknown, entirely new temporalization of cosmic space indicates several significant shifts within the genre of proto-science fiction and paves the way for a transition in which modern time travel eventually superseded early modern space travels at the end of the eighteenth century.

2 Francis Godwin's *The Man in the Moone* (1638)

Within the spectrum of early modern exoplanetary anthropology, Godwin's *The Man in the Moone*[15] represents the concept of a recursive anthropology, originally invented by Kepler in his *Somnium sive astronomia lunaris* (1634). While in *Somnium* both lunar species are intelligent but decidedly inferior, Godwin draws the relationship between lunars and humans on a much finer scale. According to the first-person narrator, an adventurous Spaniard named Gonsales,[16] who involuntarily lands on the moon by means of a self-invented goose-flight plane, the Lunarians are subdivided into two species: the so-called "true Lunars" and the "bruite beasts" (Godwin 2009, 103). In contrast to the latter, about whom Gonsales says nothing further, the true Lunarians are an equal, if not superior species to humans, both in body size and lifespan. They come in three different castes: a small-bodied lower class, 10 or 12 feet tall, with a life expectancy of just under 1,000 moons or 80 Earth years and therefore very similar to humans; a middle class of 20 feet, living up to 3,000 moons or 240 Earth years; and finally, the lunar

---

15 Godwin (1638). The title of the first edition names "Domingo Gonsales. The Speedy messenger" as author, whereas the second edition of 1657 indicates the initials "F.G. B. *of* H." without spelling out Francis Godwin's name.
16 For a detailed analysis of the main character, cf. Campbell (2011).

upper class, whose members are at least 27 feet tall and live over 30,000 moons or 1,000 Earth years (Godwin 2009, 101–103).[17]

Initially, the lunars are descendants from Earth, since their primary ancestor – so Gonsales reports – married the heiress of the Lunar kingdom 3077 Earth years ago.[18] The fictional interplanetary genealogy that turns the Lunarians into ancestors of humankind and Gonsales's voyage to the Moon into an encounter with the (biblical) past is verified by a precise dating of a common genealogical origin. Godwin's interstellar genealogy harmonizes the population of the Moon with the Exodus of the people of Israel from Egypt in an exact temporal consistency. In early modern biblical chronology, the latter was dated to the time around 1450 BC.[19] As Bishop of the Anglican Church, Francis Godwin clearly was familiar with this dating, as were most of his readers. It is thus not at all random that the beginning of the Lunar monarchy 3077 years ago concurs with the year 1447 BC.[20] This dating simultaneously answers the pressing question of the descent of the Lunarians, because referring to the biblical Exodus, they become one of the lost tribes of Israel and therefore they are identified as Christians from the very beginning of their existence on.

Among other things, they owe their Christian beliefs and near perfection to their semi-terrestrial origin, which shows itself above all in their high morality: In accordance to their superior corporal appearance they live in faithful monogamy, abhor vice, know neither war nor violence and – since there are no crimes – have no need of lawyers nor courts, just like they do not need any medicine, since they enjoy the best of health throughout their lives; even their deaths are peaceful. They just exhale their soul like a burned-down candle. Governed by a secular upper-class ruler, who directs the destiny of the lunar unitary state with the help of 696 princes (Godwin 2009, 100–101), the lunar society, which encompasses the entire moon, represents an almost ideal peace-loving monarchy. Its

---

17 Cf. also Poole's erudite introduction in Godwin (2009), 11–64.

18 Cf. also Merchant (1955).

19 Depending on the beginning of creation, the beginning of the Exodus fell between 1470 and 1430 BC. For example, the Anglican bishop James Ussher, who was one of the most influential Anglican Bible chronologists of the 17th century, in the *Annals of the Old Testament* dated the creation of the world to October 23, 4004 BC, whereupon, according to complicated calculations, the beginning of the Exodus from Egypt began in 1440 BC; cf. Ussher (1650).

20 The origin of *The Man in the Moone* can only be inferred indirectly due to a lack of historical evidence. Since Godwin takes up cosmological insights that only emerged after 1620, Grant McColley dates the work to the early 1620s; cf. McColley (1936), 387. The idea of birds flying to the moon for hibernation goes back to Francis Bacon's *Sylva Sylvarum, or Natural History*, which only appeared posthumously in 1627, so that 1628/1630 is the most plausible date of origin; cf. Godwin (2009), 27.

near perfection is completed by the fact that the Lunarians know only one language, a tonal sound system similar to High Chinese (Godwin 2009, 109).[21] Communicative misunderstandings from which hostile actions between species or nations could arise, as in early modern Europe, are thus excluded from the outset, neglecting the fact that, given the utopian quality of an alien culture that knows neither work nor property and in which all needs are satisfied by the equitable distribution of what is available, there is no deeper reason for violent conflict in the first place.

In addition to the invention of a literary anthropology based on the fictional autopsy of an alien and at the same time familiar extra-terrestrial world, the significance of *The Man in the Moone* for literary enlightenment owes to yet another moment, namely the invention of the First Encounter as the unexpected and surprising appearance of the alien stranger:

> I had never seene them [the geese; H.S.] to eat any manner of greene meate whatsoever. Whereupon stepping to the shrub, I put a leafe of it between my teeth: I cannot expresse the pleasure I found in the taste thereof; such it was I am sure, as if I had not with great discretion moderated my appetite, I had surely surfetted upon the same. [. . .] Scarcely had I ended this banquett, when *upon the sudden* I saw my self environed with a kind of people most Strange, both for their feature, demeanure, and apparell. Their stature was most divers but for the most part, twice the height of ours: their colour and countenance most pleasing, and their habit such, as I know not how to expresse. (Godwin 2009, 99)

In addition to its sudden immediacy, the significance of First Contact as the decisive narrative turning point in science fiction is reinforced by the fact that the first-person narrator, Gonsales, literally loses his speech at the sight of the Lunarians. For a long moment, he is deprived of the very instrument he needs most – his language, his ability to describe – and stands before the alien beings like a man born blind before color:

> For neither did I see any kind of Cloth, Silke, or other stuffe to resemble the matter of that whereof their Clothes were made; neither (which is most strange, of all other) can I devise how to describe the color of them, being in a manner all clothed alike. It was neither blacke, nor white, yellow, nor redd, greene nor blew, nor any color composed of these. But if you aske me what it was, then I must tell you, it was a color never seen in our earthly world, and therefore neither to be described unto us by any, nor to be conceived of one that never saw it. For as it were a hard matter to describe unto a man borne blind the difference betweene blew and Greene, so can I not bethinke my selfe of any meanes how to decipher unto you this Lunar color, having no affinitie with any other that ever I beheld with mine eyes. (Godwin 2009, 100)

---

21 Cf. Davies (1967).

The importance of Godwin's invention becomes clear when one looks at the modern genre development not only of science fiction, but of exploration narratives in general. Ever since *The Man in the Moone*, aliens, no matter their origin, appear unexpectedly and unannounced before the eyes of their earthly discoverers, their sudden arrival creating a striking narratological paradox, which Godwin already knew how to solve in an ingenious way: it is true that the human explorer should describe the shape of the alien beings as precisely as possible in order to give their appearance and existence the greatest possible vividness (*enargeia*) and accuracy (*energeia*), the two main criteria for rhetorical, thus narrative evidence. At the same time, however, a certain degree of alterity must be preserved to maintain an indisputable difference between terrestrial and extra-terrestrial anthropology, preventing the alien other from becoming completely one's own and thus losing its fascination. Hence, expressing the experience of radical alterity through the loss of language is an ingenious narrative move, for nothing can bear more eloquent witness to the unspeakability of the alien than the momentary elimination of language itself. However, not all literary space voyages of the early modern period make use of this narratological "trick" as skilfully as Godwin, but even in those anthropological designs that conceive inhabited worlds as satirical alienations of the terrestrial present, the First Encounter always contains the unexpected, as in Hector Savinien Cyrano's *L'autre monde ou les États et Empires de la Lune et du Soleil*, the most important French contribution to early modern science-fiction.

3 Hector Savinien Cyrano's *Les États et Empires de la Lune et du Soleil* (1657/1662)

In Cyrano's two-part novel, another first-person narrator, this time named Dyrcona, flies to the moon with the help of a rocket after several failed attempts to prove to himself and his drinking companions the existence of the moon's inhabitants.[22] After a short prelude in the Garden of Eden, Dyrcona unexpectedly finds himself confronted with a species whose nature he cannot classify. Only with a delay does he realize that the creatures are oversized animal men, "bêtes-hommes" (Cyrano 2006, 51), who walk on all fours and communicate in a language he cannot understand. Compared to Godwin's human-like Lunarians their physical appearance is

---

22 On Dyrcona's art of flying and the various aeronautic experiments he conducts cf. Siebenpfeiffer (2015). On the scientific background, cf. Alcover (1970) as well as Alcover (2006).

clearly parodically distorted.[23] Cyrano's Lunarians resemble anamorphic revenants, to take up a metaphor of Onfray (2008, 205), rather than ideal models. They populate an infinite, manifold and animated universe of atomistic materiality and material causality, and their external appearance represents, not like in Godwin's *The Man in the Moone*, a fixed anthropological order, but a perspectival puzzle in which the one who observes (and narrates) defines the anthropological relationship in question: in the eyes of the human narrator, the lunar "animal beings" embody an inferior species incomprehensible to him; in the eyes of the "bêteshommes," however, the opposite is true (cf. Lafond 1979).

The perspectival distortion to which the extra-terrestrial inhabitants owe their existence resembles early modern microscopic and telescopic images, which gave the subliminal a brilliant appearance in literature and philosophy, whether in the form of giants who regard the earth as the moon or of mites for whom the human body is the cosmos.[24] It is for this very reason that the world in Cyrano's novels is not dichotomously structured, but transversally.[25] Its narrative and semantic procedure of subversion creates an open relativity of anthropological orders in general, as yet unknown in Godwin's novel, that can be cited as the most striking, indeed the decisive element holding together the dizzying anthropological diversity that already characterizes the first part set on the moon. Apart from the initial topography of the Garden of Eden, in which the first-person narrator

---

23 Their external shape is indeed parodically distorted, but in contrast to Lucian's *Verae Historiae*, Cyrano's anthropology of the inhabitants of the moon and, to a certain extent, of the sun, conforms more closely to the paradigm of early modern social satire and its procedures of parody, persiflage, and caricature. Campbell speaks in this context of "libertine comedy" (1997, 7), a term that seems to me, at least for the first part, too weak to capture the critique inherent in Cyrano's subversion.

24 One could argue that the invention of the microscope in the early seventeenth century gave birth to a new philosophy of the subliminal in the guise of microscopically small insects such as mites, fleas, and lice, whose characteristic features lie beyond the natural perception of the human eye. The subliminal species, first and foremost the mite, in the conversation between Dyrcona and the Lunar Philosopher in *Les États et Empires de la Lune* is developed into "all-encompassing mite-ness" ("cironalité universelle"). Robert Hooke's *Micrographia* (1665) famously portrayed the mite, thereby illustrating Cyrano's "philosophy of mite-ness" (although Hooke presumably was not aware of Cyrano's novel). From Cyrano's novel, the mite migrated into the writings of Pascal and Voltaire. Pascal uses it in the *Pensées* to demonstrate infinity: as infinitely small as the mite is in comparison to man, so is man in comparison to the infinite. Voltaire in *Micromégas* uses the differences in size between the inhabitants of Sirius, Saturn and Earth, derived from the size of their planets of origin, to let his two space travellers ponder the ability to reason of such infinitely small beings as humans are (cf. Cyrano 2003, 174 and Voltaire 2002, 41).

25 Bezzola Lambert is one of the few who takes the perspective distortion into account (2002, 143).

initially lands but from which he is soon expelled, the moon is inhabited in Cyrano's work, as in Godwin's, by two hierarchically divergent species. Also like in *The Man in the Moone*, we are introduced above all to the superior species, the "grands lunaires." The great Lunarians communicate with each other through sounds and feed on the smell of food, the olfactory combination of which is individually adapted to their physiognomic constitution (cf. Cyrano 2006, 66–70). They sleep on flower beds, pay with verses instead of money, show the time by facial expressions, wear phalli as jewellery, punish *in effigie*, worship the young before the old and consume their dead in orgiastic rites to absorb and multiply their genius (cf. Cyrano 2006, 72–104).[26] Their cities are either on wheels and have a sail to move them horizontally with the help of an air current generated by bellows, or they are mounted on large screws to change their height (cf. Cyrano 2006, 121–123). They are called "grands lunaires" not only because they provide for the king residing in an absolutist Versailles-like court, as well as scholars and priests, but, like Godwin's "true Lunarians," because they are of giant-like stature, though without possessing the Christian bond and moral perfection of the latter. In contrast to Godwin, Cyrano's extra-terrestrials differ in nothing from humans: They, too, indulge their lusts without restraint and wage pointless wars at regular intervals in which eventually the straw decides who wins. Nevertheless, their natural philosophers advocate atomism, the infinity of the universe, the multiplicity of worlds, the materiality of perception and the equivalence of all things created at the same time (cf. Cyrano 2006, 116–132). They deny the immortality of the soul and resurrection after death, much to the discontent of the Lunar priests who violently insist on the superiority of their species, whose uniqueness they are prepared to affirm against all odds (Cyrano 2006, 144–153).

The satirical distortion in *Les États et Empires de la Lune* does not stop at the first-person narrator, who, due to his short stature, represents an inferior species from the point of view of the great Lunarians and whom they consequently consider not as a human (that is to say, Lunarian) being, but as an ape. Eventually, the acquisition of the lunar language initially promotes him from an ape to the level of an "homme sauvage" (a wild 'human' being), until he is declared truly human after a successful disputation with the priests. These recurring references to the corporal appearance of the Lunarians from the point of view of the narrator and vice versa still superficially cite the connection between physiognomy and morality familiar from Godwin's *The Man in the Moone*, according to which an extra-terrestrial species' nobility is determined by its appearance. But Cyrano only uses the supposed

---

26 This clearly is to be read as an ironic reference to Montaigne's famous essay "Des Cannibales" (*c.* 1580).

causality of an inner and outer constitution for further mockery. The first-person narrator, initially classified as inferior according to his physical appearance, eventually convinces the physically superior Lunarians of his intelligence, even though he holds views that are completely outdated: in the midst of the atheistic, enlightened Lunar society, much to the delight of the Lunarians he stands up for Christian, even Catholic beliefs, defends outdated natural philosophical views and vehemently advocates geo-centric Ptolemaic cosmology along with the superiority of his own species above all others. His outdated views form the basis upon which Cyrano unfolds his complex parodistic narrative, as the above-mentioned ritual consumption of the dead by the great Lunarians once again shows. The practice of consuming one's deceased, thus absorbing and multiplying their genius, clearly cites the early modern trope of the savage as cannibal and immediately thwarts colonial hierarchical perspectivization through allusions to Christ's Eucharist. But since Cyrano's Lunarians place their anthropophagy in the service of procreation, their action bears more resemblance to the Catholic rite of communion than to the colonial discursive devaluation of the Other as barbaric, inhuman, and beastly. In consequence, in this episode the inferior Others, who move to the position of the 'barbarians,' are not the Lunarians, but Christian humanity, in other words precisely the species embodied by Dyrcona. Taking the novel's narrative twists seriously, it becomes obvious that Cyrano, by caricaturing his philosophical adversaries in the literary figure of Dyrcona, is denying the anthropological order any firm system of identity and alterity. Instead, the relativity of perspectives which the novel unfolds replaces the (Christian) anthropological hierarchy by an anthropological ambiguity in which the extra-terrestrial itself gets an anamorphic quality.[27] What becomes visible is not the superiority or inferiority of one species over another, be it terrestrial or extra-terrestrial, but the awareness that anthropological orders, regardless of their origins, always are results of a specific perspective and therefore can never claim an overall truth. Whereas Godwin still negotiates the superiority or inferiority of different species, trying to give an answer to the conflict inherent in the extra-terrestrial anthropologies between Christian supremacy of humanity and its disempowerment in the face of an infinite number of inhabited worlds, Cyrano enters into a perspectival conundrum that informs the idea of an anthropological hierarchy. Instead, a constant anamorphosis of living things becomes the decisive feature of this plural-

---

27 The second part of the novel emphasises the anamorphic quality and its inherent relativity of order even more when Dyrcona, on his renewed journey through space, lands in the sun, where everything is in constant bodily metamorphosis, where, e.g., fruits first become fruiting beings and then turn into beautiful young gentlemen only to change into another creational species. The striking animation and material multiplication of the universe irrevocably dissolves every anthropological hierarchy.

ized universe. The anamorphic quality is further radicalized in the second part of the novel, when Dyrcona, on his second journey through space, lands first on sunspots and then on the sun itself. In *Les États et Empires du soleil*, the sun has become a celestial body in which everything is in constant metamorphosis, in which fruits give rise to "fruiting formations" from which a "beau grand jeune homme" (Cyrano 2006, 236–242) emerges only a few moments later, in which birds hold court as they transform into plants, and in which the elements fight against each other in the guise of reptiles (cf. Cyrano 2006, 250–274) – in other words, it is a world in which a vivification and multiplication of the cosmos as a whole becomes real, dissolving all anthropological orders. Instead, a constant transformation runs counter to any attempt to hierarchize humans and extra-terrestrials, while the simultaneity and equivalence of cosmic multiplicities take the place of anthropological systematics.

4 Eberhard Christian Kindermann's *Die Geschwinde Reise mit dem Luft=Schiff nach der obern Welt* (1744)

Published in 1744, Eberhard Christian Kindermann's *Geschwinde Reise mit dem Luft=Schiff nach der obern Welt* (2020) is the first true German-language literary space journey. It tells of a flight to the recently discovered moon of Mars. Even though Kindermann was a trained astronomer at the Saxon court observatory near Dresden, he adopts the materialism of the French Enlightenment advocated by Cyrano. At the same time, however, his space journey is deeply influenced by English and German physico-theology, even more than by French materialism, so that he conceives of the relationship between terrestrial and extra-terrestrial species not as a recursive past, like Godwin, or as an anamorphic present, like Cyrano, but instead, Kindermann anticipates the future of humanity by means of physico-theology.[28] The temporal, and even more so the genealogical relationship in Godwin's *The Man in the Moon* was characterized by a unidirectional hierarchy between the advanced lunar inhabitants and culturally less advanced humanity. In that the lineage narrated in Godwin's novel declared the Lunarians to be one of the lost tribes of Israel, whose biblical explanation was immediately comprehensible to the readership of the time, Gonsales, on his flight to the moon, in a sense travelled not just to another world, but to a past moment of (Christian) human history itself, which the narrator

---

[28] For further information about the physico-theological impact of Kindermann's novel see Siebenpfeiffer (2020).

Gonsales, upon his arrivals on the moon, comes face to face with in the guise of the Lunarians. Cyrano, on the other hand, transformed the unambiguous relationship of development and continuity between terrestrial and extra-terrestrial species, which is still compatible with conventional notions of purpose, into a complex web of relationships in which the lunar inhabitants no longer occupy the position of the cosmic Other, but have become a parodic inversion of humans. Genealogical questions therefore play as little a role in Cyrano's *Les États et Empires de la lune* as the attempt to make a halfway plausible conjecture about how the human-like, but nevertheless alien species could have arrived on the moon. Instead of sketching out a more or less convincing genealogy, Cyrano rather used the increasingly popular idea of the plurality of worlds for a clever and entertaining parody of the disparities of the contemporaneous by juxtaposing without hierarchy old and new knowledges as well as terrestrial and extra-terrestrial species at the same moment.

Finally, another eighty years later, Kindermann invents in *Geschwinde Reise mit dem Luft=Schiff nach der obern Welt* again a completely different time model, which conceptualizes the relation between humans and extra-terrestrials as a future prospect both in the sense of 'evolution' as in the sense of genealogy. After his five travellers named Visus, Tactus, Ordor, Auditus and Gustus – obvious allegories of the five senses – successfully constructed a vacuum driven airplane and crossed space with the help of Fama, the Roman mythological deity of fame as well as rumour, they land in an almost ideal world. Its perfection begins with the Martian moon itself. Rendered in the royal colors of gold, green, and blue, the Martian satellite is of "astonishing splendour and beauty," the air is filled with an exceedingly pleasant odour, tasty fruits grow on the trees, and centaurs, satyrs, and fauns frolic in the forests (Kindermann 2020, 29).[29] The perfection of the habitat is reflected in its inhabitants, whose significant difference becomes apparent to travellers at first glance:

> But what wonder arose among them when they saw four creatures walking before them, which in outward appearance resembled the people of the underworld. But VISUS perceived one and another extraordinary thing about them, [. . .] for VISUS saw this much, that they were not clothed, and yet were not naked [. . .]. (Kindermann 2020, 30)

The first contact in Kindermann's novel is initially violent, as the human travellers pretend to be gods in front of the Lunarians and kill one of the alien beings as proof of their supposed divine power. But the immediate examination of the slain Martian Lunarian reveals the perfection of the alien species, which "possessed all the limbs of men" without being of the same gross materiality. Instead, its body

---

29 Unless otherwise stated, all translations are mine.

was "of a fluid yet solidly composed crystalline essence," so that in the sunlight it "shone like a hundred suns" (Kindermann 2020, 32–34). The reference to the sun introduces the most important Enlightenment metaphor of the narrative, whose positive valence directly connects to the anthropology sketched out in the novel. Consequently, Kindermann's Martian Lunarians know neither social hierarchies nor inequalities and live in a unified society as equals among equals. And even if they are not entirely free of misdeeds and violence, these remain rare exceptions that are punished quickly and drastically. Their home is a city called "Fiat," which was created by divine nature itself, as they themselves are in direct communication with God, who gave them only two laws: "1) Fear GOD, 2) Love one another, one as the other" (Kindermann 2020, 37). Although their appearance is not entirely perfect, they are clearly of divine origin, since their very appearance indicates their god-likeness – they too are enlightened and self-luminous – and their physical material-ity is about to become a "fluid and yet firmly composed crystalline substance" (Kindermann 2020, 34).

In contrast, the human travellers are undoubtedly of distinctly lower origin. They are representatives of humankind "fallen to the lowest" (Kindermann 2020, 42) and in every respect inferior to the Lunarians. The basic physico-theological di-rection of Kindermann's extra-terrestrial anthropology, however, does not stop at the reversal of valence, but transforms the anthropological status quo into an an-thropological prognosis of prophetic quality. At the end of the story, the Martian moon dwellers hold out the prospect of an imminent transformation for humanity:

> So then, has divine wisdom permitted you to look into the cabinet of creation? Your intel-lect, which had been darkened completely by the Fall, will, as you can see, become light again; for you now understand and find the twinkling stars, each in itself, a world populated with creatures, since a certain number of them turn around a fixed point that communi-cates essence, warmth and light to them. [. . .] When you have passed through the putrefac-tion and become elastic, we shall, God willing, see each other again; then you will behold more than you have been able to behold at present due to your physical properties. So live well, and travel happily. (Kindermann 2020, 47–48)

Despite the recognizably enlightened narration, the most important innovation is the revaluation of the alien with the help of a future-oriented temporalization of the universe. Whereas the inhabited worlds previously were always seen as differ-ent worlds existing at the same time in the same homogenous space, the simultane-ity of the universe is now given a new orientation. In *Die Geschwinde Reise*, aliens and humans still share one space, but now they embody two different temporal dimensions. For what is already the present for the Martian Lunarians has yet to become reality for the humans: the extra-terrestrial present has turned into the human future. In Kindermann's novel, the present and the future are paralleled in a novel line of development that was previously unknown in the genre of science fiction. Kindermann's narrative invention fundamentally changes the anthropological matrix insofar as the natural perfection of the aliens no longer represents an earlier state as in Godwin's *The Man in the Moon*, but a futuristic one, that is yet to come. That the future of the Martians truly will become human reality is vouched for by the novel itself. In the mode of fiction, it proves that man can conquer space to gain knowledge of the divine creation, knowledge that will increase his morality until mankind itself has become crystalline. This startling cosmic reorientation toward the future is made plausible to eighteenth-century readers by the fact that Kindermann combines the idea of knowledge of God with the anthropological dimension of the extra-terrestrial in order to replace Christian eschatology with a scientific invention of a manmade future. Consequently, Kindermann's anthropological design of a future of mankind mirrored in the extra-terrestrial present ends with the quoted promise to overcome the loss of the past with the help of enlightened reason.

As I have shown from the example of the novels of Godwin, Cyrano, and Kindermann, the special achievement of the literary imaginings of the alien in the early modern period lies less in their invention than in the shaping of a temporalized anthropology that places terrestrial and extra-terrestrial species in a complicated reciprocal relationship. Godwin already has designed his "true Lunarians" as a species superior to humans, although their superiority was based on the not entirely plausible circumstance that they were granted a special grace as the lost tribe of Israel. In Cyrano's *Les États et Empires de la Lune et du Soleil*, extra-terrestrial and terrestrial anthropology expose each other in a satirical conundrum, so that the alien becomes the anamorphosis of the human and vice versa. Finally, in Kindermann's *Geschwinde Reise*, the terrestrial and extra-terrestrial worlds become intertwined not only in a spatial, but in a temporal dimension that turns the alien present into humanity's future. With this twist at the latest, the way was clear for future science fiction to move from space to time, and modernity could now begin inventing its own aliens.

Bibliography

Alcover, Madeleine. *La pensee philosophique et scientifique de Cyrano de Bergerac*. Geneva: Droz, 1970.

Alcover, Madeleine. "Commentaire." Cyrano de Bergerac. *Œuvres completes*, vol. 1: *L'autre monde ou les états et empires de la lune*. Paris: H. Champion, 2006. Clxiv–ccix.

Bezzola Lambert, Ladina. *Imagining the Unimaginable. The Poetics of Early Modern Astronomy*. Amsterdam and Atlanta: Rodopi, 2002.

Campbell, Mary Baine. "Impossible Voyages: Seventeenth-Century Space Travel and the Impulse of Ethnology." *Literature and History* 6.2 (1997): 1–17.

Campbell, Mary Baine. "Speedy Messengers: Fiction, Cryptography, Space Travel and Francis Godwin's 'The Man in the Moone'." *Yearbook of English Studies* 41.1 (2011): 190–204.

Cavendish, Margaret. *Observations upon Experimental Philosophy to which is added The Description of a New Blazing World*. London, 1666.

Cusa, Nicolaus de. *De docta ignorantia*. Eds. Ernst Hoffmann and Raymund Klibansky. Leipzig: Meiner, 1932.

Cyrano, Hector Savinien. *Œuvres completes*, vol. 1: *L'autre monde ou les états et empires de la lune*. Ed. Madeleine Alcover. Paris, H. Champion: 2006.

Davies, H. Neville. "Bishop Godwin's Lunatique Language." *Journal of the Warburg and Courtauld Institutes* 30 (1967): 296–316.

De la Roche, Charles-François Tiphaigne. *Amilec: ou la Graine d'hommes qui sert à peupler les planètes* (1753). Nouvelle édition avec des remarques amusantes. 'Somniopolis': 'Morphée', 1754.

De Villeneuve, Daniel Jost [M. de Listonai]. *Le Voyageur Philosophe dans un Pays Inconnu aux Habitants de la Terre*. Amsterdam: [s.n.], 1761.

Defoe, Daniel. *A New Journey to the World in the Moon*. London: C. Corbett, 1741.

Godwin, Francis [as Gonsales, Domingo]. *The Man in the Moone or a Discourse of a Voyage thither. The Speedy Messenger*. London: [s. n.], 1638.

Godwin, Francis. *The Man in the Moone*. Ed. William Poole. Peterborough: Broadview, 2009.

Hertel, Johan Jacob. *Jonas Lostwaters eines Holländischen Schiffsbarbiers Reise nach Mikroskopeuropien, einem neuerer Zeit entdeckten Weltkörper*. Glückstadt: Johann Jacob Babst, 1758.

Hopkins, Jasper. *Nicholas of Cusa "On Learned Ignorance"*. 2nd ed. Minneapolis: A.J. Banning Press, 1985.

Huygens, Christiaan. *Oeuvres complètes*. Amsterdam: Swets & Zeitlinger, 1967–1970.

Huygens, Christiaan. *Systema Satvrnivm, sive De causis mirandorum Satvrni Phœnomenôn, Et Comite ejus Planeta Novo*. Hagae-Comitis: Vlacq, 1659.

Huygens, Christiaan. *The celestial worlds discover'd, or, Conjectures concerning the inhabitants, plants and productions of the worlds in the planets*. London: Timothy Childe, 1698[b].

Huygens, Christian. *De wereldbeschouwer, of Gissingen over de hemelsche aardklooten*. Rotterdam: B. Bos, 1699.

Huygens, Christian. *Herrn Christian Hugens etc. Cosmotheoros Oder Welt-betrachtende Muthmassungen von denen himmlischen Erd-Kugeln und deren Schmuck*. Leipzig: Lanckisch, 1703.

Huygens, Christian. *Kosmotheóros, Sive De Terris Coelestibus, earumque ornatu, Conjecturœ*. Hagae-Comitum: Moetjens, 1698[a].

Huygens, Christian. *Traité de la pluralité des mondes*. Paris: J. Moreau, 1702.

Jonson, Ben. "News from the new Worlds Discovered in the Moon: A masque, as it was presented at court before King James, 1620." Ed. James Knowles. *The Cambridge Edition of the Works of Ben Jonson: 5. 1616–1625*. Eds. David Bevington, Martin Butler, and Ian Donaldson. Cambridge: Cambridge University Press, 2012. 423–444.

Kepler, Johannes. *Somnium, seu opus posthumum de astronomia lunari*. Frankfurt, 1634.

Kindermann, Eberhard Christian. *Die Geschwinde Reise auf dem Luft=Schiff nach der obern Welt* (1744). Ed. and comm. Hania Siebenpfeiffer. 2nd, augmented ed. Hannover: Wehrhahn, 2020.

Lafond, Jean. "Le Monde à l'envers dans "Les Etats et empires de la lune" de Cyrano de Bergerac." *L'image du monde renverse et ses représentations littéraires et paralittéraires de la fin du XVIe siècle au milieu du XVIIe*. Eds. Jean Lafond and Augustin Redondo. Paris: J. Vrin, 1979. 129–139.

McColley, Grant. "The Seventeenth-Century Doctrine of a Plurality of Worlds." *Annals of Science* 1.4 (1936): 385–430.

McDermot, Murtagh. *A Trip to the Moon. Containing some observations and reflections, made by him during his stay in that planet, upon the manners of the inhabitants.* Dublin: Christopher Dickson; London: J. Roberts, 1728.

Merchant, W. M. "Bishop Francis Godwin. Historian and Novelist." *Journal of the Historical Society of the Church of Wales* 5 (1955): 45–51.

Omodeo, Pietro Daniel. "Minimum und Atom. Eine Begriffserweiterung in Brunos Rezeption des Cusanus." *Die Modernitäten des Nikolaus von Kues. Debatten und Rezeptionen.* Ed. Tom Müller. Bielefeld: transcript, 2014. 289–308.

Onfray, Michel. *Contre-histoire de la philosophie,* vol. 3: *Les libertins baroques.* Paris: B. Grasset, 2008.

Pascal, Blaise. *Pensées.* Ed. Philippe Sellier. Paris: Pocket, 2003.

Pasqua, Hervé, ed. *Infini et altérité dans l'oeuvre de Nicolas de Cues (1401–1464).* Louvain-La-Neuve and Leuven: Peeters, 2017.

Reeves, Eileen and Albert Van Helden. "Turning the Telescope to the Sun. Thomas Harriot and Johannes and David Fabricius." Galileo Galilei, Christoph Scheiner. *On Sunspots.* Ed. and trans. Eileen Reeves. Chicago: University of Chicago Press, 2010. 25–36.

Siebenpfeiffer, Hania. "'un petit ciel de pourpre émaillé d'or' – Die Faszination optischer Medien in der frühen Neuzeit." *Zeitschrift für Germanistik* NF 25.2 (2015): 268–286.

Siebenpfeiffer, Hania. "*Die Geschwinde Reise* im Kontext frühneuzeitlicher Science-Fiction." Eberhard Christian Kindermann. *Die Geschwinde Reise auf dem Luft=Schiff nach der obern Welt* (1744). Ed. and comm. Hania Siebenpfeiffer. 2nd, augmented ed. Hannover: Wehrhahn, 2020. 101–124.

Siebenpfeiffer, Hania. *Die literarische Eroberung des Alls. Literatur und Astronomie (1593–1771).* Göttingen: Wallstein, 2025.

Ussher, James. *Annales Veteris Testamenti. A prima mundi origine deducti: una cum rerum Asiaticarum et Aegyptiacarum chronico, a temporis historici principio usque ad Maccabaicorum initia producto.* Londinum: Bedell, 1650.

Van Helden, Albert. "Huygens and the Astronomers." *Studies on Christiaan Huygens.* Eds. H. J. M. Bos, M. J. S. Rudwik, H. A. M. Snelders and R. P. W. Visser. Lisse: Swets & Zeitlinger, 1980. 147–166.

Venator, Balthasar. *Kurtze and Kurtzweilige Beschreibung der zuvor unerhörten Reise [. . .] in die neue Ober-Welt des Monds.* Nürnberg: [s. n.], 1660.

Voltaire [François-Marie Arouets]. *Micromégas. L'ingénu.* Ed. Jacques Van den Heuvel and Frédéric Deloffe. Paris: Gallimard, 2002.

Alexander Honold

Celestial Education

The Formative Impact of Astronomy on the German *Bildungsroman*: Goethe, Hölderlin, and Jean Paul

Abstract: In German-language literature around 1800, the *Bildungsroman* was one of the dominating genres. When sketching the adventurous trajectories of the problematic subject on its path towards self-development, authors like Goethe, Hölderlin, and Jean Paul make intense use of the astronomic patterns describing the celestial bodies in their movements. Especially, the planets and their orbits around the sun were conceived of as an educational model that gave evidence of how oppositional driving forces could be taken into a balanced, self-reproducing mechanism (Hölderlin, *Hyperion*). Thus, in the literary reception of astronomical knowledge in that period, notions like revolution, eccentric paths, and kinetic energy were the main topics, closely related to a dynamic understanding of the individual subject in its development.

1

Around 1800, astronomy represented in the realm of cultural knowledge both an empirical-mathematical discipline and a magical practice. Astral scholarship stood at the threshold of scientific specialization but had not yet taken the step of relegating unscientific questions of cultural semantics to the field of astrology, already stigmatized as obscure. Astronomers held leading positions during the French Revolution or were the founders and leaders of scientific associations; conversely, it was not at all unusual for artists, musicians, and writers to do some astronomy on the side. Authors like Hölderlin, Jean Paul, Novalis and Kleist tackled astronomical issues in numerous passages of their works and used them as exemplary sources for poetic images; not to mention Goethe's enlightened astral devotion or the linguistic and natural history research programs of the two Humboldts. As is well known, astronomical knowledge also plays a supporting conceptual role in Kant's determination of the position of the critical age, in Schiller's schematization of historical thinking and in Hegel's systemic philosophy.[1]

---

1 For a composite overview of the literary and cultural significance of astronomical knowledge around 1800, cf. Honold (2005), Hunfeld (2004), Briese (1998).

Viewed within the history of knowledge, astronomy, in the decades around 1800, undergoes an epistemic transition. Already, as in other areas of natural knowledge, methodically guided rules of data collection and their analysis apply, but still the starry sky and its wandering planets can be understood as a large canvas upon which mythological figures and story stock can be projected. But while, for example, in analytical chemistry with and after Lavoisier, the description of reaction processes led to the dissection of those supposed original substances which had represented the basic framework of the ancient elemental theory and thus made for a radical and complete paradigm shift whence there was no return (Stengers 1994), the field of astronomical knowledge retained the residues of mythological nomenclature that seem to have been better reconciled with the progress of empirical observation techniques.

In the following considerations, I would like to elaborate on the cultural significance of astronomy in this epochal transitional period around 1800, primarily in the field of literature. I will focus on an area that was particularly decisive for the genre-poetics development of both the German-language and international novel during this time: the narratives of "apprenticeship" and educational trajectories. The so-called *Bildungsroman*, which first emerged as a special development within German-language literature with Wieland's *Geschichte des Agathon* and Goethe's *Wilhelm Meister* novels as the founding texts, has as its subject the gradual development of an individual who progressively expands his abilities, character traits, and behavior as he interacts with his environment and thereby learns not to reject the realities of the external world, but to incorporate them into the development of his own personality, meeting the challenges to his own thinking and acting. To educate oneself successively, to develop one's talents and to pursue self-imposed goals, these values, taken from the models established by Wieland and Goethe, become the most important characteristics of a self-confident and capable personality. Whereas in other European literary traditions (Spain, England, France), the picaresque novel and the comedy of the debauched art of storytelling developed along strong lines of tradition, and in the nineteenth century, the social novel in all its varieties also immersed itself in diverse social milieus and atmospheres (France, England, Russia), in the German-language literary landscape one can observe an eminent interest in questions of upbringing and education that continues to echo far into the modern age.

The career of an individual character in a novel, be it Goethe's Wilhelm Meister, Gottfried Keller's Grüner Heinrich or Robert Musil's Man without qualities, develops within the respective, and extensive, plot through a constant back and forth between intrinsic and extrinsic factors, inner driving forces incessantly confronting the environment in an interplay that forces shape upon the developing personality. It may be surprising if I here forward the thesis that for this narrative model of the subjective educational path, not only pedagogical ideas in the narrower sense play an active role, but also an intensive orientation towards astronomical phenomena and findings can be observed in many of these literary personality designs. For some authors, this may also have to do with a specific interest in current discoveries of celestial science, as is attested above all for Hölderlin and Jean Paul. But beyond that, contending with astronomical models and contexts also proves a fundamental form of self-understanding for the poets and thinkers of German idealism around 1800, and this can particularly be shown in the narrated educational paths of the literary protagonists around 1800. It was, according to my assumption, just the pedagogical charge of the astronomical discourse at that time which was partly responsible for the popularity and applicability of the phenomena of celestial science to the problem of the individual personality based on astronomical models.

2

Apparently, despite the enlightened age or precisely because of it, in the second half of the eighteenth century the question of the correct or appropriate development of one's own personality continued to be under the necessary assistance of heavenly powers. However, these protective powers could no longer go hand in hand with a Christian conception of God and the certainty of salvation derived from it. By the catastrophic earthquake in Lisbon in 1755, at the latest, which had a lasting impact on the European public and claimed hundreds of thousands of lives as it turned a shining port city into a field of rubble, the belief in a benevolent God had to be regarded as fundamentally shaken. With the theodicy problem raised by Leibniz, Voltaire, Rousseau and others, the deep rift between metaphysical hopes and earthy experiences had become obvious, however one tried to reconcile faith and knowledge; historical contingency had taken the place of divine providence. But what could a young person hope for or expect from his initial career when neither divine providence nor the far-reaching protection of an influential family father could secure undisturbed development in conformity with tradition? Goethe's *Wilhelm Meister* novel narrates the educational career of its main character against the background of a structural weakness not only of the divine but also of the familial fatherly authority. In its place, a so-called "tower society" (*Turmgesellschaft*) acts as a caring protective power, watching over the main character every step from the background, including his missteps and his challenges, which, for this reason, never intensify into such serious crises that they could throw the hero completely off his course.

Without delving into the complex plot of these novels in more detail here, I would like to emphasize at least the name of this providential institution, the tower society, because a tower connotes, in remarkable iconic condensation, first, the arcane knowledge of a secret society or lodge (as with the Freemasons); second, the upwardly oriented (and symbolically transcendence seeking) verticality, as it is known from sacred buildings; and thirdly, the impression of an astronomical observer's post that arises from this vertical orientation. From the tower, all the movements of the protagonist laboriously searching for his way, can be viewed, as it were, from on high; there, the threads come together, and the obstacles, errors, and dangers recognized at the outset are resolved.

The model of the tower society corresponds to the successive progress of the novel's character throughout the trials and tribulations of life with a simultaneous image that both absorbs linear time and cancels it out in a kind of total representation. In this way, the idea of the tower bears a striking resemblance to the devices of the amateur astronomer Makarie, who in the second Wilhelm Meister novel transfers all her interactions into the planetary and sidereal realm. For this purpose, she uses a device known from traditional astronomy, an armillary sphere. With the help of her armillary sphere, she arranges all the personal relationships of the novel around herself, the figure of Makarie; she also uses the armillary sphere to restore that human illusion, lost since Copernicus, of being at the virtual center of the universe. This poetically restituted, but fictional, center is described in the novel as follows:

> Makarie befindet sich zu unserm Sonnensystem in einem Verhältnis, welches man auszusprechen kaum wagen darf. Im Geiste, der Seele, der Einbildungskraft hegt sie, schaut sie es nicht nur, sondern sie macht gleichsam einen Teil desselben; sie sieht sich in jenen himmlischen Kreisen mit fortgezogen, aber auf eine ganz eigene Art; sie wandelt seit ihrer Kindheit um die Sonne, und zwar, wie nun entdeckt ist, in einer Spirale, sich immer mehr vom Mittelpunkt entfernend und nach den äußeren Regionen hinkreisend. (Goethe 1989, 734–736)

> [Makaria is found to be in a relation to our solar system which one may hardly venture to express. In the spirit, the soul, the imagination, she cherishes it; she not only contemplates it, but forms as it were a part of it. She sees herself drawn onward in those heavenly orbits, but in a manner quite peculiar; she has revolved round the sun since her childhood, and in fact, as is now discerned, in a spiral continually receding from the central point and circling towards the outer regions. (Goethe 1885, 426)]

The studious Makarie indulges her astronomical interests in the application of that constellatory knowledge gained from the contemplation of her armillary sphere to the familial, cordial, and social relationships unfolding around her. It is no accident that Makarie's astronomical studies, as well as the mechanical structure of the armillary sphere itself, are set in a tower room, connecting them to the metanarrative observation, control, and intervention possibilities of tower society.

Already in the mid-1790s, when Goethe was busy with the completion of the first *Wilhelm Meister* volume, he had, due to his intensified cooperation with Friedrich Schiller, also become more closely acquainted with the latter's extensive Wallenstein historical drama, in which a protagonist strongly occupied by astronomical questions also makes an appearance. At the time of Wallenstein and his court astrologer, astronomy and astrology were still intertwined in a way that could hardly be disentangled. Thus it is a rather ambivalent moment when, before the most important strategic maneuver of his life, the great commander Wallenstein procrastinates by futilely consulting his astronomical tower cabinet to ask the stars for their advice.

Schiller's Wallenstein tries in vain to determine from the heavenly signs whether his future is to be happy or fatal; but if he feels encouraged by astrological omens to make a bold switch to the opposing military opposite, then, as Schiller makes clear, this only results from his own forced projection, which lets the commander see in the book of the heavens what he wants to see – namely a god of war whose happiness can be literally forced upon him by political rule and marriage policy:

> Glückseliger Aspekt! So stellt sich endlich
> Die große Drei verhängnisvoll zusammen,
> Und beide Segenssterne, Jupiter
> Und Venus, nehmen den verderblichen,
> Den tück'schen Mars in ihre Mitte, zwingen
> Den alten Schadenstifter mir zu dienen.
> (Schiller 2000, 155)[2]

> [Auspicious aspect! fateful in conjunction,
> At length the mighty three corradiate;
> And the two stars of blessing, Jupiter
> And Venus, take between them the malignant
> Slily-malicious Mars, and thus compel
> Into my service that old mischief-founder (. . .).
> (Schiller 1906, 153)]

The whole tradition of the planetary allegorism and its mythological origin is recalled in Wallenstein's fatal fallacy. Mars as the red, warlike planet was such a difficult, stubborn celestial body only because the old circular models of astronomy could not sufficiently explain its looping and putatively regressive backward

---

2 Cf. Alt (2008, 157–158).

movements, given that this planet's orbit was extremely elliptical, something Johannes Kepler, working together with Tycho Brahe in Prague, solved exactly during Wallenstein's time. And when Kepler then drew the conclusion suggested by Brahe's observational data and calculation models that the planets of the solar system did not move on circular orbits around the central star, but elliptically, the planet Mars was indeed his most important point of reference – a truly groundbreaking discovery, to which Schiller's emphatic emphasis on this planet in *Wallenstein* implicitly alludes.

If the authors of the German classical period did not completely and consistently abandon the images and forms of expression inherited from mythical and astrological traditions, this was, on the one hand, due to a better rhetorical grasp on these metaphorical concepts, which ultimately persists to this day and still allows us to speak of 'sunrise' and 'sunset.' On the other hand, however, the classics could only work out their drafts of the self-designed, free development of man and society, shaped as they were by bold idealistic values, with the help of some positive basic assumptions and cultural foundations. Among these foundations, especially after the aforementioned profound shaking of the former trust in the goodness of God and the happy course of world history, was the certainty, not yet damaged in the same way, that the creation surrounding the man-made world was well-formed, regular and significant. The aspect of "meaningfulness" (*Bedeutsamkeit*), named by Hans Blumenberg as a central category of the persistence of mythical ideas (cf. Heidenreich 2014), may indeed have played a decisive role, at least for Goethe; meaningfulness could only be established where a close, dialogical relationship could develop between the realm of an individual life in its factual contingencies and the objective surrounding world. The whole world becomes the responding or resonating space of the ego: the emphatic conception of the subject of German idealism and its idea of education aims at nothing less.

At no point did Goethe more succinctly express his insistence on the significance of and strong interest in astronomical thinking than in the spectacular description of the moment of his birth, which is punctuated with all the registers of astrological and astronomical phenomena. In the autobiographical retrospective *Truth and Poetry*, the poet reenacts the auspices of his thoroughly difficult and perilous entrance into life, endowing day and hour with a wealth of meaningful astrological concomitants:

> Am 28. August 1749, Mittags mit dem Glockenschlage zwölf, kam ich in Frankfurt am Main auf die Welt. Die Konstellation war glücklich, die Sonne stand im Zeichen der Jungfrau, und kulminierte für den Tag; Jupiter und Venus blickten sie freundlich an, Merkur nicht widerwärtig; Saturn und Mars verhielten sich gleichgültig; nur der Mond, der so eben voll ward, übte die Kraft seines Gegenscheins um so mehr, als zugleich seine Planetenstunde einge-

treten war. Er widersetzte sich daher meiner Geburt, die nicht eher erfolgen konnte, als bis diese Stunde vorübergegangen. (Goethe 1986, 15)

[On the 28th of August, 1749, at mid-day, as the clock struck twelve, I came into the world, at Frankfort-on-the-Maine. My horoscope was propitious: the sun stood in the sign of the Virgin, and had culminated for the day; Jupiter and Venus looked on him with a friendly eye, and Mercury not adversely; while Saturn and Mars kept themselves indifferent; the Moon alone, just full, exerted the power of her reflection all the more, as she had then reached her planetary hour. She opposed herself, therefore, to my birth, which could not be accomplished until this hour was passed. (Goethe 1897, 1)]

This lush and baroque description obviously recalls the characteristic style of traditional so-called *nativities*, birth horoscopes often rendered only for princely persons under the auspices of the state. With Goethe's detailed self-portrayal of his own horoscope, the bourgeois personality comes into view as a new guiding value, its founding date and solemnly-celebrated anniversary will henceforth be his own birthday; the actual date of his birth, not the name day determined by the calendar of the saints.

Goethe thus erected a literary monument to the initial moment of bourgeois subjectivity, in whose rhetorical shaping mythical astrology and enlightened self-confidence converge in an exciting way. The fact that he, the poet beginning his life's journey, no longer traces himself from a familial genealogy nor from the social position given to him constitutes an unconditional claim to freedom for the bourgeois subject, now powerful in his own right. By calling upon the causally uninvolved forces of the heavens instead of the powers of tradition, the subject, forging his own precipitous career path, turns them into solemn-symbolic witnesses of this founding act. Planets and celestial bodies therefore no longer function as influences in the sense of magical correspondences and guides of this earthly course of life but act in an emotionally reinforcing way as a cosmic sphere of resonance.

In this sense, Goethe's autobiographical self-interpretation of an epochally significant life path originating from a "favorable constellation" (or, in Oxenford's translation quoted above, a 'propitious horoscope') creates the cosmic-symbolic frame, which the narrated educational course in his novel still needed in order to map his protagonist's winding paths through various social spheres of activity, operating under a celestial orientation guiding him, as the tower society had been able to offer only in a discreet background manner.

3

Goethe's autobiographical self-justification through the consecrations of astronomical choreography was highly consequential for German literature in the nineteenth century and remains so for modernity. Biographies, autobiographies and fictional life stories repeatedly reference the auspices of some birth with symbolic calculation, but always with a certain awareness that no magical powers or mysterious forces of predestination were at work, rather in each case an individual life dared to relate itself, in a self-responsible manner, to the thought patterns of astrological-mythical interpretations of fate. In this context, Joseph von Eichendorff's novel *Ahnung und Gegenwart* (pub. 1815) deserves a special mention, because here we find a deliberate counterfactual of Goethe's narrative scheme, which makes all astronomical accompanying circumstances, which appear so positive in Goethe's work, now simply appear as the opposite.

Even before Goethe's significant revival of astronomical symbolism was carried out, it had been stated in the work of the political philosopher Jean-Jacques Rousseau that familiarity with astronomical contexts took on an important role in the processing of intellectual formation and the shaping of human self-awareness. In the age of science-based enlightenment, astronomy was a domain for the practice of critical competence – from Copernicus to Kepler and Newton to Kant and Laplace. Nowhere was the split between eye and judgment greater than when looking at celestial phenomena; and nowhere else was the plausibility of abstract models of thought more needed to subsequently reconcile observation and thinking.

In Rousseau's seminal educational novel *Émile ou de l'éducation* (1762), the exemplary education of a pupil named Émile by an ideal pedagogue (probably none other than Rousseau himself) is sketched out in a kind of *Bildungsroman avant la lettre*. Thereby, the pedagogue should not presuppose or assert anything the pupil cannot bring directly before his eyes and be so seen and understood. The phenomena of inner and outer nature in their physical interrelationships set the rhythm and the scope of the pupil's progress; Émile's teacher confines himself to that Socratic midwifery which gives the necessary jump-start to the pupil's understanding of nature, as it unfolds as if by itself.

To bring about important insights unobtrusively, the teacher discreetly employs the wonderful laws of astronomy. The dates of the observations guided by the teacher are anything but arbitrary. They are watching the sunrise on the summer solstice and its winter counterpart immediately before Christmas (Rousseau 1969, 433). The aim is to anchor the symmetry and turning points of the annual cycle in the conscious experience of the young observer and to stimulate curiosity and reflection by means of their difference. Spring and autumn, on the other hand, less striking than the extremes of the sun's lowest and highest points, simultaneously remind us of both, pointing forward and backward at the same time. Especially within the scenery of spring, Rousseau notes, imagination is prepared to follow the seasonal course of the year, and unites manyfold dimensions of time in just one single moment:

> C'est qu'au spectacle du printemps l'imagination joint celui des saisons qui le doivent suivre; à ces tendres bourgeons que l'œil apperçoit elle ajoûte les fleurs, les fruits, les ombrages [. . .]. Elle réunit en un point des tems qui se doivent succéder. (Rousseau 1969, 418)

> [Because imagination adds to the sight of spring the image of the seasons which are yet to come; the eye sees the tender shoot, the mind's eye beholds its flowers, fruit, and foliage (. . .). It blends successive stages into one moment's experience. (Rousseau 1955, 123)]

In an act of symbolic significance, Rousseau places his pupil in the middle of the cosmic annual cycle of the sun. Nothing else is needed to know the shape of the earth but to orient the gaze towards the course of the sun (Rousseau 1969, 430). With a remarkable effort of direction, threaded through a harmless morning walk, the pupil witnesses the magnificent spectacle of sunrise, this apotheosis of the enlightenment (French *éclairage*) driving away the shadows. Rousseau's astronomical theater of nature stages enlightenment by means of aesthetic absolutism: first the gradual gathering of the court, then the flashing appearance of the central ruling authority. Announced by a nameless crowd of atmospheric harbingers, the shining star makes its great, eventful leap into existence – just as we have seen in Goethe's later birth of the individual.

But Rousseau's Émile, brought up in skeptical sobriety, is not one to be dazzled by this play of light. What the pupil notices, and also should notice, is an inconsistency. Brought under the earth in the evening sky, the sun rises miraculously in the morning on the opposite side. An enveloping mystery, hiding all the power of nature to transform death and destruction into rebirth. The restriction of Émile's education to empirical observation and inductive learning processes means his astronomical instruction initially operates within the framework of a geocentric world view, where the sun circles around the earth. In fact, that "experience" he lacks in order to appreciate the spectacle of the sunrise, is merely the flipside of an observation the child is very well able to make. The pupil follows the arc of the sun's course as it curves in a semicircle around his own point of view. Now he must try to envision the latent second half of the circle, using his imagination to add the hidden to the visible first half. To follow the arc further from the evening to the morning would mean nothing less than a wandering around the globe.

Celestial science became a touchstone of pedagogical ambition because of a unique quality: its anthropological unsuitability. The relation between the astronomically observable and the humanly comprehensible involves, in several respects, glaring disproportions. The spatial extension and temporal dimension of astronomical cycles exceeds the span of the human perception considerably; to grasp them in their systematics, a human being is, moreover, an unfavorably positioned observer. The epistemological critical sensorium of the Enlightenment was not the first to address these discrepancies: the forefather of pedagogical philosophy himself, Plato, already thought through them in a similar manner. In a passage from the *Timaeus*, Plato presents an argument that later became relevant for didactic applications: God had gifted man with the visual sense in order to make him aware of the regularity of celestial orbits and to mimetically establish the independent course of his own mindful considerations (*Timaios*, 47c; Cornford 1997, 157–158). Heaven provides the human mind with the trajectory to think *further*. Despite all these discrepancies, an affinity arises between the two that keeps knowledge from failing after all. It is these *revolutions* of the celestial bodies that bring men face to face with the basic concepts of space and time, of nature and history.

4

The exemplary astronomical instruction reported by Rousseau in his fictionalized novel of education was not without its reverberations among the poets of the subsequent generation, who in turn witnessed the political awakening of the revolutionary years in France, either actively or as enthusiastic spectators. Friedrich Hölderlin, in particular, saw his own present as an age of epochal change, both politically and astronomically:

> Unter den Sternen ergehet sich
> Mein Geist, die Gefilde des Uranus
> Überhin schwebt er und sinnt; einsam ist
> Und gewagt, ehernen Tritt heischet die Bahn.
> (Hölderlin 1992, 71)

> [Amongst the stars expands
> My spirit, the domain of Uranus
> Exceeding, it roams and contemplates; lonely
> And daring, with iron steps broaches its orbit. (my transl.)]

For Hölderlin, the great date 1789 relates to a heroic figure of astronomy. Johannes Kepler is his man of the hour, a star-seeker and celestial interpreter. Why Kepler? The reason for recalling this historical role model was perhaps an external one: exactly two hundred years earlier, Kepler, who, like Hölderlin, was born in Württemberg, had been admitted to the Tübingen seminary to study the liberal arts, and later theology. Whether Hölderlin and the other graduates of the higher seminary school of Maulbronn celebrated this anniversary in their first year at Tübingen with some event or festive speeches is not certain, but it can be assumed. The memory of Kepler and the view of the planetary orbits discovered by him opens a space, the vastness of which transcends the limitations of the clerical erudition at Tübingen. To walk "amongst the stars" means to be intended by them, to be subject to their force field and their significance.

Hölderlin's own career evinces almost from the beginning an orientation towards astronomical goals and guiding principles. The year 1789 gave rise to a comprehensive determination of his position insofar as Hölderlin had moved into the Tübingen *Stift* the previous October to study philosophy and theology at the same time as Hegel, the beginning of a wonderful constellation in which Kepler's astronomy also played a role.[3] Kepler's insight that the planets do not move in circular orbits but in ellipses around the sun became relevant for Hölderlin in those very years when he began to occupy himself more intensively in literary terms with the narrative scheme of the *Bildungsroman* and the development of personality. The fruit of these studies is the epistolary novel about the enthusiast Hyperion, who, living in modern Greece, suffers under the depressing political conditions and longs for the freedom and beauty of ancient life. This Hyperion behaves eccentrically as he unhappily moves through and suffers in the present, all the while envisioning a utopian refuge beyond. At the same time, its author, Hölderlin, describes such an eccentric nature as an almost necessary form of traversing the human developmental trajectory, much like the elliptical planetary orbits in Kepler's model: "Wir durchlaufen alle eine exzentrische Bahn, und es ist kein andrer Weg möglich von der Kindheit zur Vollendung" (Hölderlin 1992, 558) ["We all run an eccentric course, and no other path is possible from childhood to perfection" (my transl.)].

Hölderlin forged another affinity in the Tübingen years, as already mentioned, with the deceased philosopher Rousseau, from whose *Contract social* (1762) Hölderlin borrows a programmatic sentence as a motto for his Tübingen 'Hymn to Humanity': "*Les bornes du possible dans les choses morales sont moins étroites, que nous ne pensons*" (Hölderlin 1992, 120). A sentence that spurs freedom of thought

---

3 The Tübingen student community is a prototypical expression of the practice of "symphilosophizing," as the early Romantics called it. For Hölderlin's time in Jena, Frankfurt, and Homburg, the association of like-minded thinkers and friends working on common problems is also a significant factor. The continuing philosophical conversation between Hölderlin and Hegel into the Frankfurt and Homburg periods has been carefully traced by both Christoph Jamme (1983) and Panajotis Kondylis (1979).

and action: the limits of what is socially possible are drawn 'less tightly' than we imagine, and they are not to be measured with straightforward guidelines. Rousseau soon assumed a preeminent position in the constellation of Hölderlin's self-chosen guiding stars. In the novel, Hyperion's first teacher, Adamas, possess distinct traits that connect him with Rousseau (Link 1999, 179–181). As in Rousseau's *Émile*, friendly instruction by the master is preferred to the rigid corset of institutional "schools," which Hyperion vehemently berates in retrospect.[4]

Understood in the geometric sense, Adamas' formation of Hyperion is "encyclopedic," bringing the adept into his eccentric orbit before breaking out the *cyclos* of the Greek island world. Adamas travels with the boy to the farthest corners of the Greek world, to Mount Athos and the Hellespont, to Rhodes and the southern tip of the Peloponnese. He introduces Hyperion to the kind of astrospherical reading that Rousseau also allows Émile and curbs his temperament with a love of geometry. The Pythagorean sense for order and the beauty of proportion also underlies the central initiation ritual of the Adamas period when they visit the island of the sun god, the climax of all prior educational experiences. Almost verbatim, Hölderlin follows the description of a contemporary travelogue, Choiseul's *Voyage Pittoresque*, where even the marble steps are mentioned.[5] The orchestration of the plot with topographical details can be seen as a clever creation of an *"effet de réel"* (Roland Barthes), but the marble staircase also serves a dramaturgical function. As he ascends, the light brightens, step by step, around the pupil until finally, at the top, a spectacle familiar from Rousseau begins. There, standing side by side, teacher and pupil await the sunrise:

> Es dämmerte noch, da wir schon oben waren. Jezt kam er herauf in seiner ewigen Jugend, der alte Sonnengott, zufrieden und mühelos, wie immer, flog der unsterbliche Titan mit seinen tausend eignen Freuden herauf [. . .]. Sei, wie dieser! rief mir Adamas zu, ergriff mich bei der Hand und hielt sie dem Gott entgegen, und mir war, als trügen uns die Morgenwinde mit sich fort, und brächten uns in's Geleite des heiligen Wesens, das nun hinaufstieg

---

4 "Ach! wär' ich nie in eure Schulen gegangen. Die Wissenschaft, der ich in den Schacht hinunter folgte, von der ich, jugendlich thöricht, die Bestätigung meiner reinen Freude erwartete, die hat mir alles verdorben" (Hölderlin 1992, 615) ["Oh, had I never entered your schools! Science, whom I followed into its mines, hoping with youthful foolishness for confirmation of my purest joy, has spoilt it all for me" (my transl.)].

5 "If one continues to climb upward, one comes, by a path hewn in granite, to Mount Cynthus; old marble steps help fully up to the summit" (Choiseul-Gouffier 1780/1782, vol. 1.1, 135). Hölderlin also used Richard Chandler's travel descriptions (1775/1776, transl. as *Reisen in Klein Asien* [Leipzig, 1776] and *Reisen in Griechenland* [Leipzig 1777]). For an overview of the models and sources for Hölderlin's Greek topography, cf. Volke (1984/1985). On the significance of the paradigm of the *Voyage Pittoresque* for Hölderlin's 'archaeological view' of Greek antiquity, cf. Honold (2002).

auf den Gipfel des Himmels, freundlich und groß, und wunderbar mit seiner Kraft und seinem Geist die Welt und uns erfüllte. (Hölderlin 1992, 621)

["It was still dawning as we reached the top. Now he arose in his eternal youthfulness, the eternal sun-god, content and effortlessly as ever, the immortal titan arising with his thousand proper joys. Be thus, Adamas called to me, seizing my hand and extending it towards the god, and it seemed as though the morning winds would carry us with them towards the host of the holy being that now ascended the peak of heaven, friendly and great, filling the world and ourselves with its power and its spirit." (my transl.)]

"Be thus!" The scene culminates in the half-secular, half-mythical variant of the Christian baptismal ritual, committing the hero of the novel to the name and course of the sun-god. Stars and men share names whose repeated invocation[6] instigates a communion between the two that (following a concept unfolded in Plato's *Symposium*) can be called symbolic. Thus, for example, the two main stars of the Twins are, according to their mythological meaning as the constellation of the Dioscuri, the heavenly half of the eternally walking pair of brothers; but for those on earth, these stars are to be read as letters inscribing the heroic brothers' names onto the heavens ("Buchstaben, womit der Nahme der Heldenbrüder am Himmel geschrieben ist," Hölderlin 1992, 61). Celestial names are programmatic abbreviations of earthly careers, also and especially for Hyperion, whose astronomical baptism on Cynthus will determine his life. Interpreters have taken this initiation as Hyperion's calling to be a poet,[7] but it is primarily an astronomical form of self-foundation, comparable to how the constellations imprinted Goethe's hour of birth.

5

Astronomical facts and models of thinking also play a complex role in Jean Paul's novels and stories, combining several functions. First, the astronomical frame of reference gains importance for his thinking in relation to the body-soul problem or body-mind duality; here astronomical phenomena create space for the imaginability of a manifoldly theorizable, so-called "second world" (Hunfeld 2004, 143).

---

6 "Oft wenn über mir die Gestirne aufgiengen, nannt' ich ihre Nahmen, die Nahmen der Heroën, die einst auf Erden lebten" (Hölderlin 1992, 533) ["Often, when the stars rose above me, I called their names, the names of the heroes who once dwelt on Earth" (my transl.)].
7 According to Binder (1963), one must conclude from the scene that Hyperion is destined to be a poet and that this is the ultimate meaning of his name. This is supported by the repeatedly suggested identity of the sun-god and Apollo, the god of the poets, and the assignment of both to the sphere of the protagonist.

It is this motif complex of the 'second world' and the associated aesthetic maneuvers of rapture or perspective shift with which Jean Paul responds to the philosophical problem of *commercium mentis et corporis*. What interests him most about the second world is its unavailability or inaccessibility, as well as the possibilities of reversing and questioning earthly 'centrisms.' A central text for these problems is the idyllic tale of the *Campaner Thal* (1797), together with its doubled continuations in the *Catechism of Woodcuts* (1797) and in the late work *Selina* (1823).

Secondly, Jean Paul takes concrete facts from astronomical studies and calculation methods, which refer to the kinetics of planetary and cometary orbits and the energetic filling of the universe. He adopted, as documented in his excerpted notebooks, broadly and intensively the cosmological and astronomical theory debates of the time, from Kant's *Allgemeine Naturgeschichte und Theorie des Himmels* (1755) to Johann Heinrich Lambert's *Cosmologische Briefe über die Einrichtung des Weltbaues* (1761) to William Herschel's *Abhandlungen über den Bau des Himmels* (1785) (Esselborn 1989; Hunfeld 2004, 103).[8] But Jean Paul also quickly and expertly absorbs new empirical discoveries and methodological advances of contemporary celestial science; for example, he observationally follows the recommendations of Rösler's *Handbuch der praktischen Astronomie* from 1788 and consults astronomical yearbooks, journals, and calendars for concrete celestial phenomena (Brosche 2004, 215–225). Together with Euler, Herschel and other experts, Jean Paul emphasizes that the universe, despite its immense dimensions, is not empty, but filled with radiation-generating energy, active forces and diffusing matter. His numerous recourses to Newtonian mechanics and Kepler's laws of planetary orbits also evince his expertise. The fact that the planets move on an elliptical path around the sun, and that this elliptical curve cycle is aligned on two focal points (rather than one, like a circle), is already reflected in German literature around 1790 in Hölderlin's epistolary novel *Hyperion*, but this knowledge also finds a literary expression in the sun-like upward striving of the young men in Jean Paul's novels (Honold 2001, 309–333). The permanent balance maintained by centripetal and centrifugal forces appears in Jean Paul's work in an unexpected place as a sort of dynamic model of forces. When he discusses the disposition of characters and their contribution to the plot in his poetic primer, *Vorschule der Ästhetik* (1804), the author sketches the basic bipolar nature of a literary character according to the pattern of the physical parallelogram of forces as applied to planetary orbits.[9]

---

8 The sources are documented in detail in Müller (1988).
9 Cf. Jean Paul (1960a).

Thirdly, and of particular literary consequence, the entire occidental tradition of knowledge with regard to the correspondence of astronomical and terrestrial planes of action is expressed in a poetics of natural ciphers specifically elaborated by Jean Paul. The author pursues the view that the knowledge of the non-authentic or merely figurative within astronomical phenomena need by no means prevent a compassionate observer of the universe from continuing to integrate celestial phenomena into his life as signs of meaning and destiny. Thus, astronomical ciphers become for Jean Paul a model case for the reflected practice of interpretation (Schäfer 2002). Like the hieroglyphic writing of the universe, any sign-like structure ultimately represents only a heuristic 'as if' construction that treats the reality of external and internal nature as creation – and thus as a readable text. For millennia, people have seen images, figures and temporal signs in the universe and derived from them a semiotics of readability, which is of quite current interest for Jean Paul insofar as it allows him to take a middle position between mere superstition (astrology) and mere unbelief (materialistic agnosticism).

Like Friedrich Hölderlin, who was born in Laufen am Neckar on 20 March 1770, the somewhat older Jean Paul, whose civil name was Friedrich Richter and who was born on 21 March 1763 in the Upper Franconian town of Wunsiedel, is one of the "celestial children" of German literary history. For these authors, whose work is imbued with astronomical phenomena and who took a specific interest in celestial cycles, it was of particular advantage that both their births fell on the vernal equinox, which begins the year of the Zodiac in the sign of Pisces.

In the first lecture of his *Selberlebensbeschreibung* (posth. pub. 1826), which was how he translated 'autobiography' into German, Jean Paul, in the first paragraph, immediately lays out the year of his birth – 1763 –, then the month – March – and the ordinal number of the corresponding day – that is, the twenty-first –, and finally the exact time "at 1 ½ o'clock in the morning," in order to then draw a semantically far-reaching conclusion from this funnel-shaped focusing on a singular point in time. Not only in Goethe's autobiography, then, do we encounter the narrative maneuver of an astrological-astronomical self-justification via an auspicious horoscope. When regarded as events upon the celestial clock, both the birthday and hour of birth are associated with a special astronomical position of the stars and planets, both in relation to each other as well as to the earthly observer's position, and from this specific constellation the tradition of astrological prognostication determined the so-called *nativity*, namely how the designated hour of birth fatefully impacted the subsequent life of the person born under this constellation provided by astro-calendrical time. Since at least the baroque era and its calendars, literature was also busy parsing the pre-significance of this moment of birth and its astronomical constellation and narrating it as an object of artful staging.

Jean Paul's *Selberlebensbeschreibung* is no exception in this respect; he does not miss the opportunity to capitalize narratively upon the peculiarity of the astronomical equinoctial situation of his biographical origin and create his own nativity.

Jean Paul's *Selberlebensbeschreibung* places the date and hour of his own birth so demonstratively at the beginning of his description that one must resist involuntarily holding these records, written in 1818, next to Goethe's autobiographical work (published a mere ten years later), in which day and hour are designated and interpreted in the classical astrological manner. For Jean Paul, however, the detail that "crowns everything," which he adds to the chronographical circumstances of his birth, is the simple but significant fact "daß der Anfang seines Lebens zugleich der des damaligen Lenzes war" (Jean Paul 1960b, 1039) ["that the beginning of his life was also the beginning of that year's spring" (my transl.)]. In this double determination of a seasonal as well as biographical beginning, a core idea of astrological speculation comes to the fore: that of coincidence. Events that take place independently of each other, but at the same time, must, so the idea goes, because of their simultaneity, also have some contextual interaction, or at least some common ground of reference, against the backdrop of which they can only really be perceived by observers as coincidence, that is, as a form of contingency. In this case, the narrating self-biographer explicitly becomes the observer, who makes the coincidence of the matter explicit and thus brings it into the spotlight of our attention – admittedly not without a due measure of narrative self-irony.[10]

With matter-of-fact seriousness, even a certain celestial solemnity, the writer deals in his literary work with how human thinking and feeling aligns with the star-spangled sky, especially at the point when an educational initiation akin to Rousseau's natural philosophy takes place with the description of the cosmic sphere of reference that also serves to imprint the protagonist. This happens in a particularly prominent place in Jean Paul's great novel *Titan* (1800–1803), the work in which Jean Paul's ambitions for 'classicism' are presented with the greatest exuberance.

The motifs of the sun-cult and the fiery figures of light form the atmospheric foundation for this enormous work, which in its scope and high tone alone testi-

---

10 "Den [. . .] Einfall, daß ich und der Frühling zugleich angefangen, hab' ich in Gesprächen wohl schon hundert Male vorgebracht; aber ich brenn' ihn hier absichtlich wie einen Ehrenkanonenschuß zum 101ten Male ab, bloß damit ich mich durch den Abdruck außer Stande setze, einen durch den Preßbengel schon an die ganze Welt herumgerichteten Bonmot-Bonbon von neuem aufzutragen." (Jean Paul 1960b, 1039) ["I must have mentioned the idea that spring and I took our beginning simultaneously in a hundred conversations, but here, I purposely launch it, like a cannon salute, for the one-hundred-and-first time, if only to prevent myself from re-using in the future a *bon-mot* made common in print to the whole world" (my transl.)].

fies to the author's aesthetic aspirations. Substantial parts of this work were written during his stay in Weimar, under the impression of the court of the Muses and its great poetic figures, Goethe and Schiller, Herder and Wieland. Based on the models of these "Himmelsstürmer der Epoche" ["period's reachers for the stars"], this work had set out to dissect the period's genius in a novel ("das Geniewesen der Epoche in einem Roman auszufalten", Pfotenhauer 2013, 237). However, Jean Paul's novel title does not primarily allude to the mythological Titans in the plural – those elemental rebels against the Olympians – but to 'the' Titan in the singular, that solar youth who bears the name Hyperion, also taken up by Friedrich Hölderlin.

Jean Paul's *Titan* and Hölderlin's *Hyperion*, written at the same time in the 1790s, are high-register glorifications of a young man under a southern light, who, like the sun-god of antiquity, precipitously traces his rapturous course across the firmament. In Hölderlin's work, the hero is a modern Greek who dreams of ancient Hellas in elegiac letters and throws himself energetically into the political struggle for freedom aiming to become a hero the old-fashioned way. In Jean Paul's novel, Albano de Cesara, a descendant of a Spanish count transplanted as a youth to Germany for years of strict education, now, at the age of twenty, meets his father for the first time. As a half-orphan, Albano had come under the "pedagogical artisan gardener" (*pädagogischen Kunstgärtner*); with him, as with the pupil in Jean Paul's earlier novel *The Invisible Lodge* (1793), rather drastic experiments had been undertaken,[11] consisting of years in isolation and turning away from the world. The alienation from his biological parents was intended to facilitate the reorientation of the subject with the help of calculated educational programs.

Only through the detachment from genealogical origins, one of the recurring principles of Jean Paul's educational models, could the counterforce of development effectively create a 'second world' at all, could a cultural recoding of the identity and destiny of the hero succeed. Whereas the *Invisible Lodge* allows its subject to become a novelist at the age of thirteen, the heroic career of Count Albano in *Titan* does not begin until the advanced age of twenty. The beginning of the novel makes it clear that here, too, the world of childhood and origin must be overwritten or reformatted by subjecting the protagonist to a kind of initiation ritual that takes place surrounded by dazzling water and mountain panorama of Isola Bella in Lake Maggiore, following the natural choreography of the sun. There, on the Borromean Islands, where the hero had spent a short part of his childhood, is he finally to meet Gaspard, his father.

---

11 Cf. Pethes (2014).

On the eve of this encounter, eagerly awaited by Albano, the hero, full of melancholy longing, watches the setting sun. This departure makes him painfully aware once again of the unoccupied or alienated place of his father. This may resonate with a memory of a similar scene in Rousseau's *Émile*. At the hour before sunrise ("Stunde vor Sonnenuntergang") Albano is driven "hinaus ans Ufer des Lago [. . .]. Hier stand der Jüngling, das beseelte Angesicht voll Abendrot, mit edeln Bewegungen des Herzens und seufzte nach dem verhüllten Vater, der ihm bisher mit Sonnenkraft, wie hinter einer Nebelbank, den Tag des Lebens warm und licht gemacht" (Jean Paul 1960c, 16) ["he must away to the shore of the Lago (. . .). Here stood the noble youth, his inspired countenance full of the evening glow, with exalted emotions of the heart, sighing for his veiled father, who, hitherto, with an influence like that of the sun behind a bank of clouds, had made the day of his life warm and light" (Jean Paul 1863, 6)].

The power which the alleged father still exerts over the son, even from a distance, is compared to the effect of sunbeams penetrating the veils of fog (in Brooks' translation, a 'bank of clouds'). This can even be a protective mechanism, recognizing this father only "veiled" respectively by his effects instead of facing him directly. The commandment not to look directly at the sun is already in ancient times a much-used *topos* for the attitude of an astronomical observational culture that ties its viewing conditions to the eclipse. For there, where the overlarge brightness of the sun superimposes and extinguishes all starlight behind it, the physical condition of all seeing turns into its opposite. On the other hand, only the absence of the central star and its optical superiority makes it possible to perceive the multitude of stars and planets in their subtleties on the dark side of the world with sensitively dilated eyes. The fact that in daylight the stars are eclipsed and shine only in the darkness belongs to the early teachings of astronomical instruction, as it was staged from Plato to Rousseau with the celestial bodies as visual objects in literature many times.

The rising of the light in the east, glorified by Herder as the primal scene of human knowledge in general (Herder 1883, 224), also forms, in the opening of *Titan*, a counterpart that effectively responds to the elegiac mood of doom. Here, one can speak of an astronomically grounded scene of 'imprinting' insofar as the dramaturgy of the narrative orients itself precisely to the choreography of the sun's path, in that it lets the hero and his companions undertake the crossing to the beautiful island in the softened darkness of the moonlight, so he may, with astonishment, witness the growing day, the first day of his new life, on a hill that opens to wide vistas.

As in the *Invisible Lodge* with its solemn exit from the cave, so also in *Titan*, the encounter with the father, this second birth, is completely drawn from the contrast of shadow giving in to light. To preserve the night sensitivity of his eyes until the last moment and to expose himself without transition to the epiphany of a sun shining as if anew, Albano (after Latin *albus*, 'white,' the 'pale one') keeps a light-swallowing bandage around his eyes as he ascends the island's height, although he already hears lively signs of revival from his friends, Schoppe and Dian.[12] This is, of course, not yet the initiated re-encounter with the distant father, but this ascent to encounter the sun far surpasses the family reunion, for which the sun encounter prepares atmospherically, in terms of aesthetic impact. For Albano, the beginning of this solar day becomes a second moment of birth, as it goes hand in hand with the emergence of the outer world surrounding him, an un-paralleled landscape panorama, to which the first morning rays lend the redness and freshness of a creation that has just burst forth.

Albano no longer needs to look back to his own origin, because he becomes a witness of a far more comprehensive process of creation, which his wandering eye takes in and fully absorbs. The high, rapturous tone of Jean Paul's early novels, which, following his own schematization, are modeled after the Italian school, has found its exemplary form in enthusiastic passages, in which something like the pathos of creation actually comes to life:[13] it is the elemental forces of nature

---

12 "Der verhangne Träumer hörte, als sie mit ihm die zehn Terrassen des Gartens hinaufgingen, neben sich den einatmenden Seufzer des Freudenschauers und alle schnelle Gebete des Staunens; aber er behielt standhaft die Binde und stieg blind von Terrasse zu Terrasse, von Orangendüften durchzogen, von höhern freiern Winden erfrischt, von Lorbeerzweigen umflattert – und als sie endlich die höchste Terrasse erstiegen hatten, unter der der See 60 Ellen tief seine grünen Wellen schlägt, so sagte Schoppe: 'Jetzt! jetzt!' – Aber Cesara sagte: 'Nein! Erst die Sonne!' Und der Morgenwind ward die Sonne leuchtend durchs dunkle Gezweit empor, und sie flammte frei auf den Gipfeln – und Dian zerriß kräftig die Binde und sagte: 'Schau umher!' – 'O Gott!' rief er selig erschrocken, als alle Türen des neuen Himmels aufsprangen und der Olymp der Natur mit seinen tausend ruhenden Göttern um ihn stand" (Jean Paul 1960c, 21–22) ["The veiled dreamer heard, as they ascended with him the ten terraces of the garden, the deep-drawn sigh and shudder of joy close beside him, and all the quick entreaties of astonishment; but he held the bandage fast, and went blindfold from terrace to terrace, thrilled with orange-fragrance, refreshed by higher, freer breezes, fanned by laurel-foliage, – and when they had gained at last the highest terrace, and looked down upon the lake, heaving its green waters sixty ells below, then Schoppe cried, 'Now! now!' But Cesara said, 'No! the sun first!' and at that moment the morning wind flung up the sunlight gleaming through the dark twigs, and it flamed free on the summits, – and Dian snatched off the bandage, and said, 'Look round!' 'O God!' cried he with a shriek of ecstasy, as all the gates of the new heaven flew open, and the Olympus of nature, with its thousand reposing gods, stood around him" (Jean Paul 1863, 12)].

13 "Welch eine Welt! Die Alpen standen wie verbrüderte Riesen der Vorwelt fern in der Vergangenheit verbunden beisammen und hielten hoch der Sonne die glänzenden Schilde der Eisberge entgegen – [. . .] und zu ihren Füßen lagen Hügel und Weinberge – und zwischen den Gewölben aus Reben spielten die Morgenwinde mit Kaskaden wie mit wassertaftnen Bändern – und an den Bändern hing der überfüllte Wasserspiegel des Sees von den Bergen nieder, und sie flatterten in

and life that begin to stir in this landscape painting; the light warms the icy glaciers of the high mountains and thereby enlivens the water as it flows away, vineyards and chestnuts suggest the blessings of a lush cultivated landscape. As a mirror, the surface of the water is a repetition of what is happening on high, and so the earthly and heavenly spheres seem in reciprocal recognition to double each other.

The sacred marriage ("hieros gamos") of heaven and earth, of which the ancient creation myths already tell – for example in Hesiod's *Cosmogony* or in Ovid's *Metamorphoses* – also underlies this cheerful day of creation, with which *Titan* begins. The 'Italian' tone of the novel invokes the astronomical order of being whenever a staging for great moments is needed (Hunfeld 2004, 140). But the phenomena of space are more than merely expedient resonance amplifiers for moments of the sublime. What the celestial signs bear witness to is the almost magical communication between self and world. In the spectacular sunrise of the beginning, the author stages a particularly meaningfully orchestrated synopsis of natural signs, which together in turn produce a chronological and temporal picture significant for the individual educational event, that is, a horoscope in the cosmological sense. And as the one personally meant by this auratic world, Albano moves into focus; just as the moment itself becomes the hour of a second birth, so this new self-foundation of the hero in turn indicates a second world of creation still in the process of becoming.

Bibliography

Alt, Peter-André. *Klassische Endspiele. Das Theater Goethes und Schillers.* Munich: C. H. Beck, 2008.

Binder, Wolfgang. "Hölderlins Namenssymbolik." *Hölderlin-Jahrbuch 12* (1961/1962). Eds. Wolfgang Bilder and Alfred Kelletat. Tübingen: J. C. B. Mohr (Paul Siebeck), 1963. 95–204.

Briese, Olaf. *Die Macht der Metaphern. Blitz, Erdbeben und Kometen im Gefüge der Aufklärung.* Stuttgart and Weimar. Metzler, 1998.

Brosche, Peter. "Jean Paul unter dem Himmel der Astronomen." *Jahrbuch der Jean-Paul-Gesellschaft* 39 (2004): 215–225.

---

den Spiegel, und ein Laubwerk aus Kastanienwäldern faßte ihn ein." (Jean Paul 1960c, 22) ["What a world! There stood the Alps, like brother giants of the Old World, linked together, far away in the past, holding high up over against the sun the shining shields of the glaciers (. . .) and at their feet lay hills and vineyards, and through the aisles and arches of grape-clusters the morning winds played with cascades as with watered-silk ribbons, and the liquid brimming mirror of the lake hung down by the ribbons from the mountains, and they fluttered down into the mirror, and a carved work of chestnut woods formed its frame" (Jean Paul 1863, 12–13)].

Chandler, Richard. *Travels in Asia Minor, and Greece or An Account of a Tour, Made at the Expense of the Society of Dilettanti*. Oxford 1775/1776.

Choiseul-Gouffier, Marie Gabriel Florente Auguste. *Voyage pittoresque de la Grèce*. Paris 1780/1782.

Cornford, Francis M. *Plato's Cosmology. The* Timaeus *of Plato*. Indianapolis: Hackett, 1997.

Esselborn, Hans. *Das Universum der Bilder. Die Naturwissenschaft in den Schriften Jean Pauls*. Tübingen: Niemeyer, 1989.

Goethe, Johann Wolfgang von. *Wilhelm Meister's Travels*. Ed. and trans. Edward Bell. London: George Bell and Sons, 1885.

Goethe, Johann Wolfgang von. *The Autobiography of Goethe. Truth and Poetry: From My Own Life*. Trans. John Oxenford. Vol 1., rev. ed. London: George Bell and Sons, 1897.

Goethe, Johann Wolfgang von. *Aus meinem Leben. Dichtung und Wahrheit*. Ed. Klaus-Detlef Müller. *Sämtliche Werke. Briefe, Tagebücher und Gespräche, I. Abteilung*, vol. XIV. Frankfurt am Main: Deutscher Klassiker Verlag, 1986.

Goethe, Johann Wolfgang von. *Wilhelm Meisters Wanderjahre. Sämtliche Werke. Briefe, Tagebücher und Gespräche, I. Abteilung*, vol. 10. Eds. Gerhard Neumann and Hans-Georg Dewitz. Frankfurt am Main: Deutscher Klassiker Verlag, 1989.

Heidenreich, Felix. "Bedeutsamkeit." *Blumenberg lesen. Ein Glossar*. Eds. Robert Buch and Daniel Weidner. Berlin: Suhrkamp 2014. 43–56.

Herder, Johann Gottfried. *Aelteste Urkunde des Menschengeschlechts*. Erster Band (1774). *Sämmtliche Werke*, vol. 6. Ed. Bernhard Suphan. Berlin: Weidmannsche Buchhandlung, 1883.

Hölderlin, Friedrich. *Sämtliche Werke und Briefe*. Ed. Michael Knaupp. 3 Vols. Munich: Hanser, 1992.

Honold, Alexander. "Krumme Linie, exzentrische Bahn: Hölderlin und die Astronomie." *Erschriebene Natur. Internationale Perspektiven auf Texte des 18. Jahrhunderts*. Ed. Michael Scheffel. Bern: Peter Lang, 2001. 309–333.

Honold, Alexander. *Hölderlins Kalender. Astronomie und Revolution um 1800*. Berlin: Vorwerk 8, 2005.

Honold, Alexander. *Nach Olympia. Hölderlin und die Erfindung der Antike*. Berlin: Vorwerk 8, 2002.

Hunfeld, Barbara. *Der Blick ins All: Reflexionen des Kosmos der Zeichen bei Brockes, Jean Paul, Goethe und Stifter*. Tübingen: Niemeyer, 2004.

Jamme, Christoph. *"Ein ungelehrtes Buch." Die philosophische Gemeinschaft zwischen Hölderlin und Hegel in Frankfurt 1797–1800*. Bonn: Bouvier, 1983.

Jean Paul. *Titan: A Romance*. Trans. Charles T. Brooks. London: Trübner & Co., 1863.

Jean Paul. *Vorschule der Ästhetik. Sämtliche Werke*. Ed. Norbert Miller. Munich: Hanser 1960[a].

Jean Paul. *Selberlebensbeschreibung; Sämtliche Werke*. Ed. Norbert Miller. Munich: Hanser 1960[b].

Jean Paul. *Titan. Sämtliche Werke*. Ed. Norbert Miller. Munich: Hanser 1960[c].

Kondylis, Panajotis. *Die Entstehung der Dialektik. An Analysis of the Intellectual Development of Hölderlin, Schelling and Hegel up to 1802*. Stuttgart: Klett-Cotta, 1979.

Link, Jürgen. *Hölderlin-Rousseau. Inventive Rückkehr*. Wiesbaden: Westdeutscher Verlag, 1999.

Müller, Götz. *Jean Pauls Exzerpte*. Würzburg: Königshausen & Neumann, 1988.

Pethes, Nicolas: "Telling Cases: Writing Against Genre in Medicine and Literature." *Literature and Medicine* 32.1 (2014): 24–45.

Pfotenhauer, Helmut. *Jean Paul. Das Leben als Schreiben*. Munich: Hanser, 2013.

Rousseau, Jean-Jacques. *Émile*. Trans. Barbara Foxley. London: Dent, 1955.

Rousseau, Jean-Jacques. *Émile ou de l'éducation. Œuvres complètes*. Eds. Bernard Gagnebin and Marcel Raymond. Vol. 4. Paris: Gallimard, 1969.

Rousseau, Jean-Jacques. *Émile oder Über die Erziehung* (1762). Ed. Martin Rang, transl. Eleonore Sckommodau. Stuttgart: Reclam, 1963.

Schäfer, Armin. "Jean Pauls monströses Schreiben." *Jahrbuch der Jean-Paul-Gesellschaft* 37 (2002): 218–234.

Schiller, Friedrich. *Dramatic Works of Friedrich Schiller. Wallenstein and Wilhelm Tell*. Trans. S. T. Coleridge, J. Churchill, and Theodore Martin. London: George Bell and Sons, 1906.

Schiller, Friedrich. *Werke und Briefe. Bd. 4: Wallenstein*. Ed. Frithjof Stock. Frankfurt am Main: Deutscher Klassiker Verlag, 2000.

Stengers, Isabelle. "Die doppelsinnige Affinität: Der newtonsche Traum der Chemie im achtzehnten Jahrhundert." *Elemente einer Geschichte der Wissenschaften*. Ed. Michel Serres. Frankfurt am Main: Suhrkamp, 1994. 527–567.

Volke, Werner. "'O Lacedämons heiliger Schutt!' Hölderlins Griechenland: Imaginierte Realien – Realisierte Imagination." *Historisches Jahrbuch* 24 (1984/1985): 63–86.

Reto Rössler

The End of 'Heavenly Writing', or: *Speech of the Dead Christ down from the Universe That There Is No God* (1796)

Jean Paul's Cosmopoetics as Paratextual Prose

Abstract: Jean Paul Richter's *Speech of the Dead Christ* (1796) can be read as a literary thought experiment and metapoetic reflection on Enlightenment cosmology and its shifting conceptual metaphors. In close formal-aesthetic relation to one of the most influential early modern fictions of the Copernican cosmos, Johannes Kepler's *Somnium*, Richter's dream narrative reveals the contingent downside of the harmonious world model and its nihilistic consequences on at least three levels: *Cosmologically*, the *Speech* marks the decline of the metaphor of the 'readability' of the world around 1800, concerning *anthropology*, it shows man thrown back on himself in an infinite, empty, and chaotic cosmos. Contrary to these impositions, as an artist, man again experiences a greater degree of freedom, since, *poetologically*, he now conceives of himself as a creator of possible worlds, beyond the idea of mimesis.

1 Introduction

Since the beginnings of the scientific revolution, poetic fiction has accompanied the methodically guided exploration of the Copernican cosmos, expanding the boundaries of astronomical visibility and telescopic observation. From 1600 onwards, for almost two centuries, a productive interaction developed between the 'new' natural sciences and poetry, in which the latter (with few exceptions) succeeded in balancing cosmic contingencies. If the 'plurality of worlds' and the infinite nature of the universe may have given rise to a number of cosmic anxieties, they could be alleviated through narrative and metaphorical framing. For example, the idea of the heavens as a divine writing tablet on which the omnipotence of God is revealed, harmonised the cosmos and placed man as reader in the role of deciphering the heavenly scriptures.

Using Jean Paul Richter's short narrative *Rede des toten Christus vom Weltgebäude herab, daß kein Gott sei* [*Speech of the Dead Christ Down from the Universe that there is no God*] as an example, I would like to show that the role of literary texts in the context of cosmology was not only to construct, but also to deconstruct established models of the world. Depending on the perspective and focus from which the *Speech* is viewed, it marks either a beginning or an end point in the poetics of knowledge on the threshold of the nineteenth century by overlaying the reflection of cosmological (and anthropological) transformations with questions of the poetic representability of a contingent cosmos.

As one of the few examples in German literature of a paratext (Genette 2014) that in its radiance and aesthetic significance goes far beyond the novel in which it is mounted (*Siebenkäs*; 1796), the *Speech of the Dead Christ* has served literary scholarship either as a hermeneutic key to unlock the author's work or as an experimental field for theoretical approaches (Pietzcker 1985; Andringa 1994) – and in both cases has provoked contrary readings: for some, *The Speech of the Dead Christ* marks the beginning of a materialist cosmology at the end of the eighteenth century (Esselborn 2011, 64ff.), a philosophy of nihilism (Moros 2007), or a scenography of secularised apocalypses in modernity (Agazzi 2008; Horn 2013). With reference to Jean Paul's concept of the 'second world' and the recurring motifs of transgression and ascent, others have argued that his oeuvre should rather be understood in terms of mythopoetic attempts to reconcile man and cosmos in a secularised age (Hunfeld 2004, ch. 3; Honold 2019).

However, it cannot be denied that Jean Paul, as a well-read author, was aware of the history of knowledge and literature of astronomy and cosmology like no other German writer around 1800 (Bühlmann 1996). While his contemporaries, the Classical and Romantic authors such as Hölderlin, Novalis, Kleist, Schiller or Goethe, were interested in individual aspects within this field of knowledge, Jean Paul's cosmological references are of a much greater systematic interest. For example, he read popularising descriptions of astronomy, such as Christian Ernst Wünsch's *Kosmologische Unterhaltungen für junge Freunde der Naturerkenntniß* (2 vols.; 1791–1794), Friedrich Gottlieb Rösler's *Handbuch der praktischen Astronomie für Anfänger und Liebhaber* (1788), or Georg Christoph Lichtenberg's Essays from the *Göttinger Taschen-Calender*. He made selective extracts from these, which served him as material for poetic comparisons and metaphorisations in his later novels (Müller 1988; Esselborn 1989, 203–215).

For Jean Paul cosmology and literary writing were historically and conceptually intertwined. This applies to how he received scientific texts, but it is also evident in the aesthetic production of his literary works. The often repeated and varied motif of cosmic ascents – whether through technical aircraft such as airships or in the literary form of dreams – is a clear indication that Richter was also familiar with the literary genre of fictional space travel from antiquity (Plutarch's *De facie in orbe lunae*; Lucian's *Verae Historiae*; Cicero's *Somnium Scipionis* [*De re publica*]) to the Early Modern period (Kepler's *Somnium*; Godwin's *The Man in the Moone*; Cyrano's *Voyage dans la lune*) (cf. Nicolson 1960). In his novels, these references are made explicit on several occasions, for example in *Hesperus oder 45 Hundposttage* when the protagonist recommends to his beloved that she should "learn by heart" Bernard de Fontenelle's *Entretiens sur la pluralité des mondes* in addition to the catechism (Jean Paul 1996, I.1, 546).

Beyond the two poles of interpretation outlined above, the following reading is based on the background of the literary genre of early modern space travel. In comparison with Kepler's *Somnium* and the varied technique of narrative framing in both dream narratives, I argue that the *Speech* can be read as a poetic reflection on the historicity and conceptual transformation of both the genre and the entanglement of cosmology and fiction. What 'ends' with the literary sermon of the dead Christ is neither the idea of man as a cosmic *contemplator caeli* nor the faith in the transgressive power of myth. Jean Paul's idea of prose as a digressive and paratextual poetics reacts to the insight that the heavens no longer represent a 'language' or 'writing' of nature that can be clearly deciphered, but can at best be interpreted and *re*written in ambiguous ways.

2 Reading and writing the heavens: The early Enlightenment paradigm of literary cosmology

The conceptual history of the 'readability' (and 'writability') of the stars and the heavens has a long tradition, going back to ancient philosophy and poetry (Blumenberg 1981). However, it was in the paradigm of Copernican astronomy and the beginnings of experimental natural philosophy that the metaphor was particularly popular, and it was in the Age of Enlightenment that it gained its greatest persuasive power. The 'physico-theology' that developed in parallel with the *new science* helped to popularise the latest knowledge about nature, but also to reconcile the hypotheses of the natural sciences with theological views and the idea of a harmonious nature (Michel 2008, 80): within the physico-theological paradigm, scientific observation and experimentation were seen as practices of reading the 'book of nature' and making the divine handwriting legible.

Moreover, the great productivity and potential of the metaphor of 'readability' also made the boundaries between natural science and literature permeable (Preisendanz 1994, 489). In probably no other age, before or since, has the attraction for poets to explore the cosmos with their own imaginative resources been so great, and natural philosophers as well as writers have felt so empowered by a cosmically expanded metaphysics of a 'plurality of (possible) worlds' to consider not only the factual but also the fictional as a possibility of truth or the reality of nature (Richter 1972; Baasner 1987).

An early literary reflection of this 'enlightened' understanding of science and a poetic example in which the metaphor of cosmic readability appears even in the title, can be found in the didactic poem of the Hamburg poet Barthold Heinrich Brockes. In a stanza of *Die himmlische Schrift* [*The Heavenly Scripture*], taken from his nine-volume cycle *Irdisches Vergnügen in Gott* (1721–1748), he writes:

Seh' ich den Himmel an, so kömmt mir sein Sapphir	When I look at the sky, I see its sapphire
Als eine Tafel für,	As a writing tablet
Die unermeßlich ist, auf welcher eine Schrift,	which is immeasurable, and with a writing,
Die des allmächt'gen Schöpfers Wesen,	enclosing the whole of the almighty God,
Huld, Weisheit, Macht und Majestät betrifft,	grace, wisdom, power and majesty
Im schimmernden Gestirn, in heller Pracht zu lesen.	can all be read in the shimmering star, in bright splendour.
Hilf Gott, welch eine Schrift! O! welch ein Wunder-Buch,	Help God, what a writing! O! what a wonder-book
In welchem die Gestirne Zeilen,	in which the stars are lines,
Die Lettern grösser sind, als hundert tausend Meilen,	The letters are greater than a hundred thousand miles,
Woran, in wunderbarem Schein,	Wherein, in wondrous light,
Die Puncte selbsten Sonnen seyn!	The points themselves are suns!
(Brockes 1727, 178)	(my transl.)

Seen as a heavenly tablet or "wonder-book" in which God writes down his miracles, the entire cosmic order with its phenomena appears as letters only for the reading pleasure of the human mind, if it can correctly decipher the individual letters and assemble them into to a coherent textual cosmic order.

The changing reception history of Brockes' scientific poetry thus indicates the dwindling plausibility of the idea of a harmonic cosmos. While Brockes was one of the most widely read German-speaking poets throughout his life and for a large part of the eighteenth century, it was Jean Paul who – from the perspective of the early nineteenth century – saw in Brockes' poetry of nature no more than „Beispiele von unpoetischen Repetierwerken der großen Weltuhr" ["examples of unpoetical repeating mechanisms of the world clock"] (Jean Paul 1996, I.5, 36; Müller 1983, 68–81; Esselborn 2011). In addition to 'letters', 'books' and 'writing', Jean Paul here addresses a second group of metaphors with which the natural philosophers of the time conceived of the world as well-ordered and which underwent a similar conceptual change: that of a 'clockwork' and a 'cosmic architecture' (*Weltgebäude*). While both received decisive impulses from Copernicanism and Newtonianism, only the architectural metaphor of the *Weltgebäude* developed a model-forming potential for both astronomy and poetry (Rössler 2020). Following in the footsteps of Bernard de Fontenelle's *Entretiens sur la pluralité des mondes* (1686), a text that stands exactly on the threshold between scientific and poetic writing (Guthke 1983, 202–217), poets pursued the idea that even higher levels of cosmic system formation could be imagined beyond the Copernican order of the solar system.

A conceptual break with the cosmic idea of harmony and the metaphor of the *Weltgebäude* can be seen in Immanuel Kant's early work, the cosmological treatise *Allgemeine Naturgeschichte und Theorie des Himmels* (*Universal Natural History of the Heavens* [1755]). Here, based on Newtonian mechanics, he outlines his so-called 'nebular hypothesis', according to which the planets, solar systems and ultimately the edifice of the universe were formed from a rotating cloud of gas:

> Wenn nun alle Welten und Weltordnungen dieselbe Art ihres Ursprungs erkennen [. . .], sollten nicht alle die Weltgebäude gleichermaßen eine beziehende Verfassung und systematische Verbindung unter einander angenommen haben, als die Himmelskörper unserer Sonnenwelt im kleinen, wie Saturn, Jupiter und die Erde, die für sich insonderheit Systeme sind und dennoch unter einander als Glieder in einem noch größern zusammen hängen? Wenn man in dem unermesslichen Raume, darin alle Sonnen der Milchstrasse sich gebildet haben, einen Punkt annimmt, um welchen durch ich weiß nicht was für eine Ursache die erste Bildung der Natur aus dem Chaos angefangen hat: so wird daselbst die größte Masse und ein Körper von der ungemeinsten Attraction entstanden sein [. . .]. (Kant 1900, 307)

> [Now, if all planets and planetary systems acknowledge the same sort of origin [. . .] should not the cosmic structures have acquired in a like manner an interconnecting relationship and a systematic coordination among themselves, as the celestial bodies of our solar system have on a small scale, like Saturn, Jupiter, and the Earth, which are special systems on their own and yet are linked together amongst themselves as parts in an even greater system? If we take one point in the infinite space in which all the suns of the Milky Way were developed, a point around which, for some unknown reason, the first development of nature out of chaos began, then at that location the largest mass and a body of the most exceptional power of attraction will have arisen.] (Kant 2008, 99–100)

Although Kant emphatically defends the existence of God in the preface to his work, his version of a *Weltgebäude* remains stable without the intervention of a creator – at least for a long time in human terms. The Kantian cosmos is a cosmos *in time*, a highly dynamic structure that follows the principles of self-organisation through 'attractive' and 'repulsive forces'. Taking the concept of 'natural *history*' literally, his *Weltgebäude* has a temporal beginning and end. In an infinite process of cosmic creation, every cosmic collapse and chaos is followed by a new order like a "phoenix" rising from the ashes (Kant 1902, 321; Rössler 2022).

Kant's *Natural History* is not only a good example of the productivity of the cosmographic model of the *Weltgebäude*. Reading his treatise chapter by chapter, one also gets a clear insight into how fragile the architecture is beyond the theoretical pathos formulas of its young author. It is often only a small step between cosmic harmony and emptiness, infinity or chaos, and which side prevails seems to depend on one's perspective. The enlightened cosmos is, to speak with Hans Blumenberg, a highly "ambiguous" construction (Blumenberg 1987, ch. 1), which, along with its enthusiasm for scientific progress, is accompanied by a series of cosmic fears: of infinity, endlessness, void, an absent God and – still – apocalyptic world-ends. One of the paradoxes of the development of cosmological knowledge in the eighteenth century seems to lie in the fact that the increase in empirical data, such as that provided by the Herschelian telescopes, but also philosophical achievements, such as Kant's reflections on 'cosmological paralogisms' in his first *Critique* (1781/1787), did not stabilise the *Weltgebäude* on the threshold of the nineteenth century, but profoundly destabilised it.

Epistemologically destabilising events such as these outline the discursive framework within which Jean Paul's *Speech of the Dead Christ down from the Universe that there is no God* is set. In cosmological, epistemological and affect-poetological terms, this *Speech* marks a turning point – and *stages* it. In the following, I analyse Jean Paul's short paratext against the genre background of early modern space travel and read it as a *cosmopoetic* deconstruction scene whose dream narrative combines reflections on epistemic transformation with those on the forms and modes of its aesthetic representation.

3 The literary dream as a cosmic construction and deconstruction scene: Kepler and Jean Paul

Jean Paul's interest in natural philosophy, metaphysics, and astronomy dates back to his youth. Thanks to the ideal support of Pastor Erhard Friedrich Vogel, Jean Paul had already begun at an early age to acquire knowledge from various disciplines, the classical humanist educational canon as well as the latest scientific findings, through reading and writing, the latter in the form of various excerpts. Some 12,000 manuscript pages and more than 100,000 entries between 1778 and 1825 reveal an author who – literally – *wrote* his world. We know from his extracts that he read, among others, Kant's *Allgemeine Naturgeschichte*, Johann Heinrich Lambert's *Cosmologische Briefe über die Einrichtung des Weltbaus* (1761) and the astronomical essays by Georg Christoph Lichtenberg, written and published in his *Göttinger Taschen Calender* from 1777 to 1799.

Astronomical and cosmological knowledge enters Jean Paul's novels at a macro and micro level. It is no coincidence that he named three of his novels after celestial bodies: *Hesperus, Titan,* and *The Comet* (Esselborn 1989, 59–77). And unlike the novels mentioned, *The Speech of the Dead Christ* refers to astronomical knowledge on a meta-level: in the double sense that it 'observes' and rhetorically participates in the ongoing epistemic deconstruction of the cosmos around 1800.

As its title already suggests, Jean Paul's *Speech* dramatises cosmological knowledge by playing with different and interwoven literary forms. In this respect, the narrative belongs to the literary tradition of fictional cosmic dreams exemplified by the above-mentioned texts from Greek and Roman antiquity by Plutarch (*De facie in orbe lunae*), Lucian (*Verae Historiae*) and Cicero (*Somnium Scipionis*) and the Copernican era by Cyrano de Bergerac (*Voyage dans la lune*), Godwin (*The Man in the Moone*), and Kepler (*Somnium*).

Looking more closely at the narrative form, there are striking parallels with Jean Paul's *Speech*, especially in the latter. Regarding *Somnium* (the text was published posthumously in 1634 by Kepler's son) the formal similarities go so far as to suggest that Jean Paul not only knew the narrative, but deliberately referred to it intertextually. According to the French literary historian Fernand Hallyn, Kepler's narrative is composed by several interconnected narrative frames as the following scheme shows:

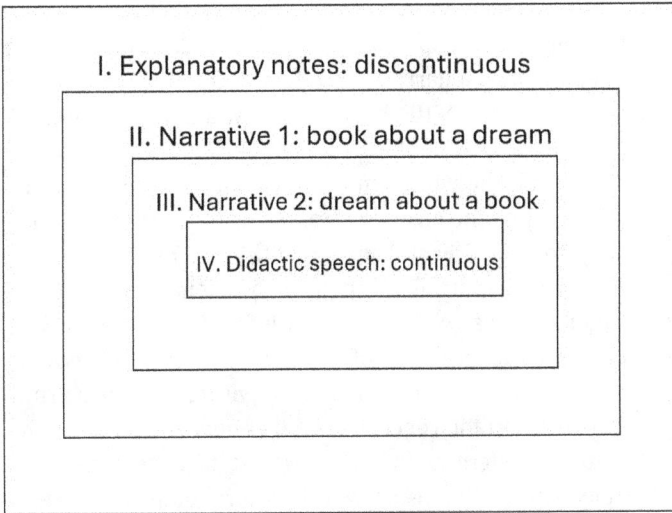

Figure 1: Narrative Scheme of Johannes Kepler's *Somnium* (Hallyn 1990, 262).

Thus *Somnium* begins with 'explanatory notes', whose elements are interwoven with several narrative frames, beginning with the first-person narrator (Kepler's alter ego) reading a book on Bohemian history about a famous magician named Libussa ('book about a dream'); he falls asleep and sees himself reading at the Frankfurt Book Fair a book about a young man, named Duracotus ('dream about a book') who, with the help of his mother's magical powers, summons a demon (Kepler 2010, 11). The demon then takes him to the moon, and here, in the inner core of the narrative, he delivers an extended speech in which – besides detailed astronomical facts, references to the times of day, the seasons and the life of the moon's inhabitants (called 'Privolvans') – he almost casually confirms the assumptions of the Copernican hypothesis: standing on the surface of the Moon, Duracotus now sees the Earth moving while feeling that he is standing on stable and immobile ground. In its multiple framing, Kepler's *Somnium* forms a thought experiment that rhetorically and narratively makes a new and uncertain astronomical knowledge plausible.

If we now look at the narrative structure of Jean Paul's *Speech of the Dead Christ*, we can see an almost identical technique of narrative framing. As in Kepler, there is a short preface which, narratively speaking, serves to link the main plot of the novel *Siebenkäs* (level N I) with the paratextual unit of the 'flower piece' of the *Speech*. Here, as in Kepler's narrative, the first-person narrator is introduced recalling the religious and metaphysical dreams of his childhood (level N II), when he falls asleep and wakes up in a dream (level N III). In the dream, he encounters the figure of the dead Christ in the cemetery, who finally (analogous to Kepler's demon) begins his nihilistic speech (level N IV), which forms the inner core of the narrative. The first-person narrator then observes the actual downfall of the *Weltgebäude* within his dream (N III), before finally waking up and joyfully worshipping God (N II).

Finally, the structural parallel between Kepler's and Jean Paul's dream narratives includes their endings, when the narrative framing is dissolved by the narrator's sudden awakening from his dream. The main difference between the two texts, however, lies in the way in which the technique of fictional framing is *epistemically* functionalised in each case. Not only Kepler in his dream text, but also the poets, astronomers, and cosmographers of the eighteenth century used, as shown, the imagination to represent the cosmos as an ordered system. Thus, when Jean Paul explicitly takes up the concept of *Weltgebäude* in the title of his paratext, he is writing, with the reference to Kepler, against the Copernican tradition, but also against idealistic-harmonising world constructions in the sense mentioned above. Instead, the *Speech of the Dead Christ* shows the 'dark side' of the enlightened cosmos by demonstrating that all cosmic ordering power is linked to the instance, and possibly the *fiction* of an omniscient, good and, above all, existing God. When the first-person narrator awakens from his dream, he wanders through an apocalyptic end-time setting:

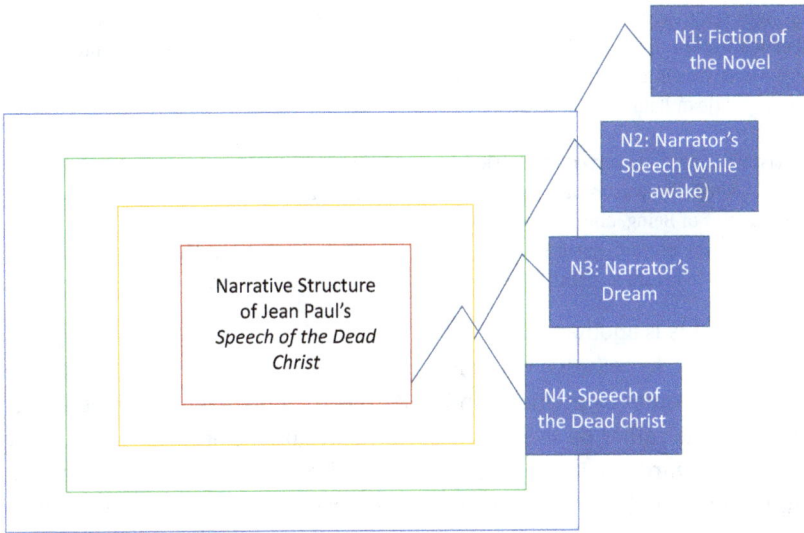

Figure 2: Narrative Frames in Jean Paul's *Speech of the Dead Christ*.

Ich lag einmal an einem Sommerabende vor der Sonne auf einem Berge und entschlief. Da träumte mir, ich erwachte auf dem Gottesacker. Die abrollenden Räder der Turmuhr, die eilf [sic] Uhr schlug, hatten mich erweckt. Ich suchte im ausgeleerten Nachthimmel die Sonne, weil ich glaubte, eine Sonnenfinsternis verhülle sie mit dem Mond. Alle Gräber waren aufgetan und die eisernen Türen des Gebeinhauses gingen unter unsichtbaren Händen auf und zu. (Jean Paul 1996, I.2, 272)

[Once on a summer evening I lay in the sun on a hillside and fell asleep. And I dreamt that I woke in the churchyard. The whirring wheels of the church clock striking eleven had wakened me. I looked for the sun in the desolate night sky, for I believed an eclipse was hiding it behind the moon. All the graves stood open and the iron doors of the charnel house opened and shut by invisible hands.] (Jean Paul 1992, 181)

The passage reactivates the pictorial and metaphorical archive of early modern cosmologies as well as the literary genre of space travel: the 'clockwork' – perhaps the most powerful allegory of a harmonious cosmos with God as the providential and wise clockmaker – here breaks into pieces, the sun as the source of all earthly life darkens, and the ethereal figure, be it an angel or an inhabitant of a distant planet (both of whom are used in the genre to share their higher insights with the space travellers) is now transformed into the figure of the dead Christ proclaiming the absence of God and the emptiness of the heavens:

Christus fuhr fort: ‚Ich ging durch die Welten, ich stieg in die Sonnen und flog mit den Milchstraßen durch die Wüsten des Himmels; aber es ist kein Gott. Ich stieg herab, soweit das Sein seine Schatten wirft, und schauete in den Abgrund und rief: ‚Vater, wo bist du?' aber ich hörte nur den ewigen Sturm, den niemand regiert, und der schimmernde Regenbogen aus Wesen stand ohne eine Sonne, die ihn schuf, über dem Abgrunde und tropfte hinunter.' (Jean Paul 1996, I.2, 273)

[Christ went on: 'I traversed the worlds, I ascended into the suns, and soared with the Milky Ways through the wastes of heaven; but there is no God. I descended to the last reaches of the shadows of Being, and I looked into the chasm and cried: 'Father, where art thou?' But I heard only the eternal storm ruled by none, and the shimmering rainbow of essence stood without sun to create it, trickling above the abyss. (Jean Paul 1992, 182)

Christ's role – this is another difference from Kepler's narrative – is not simply to prove a scientific hypothesis: neither Copernicanism or Newtonianism nor the collapse of the cosmos as predicted in Kant's *Theory of the Heavens*. The effect of his speech should rather be seen on its performative or rhetorical level. Contrary to the biblical words of Jesus on the cross (addressed to those mocking him: "Amen, I tell you, today you will be with me in paradise" [Luke 23:43]), Christ's speech is characterised by hypotactic sentence structures (sometimes close to a stream of consciousness), by a direct address to God, and finally by drastic metaphoricity and an increasing pathos: "Und als ich aufblickte zur unermeßlichen Welt nach dem göttlichen Auge, starrte sie mich mit einer leeren bodenlosen Augenhöhle an" (Jean Paul 1996, I.2, 273) ["And when I raised my eyes to the boundless world for the divine eye, it stared at me from an empty bottomless socket"] (Jean Paul 1992, 182). While retaining the sublime style of his speech (in an earlier prose version it was not Christ, but the dead Shakespeare who spoke down from the universe), the signs of cosmic affection are reversed. As soon as the instance of God is detached from the balanced model of cosmic harmony, the rhetorical gestures of 'Copernican enthusiasm', characteristic of the poetry of Brockes, Albrecht von Haller or Friedrich Gottlieb Klopstock, turn into a pathetic incantation of the cosmic downfall, ending in an *exclamation*: "und die Ewigkeit lag auf dem Chaos und zernagte es und wiederkäuete sich. – Schreiet fort, Mißtöne, zerschreiet die Schatten; denn Er ist nicht!" (Jean Paul 1996, I.2, 273) ["and Eternity lay on Chaos and gnawed it and ruminated itself. – Shriek on, ye discords, rend the shadows; for He is not!"] (Jean Paul 1992, 182).

The *Speech of the Dead Christ* is thus more than a mere philosophical speculation about an absent or non-existent God, the end of the world or the decline of the idea of cosmic harmony. It is also a dramatic narrative of decline, culminating in Christ's demolition of the *Weltgebäude* by means of a performative speech act, seen outside from space. Reading the dream in the long tradition of literary 'thought experiments' (Winkler 2016), its philosophical result can be seen in its allusion to a blind spot in Enlightenment cosmology, since within this paradigm all the attempts at stabilising the cosmos through theory could not hide the fact that the idea of cosmic harmony necessarily implied the figure of a (at least) deistic God.

Beyond these differences, one aspect that Jean Paul's *Speech* shares with the cosmopoetic fictions of the Enlightenment is its view of the relationship between poetry and knowledge as mutually dependent. The dream illuminates this relationship from the other side, showing that the transformation of a cosmological paradigm of knowledge, which was reflected in a series of literary deconstructive scenes around 1800 (in Weimar Classicism and Romanticism), had a noticeable effect on the triangular relationship between man, cosmos, and aesthetic forms in all three disciplines: cosmology, anthropology, and poetics. In its final section, the *Speech* stages a triple ending: the end of the cosmos, of man, and of the dream. Here, the "Riesenschlange der Ewigkeit" ["giant serpent of Eternity"], an allusion to Ouroboros, the cosmic serpent in ancient Egyptian mythology, crushes the *Weltgebäude* – before the narrator suddenly awakens:

> Und als ich niederfiel und ins leuchtende Weltgebäude blickte: [. . .] alles wurde eng, düster, bang – und ein unermeßlich ausgedehnter Glockenhammer sollte die letzte Stunde der Zeit schlagen und das Weltgebäude zersplittern . . . als ich erwachte.

> Meine Seele weinte vor Freude, daß sie wieder Gott anbeten konnte – und die Freude und das Weinen und der Glaube an ihn waren das Gebet. Und als ich aufstand, glimmte die Sonne tief hinter den vollen purpurnen Kornähren und ward friedlich den Widerschein ihres Abendrotes dem kleinen Monde zu [. . .], und von der ganzen Natur um mich flossen friedliche Töne aus, wie von fernen Abendglocken. (Jean Paul 1996, I.2, 275)

> [And as I sank down and gazed into the shining fabric of the Universe: [. . .] all became close, dark, and fearful – and a bell clapper, infinitely extended, was about to strike the ultimate hour of Time and shatter the Universe . . . when I wakened. My soul wept for joy that she could again worship God – and the joy, and the weeping, and the belief in Him were the prayer. And when I rose, the sun glowed deep behind the full purple ears of corn and peacefully cast the evening glow's reflection to the little moon, [. . .] and from all Nature around me flowed peaceful sounds as if from faraway vesper bells.] (Jean Paul 1992, 183)

The abrupt change of the fictional level from N3 to N2 varies the paradoxical narrative formula of the beginning only slightly: from ". . . I dreamed I woke up" to ". . . *when* I woke up" (Neumann 2003, 95–96). While the suddenness of the awakening and the narrator's joyful worship of God may appear unmotivated in such immediacy, it marks the chiastic intertwining of the two worlds: as a dreamer, the narrator experiences the finiteness of the cosmos; as an awakened person, he can believe in the resurrection all the more. On the other hand, at both the beginning and the end of the narrative, it is the full stop that conceals and reveals the transition between the two worlds as an ambiguous one. In the coda of the narrative, it is the narrator who becomes the reader of his own dream – placing the entire narrative on a meta-level. While he tends towards a more harmonising interpretation, he leaves his less faithful readers all the more uncertain. The title of the latest biography of the poet, Beatrix Langner's *Jean Paul: Meister der zweiten Welt*, is apt in that Jean Paul not only repeatedly juxtaposed the 'first' and 'second world' in his writings and novels, but his 'mastery' can also be seen in the fact that he repeatedly presented transition in terms of failure and demolition (Langner 2013). In this respect, the ambivalent ending of the *Speech of the Dead Christ* can be compared to *Des Luftschiffers Giannozzo Seebuch* (*Aeronaut Giannozzo's Voyage Diary* [1801]), where the protagonist also crashes with his balloon, or to *Das Kampaner Tal* (*The Kampaner Valley* [1797]), where the balloonists decide to return to earth on the threshold (Buschendorf 1997; Hunfeld 2004, 132–136; Eickenrodt 2006, Ch. 4).

If we now read the *ad hoc* conclusion of the *Speech* against its grain, we must conclude that it does not confirm the initial assertion of the paratext's *Vorbericht*, "[d]as Ziel dieser Dichtung ist die Entschuldigung ihrer Kühnheit" (Jean Paul 1996, I.2, 270) ["the object of this fiction is the excuse of its audacity" (Jean Paul 1992, 178)]. Instead, it undermines it. If the transition here can only be made plausible by a leap into faith, then it points rather towards the history of ideas of metaphysical destruction in the late nineteenth century, such as Kierkegaard or Nietzsche. The *Speech of the Dead Christ* thus shows both: the end of cosmologies of harmony and perfection and the end, or at least the *modern* failure, of attempts at rhetorical and narrative stabilisation of cosmic order: in the moment of awakening, the cosmological *Weltgebäude* and its various narratives frames collapse like a house of cards.

4 Jean Paul's cosmopoetics and paratextual prose

Besides Novalis, Hölderlin, or Goethe, Jean Paul is one of the authors of German Classicism and Romanticism who not only thought intensively about the aesthetic consequences of a cosmic rupture, but also made these reflections productive for his poetics of the novel on different levels. It is precisely these three levels of a cosmopoetics of the modern novel that Florian Klaeger describes in *Reading into the Stars* and presents them in the following scheme (Klaeger 2018, 79):

Cosmology:	The (auto-critical) novel:
. . . takes as its object the world from a human perspective. On a conceptual level, it speaks in *epistemological* terms about the relationship world – human, and it thus enables and provokes reflections about epistemology [C1] ('humans can (not) find out significant things about their place in the universe').	. . . takes as its object the world from a human perspective. On a 'literal' (mimetic) level, it speaks in *epistemological* terms about the relationship world – human, enabling reflections about epistemology in general [N1]. Auto-critically, it can reflect about the epistemology of the novel and the way it relates to extratextual reality [N1'].
. . . becomes the object of metaphorical signification. On a nonconceptual level, it is made to 'speak' *ontologically* about the relationship world – human [C2] ('humankind's place in the universe is (not) indicative of their importance').	. . . signifies in a non-literal, metaphoric way, making *ontological* statements about the relationship world – human [N2], and, auto-critically, about the relationship world – novel [N2'].
. . . becomes the object of metaphorology and is examined for the *anthropological* implications of its various historical ontological functions [C3] ('views about the human place in the universe can (not) tell us something about humankind').	. . . becomes the medium [N3] and object [N3'] of a meta-metaphorology, exploring the *anthropological* function of metaphoricity as a mediating instance in the relationship world – human.

For each of the three levels in the right-hand column of the scheme, there are concrete examples in Jean Paul's poetics:

1. [*N1: autocritical*]: The intertextual allusions to the early modern genre of the lunar voyages and the narrative framing of the *Speech* can be read in an *epistemological* sense as a reflection on the role of *imagination* in cosmology (including the metaphor of the 'book' from the sixteenth to the early eighteenth centuries). The reversal of the dream, on the other hand, points to its changing role as a medium for representing epistemic transformations or ruptures.

2. *N2*: On an *ontological* level (i.e., in terms of the relationship between world and novel), the narration of failed transmissions between 'first' and 'second world' shows the cosmos no longer as a harmonious, but in many ways highly contingent composition. It also points to the liminal status of literature, offering glimpses of distant planetary and possible worlds, neither of which can be inhabited.

3. *N3*: On a third and final level, the narrative serves as the medium and object of a *meta-metaphorology* that explores the role of mediating metaphoricity (as an anthropological function) between world and man. Here, *The Speech of the Dead Christ* demonstrates the decline of a crucial metaphor of science and astronomy in the eighteenth century, and stages this decline dramatically.

Looking beyond the *Speech*, Jean Paul's oeuvre offers further examples of each level:

1. *Epistemological* reflections on the role of literature can be found in his articles and essays on the phenomenon of dreaming (*Über das Träumen* [*On Dreaming*] [1796]; *Blicke in die Traumwelt* [*Views into the Dream World*] [1813]), in which he integrates his extracts from a physiological and early psychological point of view, but he also sees dreaming as intertwined with poetry when he writes: "Der Traum ist unwillkürliche Dichtkunst" (cf. Schmidt-Hannisa 2001; Schmitz-Emans 2005) ["dreams are involuntary poetry"].

2. With regard to the novel form of his last novel *Der Komet* [*The Comet*] (1820–1822), the irregular course of the celestial body – due to their unpredictable course, comets were for centuries regarded as ominous signs in the sky – serves as a model for the biography of the novel's hero, Nikolaus Marggraf, and the contingent novel form.

3. With regard to metaphoricity we can refer to Jean Paul's *Vorschule der Ästhetik* [*Preschool of Aesthetics*] (1804) and its important chapter on the aesthetic role of *Witz* (*wit*). Departing from the poetological tradition, Jean Paul does not define *wit* as a means of creating narrative coherence by levelling and smoothing out differences, but rather by relating the distant and unrelated. This technique of poetic "*metaphorisation*" is crucial to his cosmic writing, as it allows the poet to refill the cosmologically emptied, godless sky with new images, thereby creating unpredictable cross-connections between the heavenly and the earthly, bridging the abyss between the 'first' and the 'second world' (Wiethölter 1979, 130–134). This poetological means leads to the aesthetic twist that the discarded cosmological metaphor of 'celestial writing' is poetologically motivated (which can also be observed in Jean Paul's *Traum über das All* [*Dream about the Universe*] (1820). Precisely *because* cosmology has demystified the heavens, whose archive of signs and letters has been erased, poetry comes to rewrite its table with new content (Wölfel 1974, 296f.).

With regard to Jean Paul's poetics, the third level of the above model can perhaps be extended by one element. In terms of content, the *Speech* does not stop at the staging of a collapsing world structure and the separation of cosmos and subject, but it also proves to be highly reflective in terms of poetic form, transforming the double rupture into a discontinuous process of representation of prose and novel form. As we have shown, the *Speech* is characterised by a changed function of narrative framing, in which the framing no longer reveals deeper insights into a divine cosmic architecture whose language is to be deciphered, but rather unveils the missing context of the meaning of its individual celestial letters. At the end of his speech, the figure of the dead Christ tentatively addresses the crumbling stars and world structures as mines ("Bergwerke"), pit lamps ("Grubenlichter"), silver veins ("Silberadern"), coral banks ("Korallenbänke") and water balls ("Wasserkugeln"), but no coherent picture emerges (Jean Paul 1996, I.2, 274). Cosmic destruction and failed interpretation, which could just as well be understood as Christ's deconstructive reading technique, are thus also superimposed in the *Speech*.

A look at the novel *Siebenkäs* and its main plot also reveals a poetological transformation of the technique of narrative framing, which here serves as an element of modern storytelling (Erdbeer et al. 2022, 49–50). Inserted as a 'floral piece' between the second and third "Bändchen" (volume) of the novel, the narrative not only forms a multi-framed framework; with its specific position, it is no coincidence that it marks a turning point in the main plot, commenting on, interrupting, mirroring, and inverting it in a complex way. The novel's protagonist, Firmian Stanislaus Siebenkäs, is a young lawyer whose inner and very personal aspiration is to become a successful writer, but who, as the novel progresses, feels increasingly constrained by his unhappy marriage to Lenette, an ambitious milliner, striving for social advancement. Against this background, *The Speech of the Dead Christ* not only deals with the contingency of the cosmos, but also marks and highlights the contingency of Firmian's way of life. At the point where the *Speech* interrupts the plot, it seems for a moment that events might take a turn for the better: Firmian is tormented by acute money worries when he wins a shooting competition, and with the prize money he is able to avert the impending financial misery for the time being. However, this happy turn of events cannot hide the couple's escalating interpersonal problems, which are only resolved when Firmian's friend Leibgeber helps him fake his own death in order to start a new life in freedom.

In view of the novel's plot, the dream narrative of the *Speech* appears as a paratextual prolepsis, which – with ironic refraction – refers both to Firmian's growing unhappiness, his fictitious death, but also to his liberation and 'redemption' in the newly formed relationship with Natalie. In the narrative parallelism of the cosmic end of the world and the death of God with the critical scenes of a marriage, *high* and *low*, eschatological metaphysics and the banalities of everyday life, are superimposed – whereby, on closer inspection, the narrative structure is reversed: from the sequence 'fiction (of the dream)' – 'Death' – 'Life' in the *Speech* to the novel's sequence 'life' – 'death' – 'fiction (of life)'. The narrative structure of *Speech* can thus be described as a double break with the framing technique of the cosmic tradition. At the level of representation, the visions of a harmonious world structure, a wise creator and a recognisable reality are each replaced by their counterparts: the collapse of the world structure, an absent God and the opacity of the world as a whole.

A second break lies at a performative level: In Kepler's *Somnium*, the fictional frame was part of a technique of serial cognitive processing, in which hypotheses and fictions were narratively linked (Aït-Touati 2011, ch. 1; Heydenreich 2016). In the novel, by contrast, the framework of the *Speech* functions as a paratextual and thus *metapoetic* element of commentary, interruption, displacement, and rhythmisation of the narrative process.[1] In relation to its central paratext, then, the novel is characterised by a double perspective, in which the metaphysical framing on the paratextual side (which reveals the finitude of the cosmos) is countered by framing elements on the novel side that tend to reject such 'truths' and instead focus on the subjectivity and presence of the narrative process itself. This applies both to the composition of the novel and, more generally, to Jean Paul's poetics (Dembeck 2007, ch. 6): in *Siebenkäs*, the principle of paratextual framing is already announced in the preface, which not only presents the prelude to the novel as its (fictional) genesis, but also, in the end, over the course of almost thirty pages, as a series of attempts to write a good preface (Wirth 2008, 334–338).

In his *Preschool of Aesthetics*, Jean Paul undoubtedly reflected extensively, and – to speak with Ralf Simon – probably more profoundly than any other author around 1800 on the conditions and possibilities of truly modern writing, and in this context already sketched the outlines of a 'theory of prose' (Simon 2013, 211–259). In addition to *wit*, the role of humour ("Humor") is central here; contrary to tradition, Jean Paul defines humour as 'the inverted sublime' ["das umgekehrt Erhabene"] (Jean Paul 1996, I.5, 125; Müller 1983, 218–225; Dembeck 2007, 327). If we take this principle seriously as part of his poetological self-description and apply it to his cosmic dream texts (*Speech of the Dead Christ*; *Dream about the Universe*), both appear in an illusion-breaking light beyond their designation as mediating and transgressive forms. In contrast, the humorous way of writing reveals that the supposedly 'high' (*sublime*) and 'low' are not opposed as 'first' and 'second' worlds, but are always connected.

In this way, the paratexts function as prosaic reversals (*Kippfiguren*) in which the vertical movement on the content side suddenly tilts into a horizontal textual movement of the frames, which can be read as humorous digressions of the narrator as well as rhythmisations of the narrated *life* of the protagonist (Campe 2013). Jean Paul's idea of a paratextual novel prose can thus be understood as a formal-aesthetic consequence of the dream content of the narrative. The *Speech* transforms the fictional insight into a contingent and disordered cosmos into a

---

1 Monika Schmitz-Emans (2015) also ascribes a 'metapoetic' character to Jean Paul's paratexts, but sees it in the fact that – in the form of poetics sermons – they directly address the reader and thus make processes of literary communication between author and audience transparent.

(cosmo)poetics of multiple digressions (Wieland 2013), whose structural principle is that of metapoetic frames that build up complexity and create ambiguities. On the one hand, Jean Paul's novel prose demonstrates the failure of all attempts to decipher the world as a book of nature. At the same time, however, it shows the sky as a tablet on which things, signs and subjectivity can be constellated in ever new ways.

Bibliography

Agazzi, Elena. "Von der *Rede des toten Christus* bis zu den *Nachtwachen des Bonaventura*: Apokalyptische Visionen und Skepsis vor dem Weltzerfall in der Literatur." *Das Erdbeben von Lissabon und der Katastrophendiskurs im 18. Jahrhundert*. Eds. Gerhard Lauer and Thorsten Unger. Göttingen: Wallstein, 2008. 406–421.

Aït-Touati, Frédérique. *Fictions of the Cosmos. Science and Literature in the Seventeenth Century*. Transl. Susan Emanuel. Chicago: University of Chicago Press, 2011.

Andringa, Els. *Wandel der Interpretation: Kafkas ‚Vor dem Gesetz' im Spiegel der Literaturwissenschaft*. Wiesbaden: VS Verlag für Sozialwissenschaften, 1994.

Baasner, Rainer. *Das Lob der Sternkunst: Astronomie in der deutschen Aufklärung*. Göttingen: Vandenhoeck & Ruprecht, 1987.

Blumenberg, Hans. *Die Lesbarkeit der Welt*. Frankfurt am Main: Suhrkamp, 1981.

Blumenberg, Hans. *The Genesis of the Copernican World*. Transl. Robert M. Wallace. Cambridge, MA, and London: MIT Press, 1987.

Brockes, Barthold Heinrich. *Irdisches Vergnügen in Gott*. 9 vols. Hamburg: Johann Christoph Kißner, 1721–1748.

Bühlmann, Regula. *Kosmologische Dichtung zwischen Naturwissenschaft und innerem Universum: Die Astronomie in Jean Pauls „Hesperus"*. Frankfurt am Main and Bern: Peter Lang, 1996.

Buschendorf, Bernhard. "Jean Pauls *Kampaner Tal*: Ein ‚mendelssohn-platonisches Kolloquium' über die Unsterblichkeit der Seele." *Literaturwissenschaftliches Jahrbuch* 38 (1997): 63–92.

Campe, Rüdiger. "Form und Leben in der Theorie des Romans." *Vita Aesthetica: Szenarien ästhetischer Lebendigkeit*. Eds. Armen Avanessian, Winfried Menninghaus and Jan Völker. Berlin and Zurich: Diaphanes, 2013. 193–211.

Dembeck, Till. *Texte rahmen. Grenzregionen literarischer Werke im 18. Jahrhundert (Gottsched, Wieland, Moritz, Jean Paul)*. Berlin and New York: De Gruyter, 2007.

Eickenrodt, Sabine. *Augen-Spiel. Jean Pauls optische Metaphorik der Unsterblichkeit*. Göttingen: Wallstein, 2006.

Erdbeer, Robert Matthias, Florian Klaeger, and Klaus Stierstorfer. "Einleitung: Literarische Form." *Grundthemen der Literaturwissenschaft: Form*. Eds. Robert Matthias Erdbeer, Florian Klaeger, and Klaus Stierstorfer. Berlin and Boston: De Gruyter, 2022. 3–70.

Esselborn, Hans. "Der Albtraum der Leblosigkeit und Fremdbestimmtheit: Motiv und Metapher der Maschine bei Jean Paul." *Lichtenberg-Jahrbuch* (2011): 57–69.

Esselborn, Hans. *Das Universum der Bilder: Die Naturwissenschaft in den Schriften Jean Pauls*. Tübingen: Niemeyer, 1989.

Genette, Gérard. *Paratexte: Das Buch vom Beiwerk des Buches*. Transl. Dieter Hornig. Frankfurt am Main: Suhrkamp, 2014.

Guthke, Karl S. *Der Mythos der Neuzeit: Das Thema der Mehrheit der Welten in der Literatur- und Geistesgeschichte von der kopernikanischen Wende bis zur Science Fiction*. Bern and Munich: Francke, 1983.

Hallyn, Fernand. *The Poetic Structure of the World: Copernicus and Kepler*. New York: Zone Books, 1990.

Heydenreich, Aura. "Vom astronomischen Weltmodell zum literarischen Weltbild: Johannes Keplers *Somnium* zwischen faktualer Kosmographie und fiktionaler Selenographie – mit einem Kommentar zu Durs Grünbein *Cyrano oder Die Rückkehr vom Mond*." *Der Himmel als transkultureller ethischer Raum: Himmelskonstellationen im Spannungsfeld von Literatur und Wissen*. Eds. Harald Lesch, Bernd Oberdorfer, and Stephanie Waldow. Göttingen: Vandenhoeck & Ruprecht, 2016. 333–370.

Honold, Alexander. "Katechismus der Planetenkinder: Jean Pauls Astro-Poetik." *Des Sirius goldne Küsten – Astronomie und Weltraumfiktion*. Eds. Philipp Theisohn et al. Paderborn: Wilhelm Fink, 2019. 251–279.

Horn, Eva. "Die romantische Verdunklung: Weltuntergänge und die Geburt des letzten Menschen um 1800." *Abendländische Apokalyptik: Kompendium zur Genealogie der Endzeit*. Eds. Veronika Wieser et al. Berlin: Akademie, 2013. 101–124.

Hunfeld, Barbara. *Der Blick ins All: Reflexionen des Kosmos der Zeichen bei Brockes, Jean Paul, Goethe und Stifter*. Tübingen: Niemeyer, 2004.

Jean Paul. "Speech of the Dead Christ from the Universe that There Is No God." *Jean Paul. A Reader*. Ed. Timothy Casey, transl. Erika Casey. Baltimore: Johns Hopkins University Press, 1992. 179–183.

Jean Paul. *Sämtliche Werke*. 10 vols. Ed. Norbert Miller. Munich: Hanser, 1974ff. Rpt., Frankfurt: Zweitausendeins, 1996.

Kant, Immanuel. *Kant's gesammelte Schriften: 1. Abt. 1, Werke; Vorkritische Schriften; 1, 1747-1756*. Ed. Preußische Akademie der Wissenschaften. Berlin: G. Reimer, 1902.

Kant, Immanuel. *Universal Natural History and Theory of the Heavens*. Transl. Ian Johnston. Virginia: Richer Resources Publications, 2008.

Kepler, Johannes. *Der Traum, oder: Mond-Astronomie*. Ed. Beatrix Langner, transl. Hans Bungarten. Berlin: Matthes & Seitz, 2010.

Klaeger, Florian. *Reading into the Stars: Cosmopoetics in the Contemporary Novel*. Heidelberg: Universitätsverlag Winter, 2018.

Langner, Beatrix. *Jean Paul, Meister der zweiten Welt: Eine Biographie*. Munich: C.H. Beck, 2013.

Michel, Paul. *Physikotheologie: Ursprünge, Leistung und Niedergang einer Denkform*. Zurich: Gelehrte Gesellschaft in Zürich, 2008.

Moros, Zofia. *Nihilistische Gedankenexperimente in der deutschen Literatur von Jean Paul bis Georg Büchner*. Frankfurt am Main, Berlin, Bern, Vienna: Peter Lang, 2007.

Müller, Götz. *Jean Pauls Ästhetik und Naturphilosophie*. Tübingen: Niemeyer, 1983.

Müller, Götz. *Jean Pauls Exzerpte*. Würzburg: Königshausen & Neumann, 1988.

Neumann, Gerhard. "Traum und Transgression: Schicksale eines Kulturmusters. Calderón – Jean Paul – E.T.A Hoffmann – Freud." *Transgressionen. Literatur als Ethnographie*. Eds. Gerhard Neumann and Rainer Warning. Freiburg i. Br.: Rombach, 2003. 81–122.

Nicolson, Marjorie Hope. *Voyages to the Moon*. New York: Macmillan, 1960.

Pietzcker, Carl. *Einführung in die Psychoanalyse des literarischen Kunstwerks am Beispiel von Jean Pauls „Rede des toten Christus"*. Würzburg: Königshausen & Neumann, 1985.

Preisendanz, Wolfgang. "Naturwissenschaft als Provokation der Poesie. Das Beispiel Brockes." *Frühaufklärung*. Ed. Sebastian Neumeister. Munich: Wilhelm Fink, 1994. 469–494.

Rössler, Reto. *Weltgebäude. Poetologien kosmologischen Wissen der Aufklärung*. Göttingen: Wallstein, 2020.

Rössler, Reto. "Einbruch der Zeit in die Darstellung. Wissenspoetik und *deep time* in der *Allgemeinen Naturgeschichte und Theorie des Himmels*, oder: die temporale Wende beim frühen Kant." Eds. Johannes Pause and Tanja Prokic. *Zeiten der Natur: Konzeptionen der Tiefenzeit in der literarischen Moderne*. Berlin: Springer Nature, 2023. 43–59.

Richter, Karl. *Literatur und Naturwissenschaft: Eine Studie zur Lyrik der Aufklärung*. Munich: Wilhelm Fink, 1972.

Schmidt-Hannisa, Hans-Walter. ",Der Traum ist unwillkürliche Dichtkunst': Traumtheorie und Traumaufzeichnung bei Jean Paul." *Jahrbuch der Jean-Paul-Gesellschaft* 35/36 (2001): 93–113.

Schmitz-Emans, Monika. "Redselige Träume. Über Traum und Sprache bei Jean Paul im Kontext des europäischen Romans." *Traum-Diskurse der Romantik*. Eds. Peter-André Alt and Christiane Leiteritz. Berlin and New York: De Gruyter, 2005. 77–110.

Schmitz-Emans, Monika. "Religious discourse and metapoetic reflection in Jean Paul's novels. The *Rede des toten Christus*, the *Clavis Fichtiana*, and Kain's monologue in *Der Komet*." *Neohelicon* 42.2 (2015): 389–402.

Simon, Ralf. *Die Idee der Prosa. Zur Ästhetikgeschichte von Baumgarten bis Hegel, mit einem Schwerpunkt bei Jean Paul*. Boston: Brill, 2013.

Wieland, Magnus. *Vexierzüge: Jean Pauls Digressionspoetik*. Hannover: Wehrhahn, 2013.

Wiethölter, Waltraut. *Witzige Illuminationen: Studien zur Ästhetik Jean Pauls*. Tübingen: Niemeyer, 1979.

Winkler, Bernhard. "Poetische Experimentalmetaphysik. Jean Pauls Traum-Expedition in die ,2te Welt' in der *Rede des toten Christus vom Weltgebäude herab, dass kein Gott sei*." *Jahrbuch der Jean-Paul-Gesellschaft* 51 (2016): 123–140.

Wirth, Uwe. *Die Geburt des Autors aus dem Geist der Herausgeberfiktion. Editoriale Rahmung im Roman um 1800: Wieland, Goethe, Brentano, Jean Paul und E.T.A. Hoffmann*. Munich: Wilhelm Fink, 2008.

Wölfel, Kurt. ",Ein Echo, das sich selbst ins Unendliche nachhallt': Eine Betrachtung von Jean Pauls Poetik und Poesie." *Jean Paul: Wege der Forschung*. Ed. Uwe Schweikert. Darmstadt: Wissenschaftliche Buchgesellschaft, 1974. 277–313.

IV Early Modern China

Gianamar Giovannetti-Singh
Chinese Heavens in European Literatures, *c.* 1650–1700

Abstract: This chapter retraces the deployment of Chinese astral sciences in early modern European popular literatures. It suggests that Jesuit missionaries' accounts of the transition from the Ming to the Qing dynasties, which privileged "Heaven" as an agent of historical change, helped establish a trope connecting East Asia to the celestial realm in European popular culture. The chapter thus explores the many reinterpretations of the Ming-Qing transition in European literatures, focusing on the deployment of celestial rhetoric and language. It first situates the emergence of astral motifs in Jesuit accounts of Chinese cultures of knowledge by examining the central role of "Heaven" and astronomy in the imperial politics of the Ming-Qing transition in mid-seventeenth-century China. It uncovers the ways in which Jesuits took advantage of the empire's Heaven-centred politics to promote themselves as indispensable allies to China's new powerful Manchu rulers. Then, the chapter moves to Europe, where it investigates the reception and appropriation of Jesuit accounts of China in a variety of different literary genres. It retraces these texts' references to astronomy, the Heavens, and the stars, and proposes that China and its politics became intimately connected to astral language in European literatures. As such, the chapter shows that the Manchu conquest of China had a resounding impact on European literary conceptions of China, its sciences, and politics in the early modern period.

1 Introduction

Between 1644 and 1647, knowing the Chinese heavens quite literally became a matter of life or death for several Europeans. Although the night skies had long played an important role in early modern Europe, with court astrologers engaging intensely in local politics, the supreme importance of the heavens in Chinese elite culture elevated their study – and its political consequences – to another level entirely (Biagioli 1993; Cullen 2017). According to imperial China's ruling political ideology, the emperor was the Son of Heaven (*tianzi*) and required the "Mandate of Heaven" (*tianming*) to rule legitimately over the Middle Kingdom (Schäfer, Chen and Che 2020). Ever since the third century BCE, the task of generating such a mandate fell to various incarnations of the Imperial Astronomical Bureau, an institution whose labourers would watch the skies night and day, carefully noting down any unexpected celestial activity (Deane 1994). If the skies adhered to the predictions from an annually issued imperial calendar, the emperor's mandate was taken to be intact; if they deviated from it substantially, his legitimacy could be called into question. During the 1640s, China underwent a dramatic dynastic transition. The Ming dynasty's Chongzhen emperor, whose ancestors had ruled "all under Heaven" (*tianxia*) since 1368, was overthrown by the Manchus – a semi-pastoralist population from northeast Asia – who claimed to have obtained the Mandate of Heaven to set up a new dynasty, the Qing. As in previous dynastic shifts, the new claimants of the mandate invoked celestial events and "calamities from Heaven" (*tianzai*) to legitimise their political power (Elvin 1998; Janku 2009). Unlike previous dynastic revolutions, however, the different pretenders to the throne were able to exploit the Jesuits' astronomical expertise due to the missionaries' precarious status as guests in the empire.[1] To ensure they would remain welcome in China, different missionaries offered their astronomical services to different political factions, hedging their bets in a time of political upheaval.

During the tumultuous transition from the Ming to the Qing dynasties, several Jesuit missionaries, scattered across the empire, found themselves forced to offer their services as astronomers to different claimants of the Mandate of Heaven. Broadly speaking, those whose missionary work took place in northern China allied themselves with the Manchus; those in the south remained largely loyal to the Ming; and two missionaries unfortunate enough to find themselves in the western province of Sichuan became court astronomers to the brutally cruel tyrant Zhang Xianzhong (Rule 2011; Zürcher 2002). The importance of the firmament in legitimising political power in imperial China, combined with the fact that several astronomically trained Europeans were in the empire at the time and found themselves involved in celestial controversies with far-reaching political ramifications, ensured that the Chinese heavens left an indelible mark upon early modern European scholarly and popular culture, reverberating well into the eighteenth century (Lach and Van Kley 1993; Mungello 1985; Van Kley 1971; Wu 2017). This chapter, which explores different European interactions with and appropriations of Chinese celestial discourses, shows the substantial impact that the Chinese heavens had on European literatures, beyond a strictly scholarly sphere, in early modernity.

---

1 While the Ming-Qing transition was the first to be witnessed by the Jesuits, foreign, Islamicate astral knowledges also played a role in the previous dynastic shift, from the Mongol Yuan dynasty to the Ming in the 1360s. Cf. Weil (2022).

Ever since the early twentieth century, historians have discussed China's influence on different spheres of early modern European culture and society (Jacobsen 2013; Statman 2019; Statman 2023). In 1908, Henri Cordier argued that Chinese artworks and artefacts, such as paintings and porcelains, led to a European "engouement" (Cordier 1908, 769), or infatuation, with China in the eighteenth century, contributing to the craze for *chinoiserie* that captured burgher and courtly cultures alike, from Golden Age Amsterdam to Louis XIV's France to Frederick the Great's Prussia.[2] In 1932, shifting attention from visual and material cultures to intellectual history, Virgile Pinot's masterly monograph *La Chine et la formation de l'esprit philosophique en France* suggested that Chinese philosophy left a marked effect on the Enlightened "philosophical spirit" (Pinot 1932).[3] Pinot contended that debates about chronology, the relationship between religion and morality, and between morality and government were all tangibly shaped by the reports flooding into Europe from China. In 1946, turning to political and economic history, Lewis Maverick argued that Confucian moral and social philosophy heavily shaped the Physiocrats' reformist proposals for a new political economy in France. With regards to China's impact on literary cultures, Adrian Hsia (1998) examined the emergence of the construct of "China" in several different early modern European national literatures. More recently, Eun Kyung Min's superb monograph (2018) showed that the construction of a "national" literary identity in seventeenth- and eighteenth-century England was deeply imbricated with the famous quarrel between the Ancients and Moderns, which was itself heavily shaped by controversies about China's antiquity. Much of this scholarship, however, despite paying meticulous attention to the reception of all things Chinese in Europe, did little to examine the conditions *in China* that gave rise to the circulation of Chinese knowledge, practices, and artefacts to Europe.

That trend was challenged in the 1970s by Edwin Van Kley's pathbreaking work (1971), which examined how historical geopolitical events in China were understood in early modern Europe.[4] Van Kley (1973; 1976) argued that the collapse of the Ming dynasty was seen as an important event by contemporary Europeans, warranting the writing of plays, poems, and stories about it. Both erudite scholars and writers communicating to wider, more popular audiences drew extensively on eyewitness accounts of the rise of the Qing dynasty (Lach and Foss 1990). Two of the most popular descriptions of the war were the Jesuit missionary Martino Martini's *De Bello Tartarico Historia* (1654), first published by the *Officina Plantiniana* in Antwerp and republished over twenty times in ten languages by the

---

2 On the impact of Chinese luxury goods in early modern Britain, cf. Berg (2004).
3 On the Enlightened "philosophical spirit", cf. Cassirer (2009 [1932]).
4 Cf. also Wakeman (1986).

end of the century, and the Dutch merchant Johan Nieuhof's *Het Gezandtschap der Neêrlandtsche Oost-Indische Compagnie, aan den grooten Tartarischen Cham* (1665), published by Jacob van Meurs in Amsterdam (Lindgren 2016). Given the centrality of "Heaven" in the politics of dynastic transitions and the fact that Jesuits like Martini, who published accounts of their experiences across Europe, participated as astronomers in the Ming-Qing transition, European *savants* came to closely associate the Chinese empire and its political culture with the celestial (Hsia 2009; Giovannetti-Singh 2023). However, as I suggest in this chapter, the European interest in the Chinese heavens extended beyond the debates of elite scholars and into more popular texts. The loss of the Mandate of Heaven by the Ming dynasty, recounted in an almost picaresque style by Martini in *De Bello Tartarico Historia*, brought the Chinese heavens squarely into early modern popular European literatures. To use Van Kley's expression, the Manchu conquest (and its astronomical baggage) became "an alternative muse" (1976) for European literary cultures, competing against Greco-Roman antiquity. Nevertheless, given the declining credibility of judicial astrology in Europe at the time that accounts of a celestially governed China started to gain traction, the emergent stereotype that associated China to the Heavens was not always positive.[5]

The next section of this chapter describes the circumstances under which European missionaries in China came to associate the Middle Kingdom with the heavens. Then, it explores how the Chinese heavens were received in European literatures from the second half of the seventeenth century, paying particular attention to invocations of the celestial realm in romances, plays, and poems about the fall of the Ming dynasty and China more broadly. The chapter argues that whether with positive or negative connotations, China came to be closely associated with the heavens in early modern European culture, largely as a result of the far-reaching impact of Martini's account of the "Tartar War".

2 The Jesuits encounter the Chinese Heavens

Ever since the Society of Jesus first entered the Middle Kingdom in the late sixteenth century, missionaries used astronomical instruments, arguments, and their applications to engage with the class of elite Confucian scholar-officials who operated the imperial state bureaucracy (Hsia 2010). This should not strike us as

---

5 Pope Sixtus V condemned all forms of *astrologia iudicaria* in 1585, and in 1619, Johannes Kepler admonished astrologers for "portray[ing] the influences of the stars as if they were some gods, with power over heaven and earth" (Kepler 1997, 360).

particularly surprising, as the Jesuit colleges' curriculums required a thorough familiarity with Aristotle's *Physics* and several of its early modern commentaries, as part of a missionary's basic education (*Ratio Atq. Institutio* 1660). As early as 1582, Michele Ruggieri gifted a mechanical clock, which would have been used together with astronomical instruments for navigation, to the Guangzhou official Huang Yingjia, in order to gain social access to the higher echelons of the imperial polity (Zhang 2012). Perhaps the most ambitious cross-cultural astronomical collaboration in the early years of the mission was that between Matteo Ricci – who had studied at the Collegium Romanum under the famous Bamberg-born astronomer Christoph Clavius – and Xu Guangqi, a Chinese scholar-official, polymath, and Catholic convert (Jami, Engelfriet and Blue 2001). Working together, the two scholars translated Euclid's *Elements* into Chinese. Over the course of his life in China, Ricci came to view astronomy as a key means of engaging with and converting Chinese *literati*, largely due to the cultural importance of a highly sophisticated if increasingly troubled set of astral traditions in China.

Although early modern China did not have a single discipline directly translatable as "astronomy," two separate albeit connected fields involved studying the heavens: *tianwen* and *lifa* (Sun 2015). The former, loosely translated as "heavenly writing" was somewhat akin to judicial astrology in Europe and constituted interpreting the skies in terms of their relationship to human lives and terrestrial events such as earthquakes. *Lifa* ["calendrical methods"], instead, involved determining the regularity of celestial phenomena using mathematical calculations. However, as Christopher Cullen (1999) points out, these two disciplines were united through their practical value for the imperial state. While practitioners of *lifa* were required to identify regular patterns in the cosmos and produce a calendar that predicted heavenly motions accurately and reliably, thus demonstrating the emperor's harmonious relationship with Heaven, *tianwen* scholars were responsible for recording and, more importantly, *interpreting* unusual celestial events and their astrological consequences for imperial politics. When Ricci first began engaging with Chinese astral sciences, the empire's calendar was widely seen as being in crisis; several scholar-officials had noticed that the date for the Lunar New Year had drifted by one day with respect to the preceding Yuan dynasty (Elman 2005). With the Gregorian calendar reform – which had been led by Ricci's teacher, Clavius, with whom the missionary remained in contact – coming into effect in Europe in 1582, Ricci saw an opportunity for his order to ingratiate itself with the Ming regime by offering astronomical labour and expertise to help reform the Chinese calendar.

In 1605, Ricci wrote back to the Society of Jesus's headquarters in Rome, asking them to "send [him] an astrologer" to undertake further mathematical translations and calendrical calculations to gain the confidence of the Chinese emperor (Dunne

1962, 210–211). Shortly after Ricci's death in 1610, the Society sent his Flemish confrère, Nicolas Trigault, back to Europe to recruit astronomically trained missionaries and bring European books to Beijing.[6] Among the missionaries who returned to China were the Cologne astronomer Johann Adam Schall von Bell and the Swiss polymath Johann Schreck (Terrenz), who had studied medicine under Galileo in Padua and was a member of the prestigious scientific Accademia dei Lincei in Rome. Although Ricci was never commissioned to reform the Chinese calendar during his lifetime, the last ruler of the Ming dynasty, the Chongzhen emperor, finally commissioned Xu Guangqi to lead a project of calendrical reform and work with a team of Jesuits, including Schall and his Milanese colleague Giacomo Rho, in 1629. In December of that year, the Chinese and Muslim sections of the Imperial Astronomical Bureau predicted a solar eclipse less accurately than the Jesuits at court, which led Chongzhen to finally acquiesce to order a new, Jesuit-compiled calendar (Chu 2007; Väth 1991). Although Xu, Schall, Rho, and their team of calendrical labourers finished compiling the new calendar, titled *Chongzhen lishu* ["Calendrical Treatises of the Chongzhen Reign"], in 1634, the Chongzhen emperor did not publicly adopt it until 1643, as his empire was collapsing at the hands of the Manchu invaders from the northeast and internal insurrections (Chu 2007, 164).

During the civil war that engulfed the Middle Kingdom in the mid-seventeenth century, astronomy became even more important, as many different claimants of the throne argued that they had received the Mandate of Heaven to establish their own imperial dynasty. Ming Beijing first fell to the peasant rebel Li Zicheng and his army on 24 April 1644, who claimed he had received the Mandate of Heaven to establish the Shun dynasty.[7] As Martini recounted, "being by now sure of obtaining the Empire, [Li Zicheng] attributed to himself the title of emperor, and to the dynasty he thought of founding he gave the name Thienxun [Tian Shun]. Thienxun means obedient to Heaven: with this title he tried to persuade the people and their soldiers that it was the will of Heaven that he governed, and that from Heaven he had received the order to free the people from the greed of the [Chongzhen] emperor, and to eradicate the prefects devoted to him who were cruel to the people" (Martini 1654, 63). Despite defeating the Chongzhen emperor, who hanged himself when Li's troops arrived, and the latter's invocation of Heaven to legitimise his newfound political power, Li's reign did not last long. In May of the same year, the Manchu army, assisted by the former Ming general Wu Sangui, crossed the Great Wall at Shanhai Pass in May and defeated Li's troops in battle. The following month, Wu and the Manchu general Dorgon entered the Forbidden City and sought

---

6 On the Jesuit library in Beijing, cf. Golvers (2020) as well as Wu and Zhang (2020).
7 For a detailed account of the short-lived Shun dynasty, cf. Wakeman (2009).

out the astronomical expertise of the last remaining Jesuit in the city, Schall, whom Dorgon appointed director of the Imperial Astronomical Bureau (Jami 2015). Eventually, on 8 November 1644, the Manchus enthroned the six-year-old Shunzhi emperor on the dragon throne and declared to have obtained the Mandate of Heaven to establish the Qing dynasty (Wakeman 1986, 857–861). As Martini wrote, Shunzhi, who was "not childish, but mature and lofty, ascended the ancient throne of his people with great solemnity and majesty" (1654, 80–81). Having achieved harmony with Heaven, Shunzhi – son of a "barbarian" Tartar – had become both Chinese and "civilised".

Despite Shunzhi's coronation in Beijing in November 1644, fighting continued across the Middle Kingdom, with descendants of the Ming continuing to claim the Mandate of Heaven in southern China. During these tumultuous years, the Jesuit missionaries who had by this point established themselves among the Ming *literati*, found themselves forced to develop and maintain *ad hoc* coalitions with whichever armed group they happened to encounter. In 1646, Martino Martini – who had previously served as a ballistics expert in the irredentist Southern Ming court – encountered Manchu troops for the first time, and astutely decorated his mission station with a large red scroll proclaiming *"Hic habitat ex magno Occidente diuina legis Doctor"* and placed several European books, astronomical and mathematical instruments, and an image of Christ in the atrium (Collani 2016; Martini 1654, 99). When the Manchu soldiers found Martini inside, they offered him the chance to join their dynasty; unsurprisingly, the missionary agreed and began serving the Qing. Martini's display of the astronomical instruments indicates that he was aware of the importance of celestial events in Chinese dynastic politics, and the Manchu soldiers' response suggests that they sought out all the astronomical expertise they could muster to ensure that they maintained Heaven's mandate.

In 1650, Martini was summoned by his superiors to undertake astronomical work with Schall at the Imperial Astronomical Bureau in Beijing. Their collaboration did not last long, however, as Schall – who was distrusted by several other Jesuits for his relatively open attitude towards Chinese astral sciences and his non-celibate lifestyle – grew suspicious that Martini had been sent to spy on him. Martini was then sent back to Europe as a mission "procurator", tasked with recruiting further missionaries and publicising the Society of Jesus's activities in China. After a long and arduous journey, in which he was kidnapped by the Dutch Vereenigde Oostindische Compagnie (VOC) on his way to the Philippines and was imprisoned for several months in Batavia, Martini reached Bergen, Norway, on 31 August 1653, before journeying for the next five years across the Low Countries, and to Vienna, Munich, and Rome (Begheyn 2019; Cams 2020). Through his publications in a striking array of vernacular languages, bridging confessional divides, Martini brought China to Europe and, importantly, as a result of his own astronomical experiences in China, cemented the association between China and the heavens for European readerships.

3 The Chinese Heavens travel to Europe

Only a few days after he had reached Amsterdam in November 1653, Martini met the renowned cartographer and printer Joan Blaeu and began negotiating the publication of his writings about China's geography, history, arts and sciences, and the Manchu conquest (Begheyn 2012). Shortly after his first meeting with Blaeu, Martini travelled southwest to Antwerp, in the Spanish Netherlands, where he offered both his account of the Manchu war and his atlas of China to Balthasar Moretus II, director of the famous *Officina Plantiniana* printing press.[8] Moretus accepted the former, which was published as *De Bello Tartarico Historia* in 1654, but rejected the latter, for its map plates would be too expensive to engrave (Golvers 2012). Martini returned to Amsterdam, where, after discussing the particularities of his atlas of China with Blaeu, eventually published it as the *Novus Atlas Sinensis* in 1655, constituting the sixth volume of Blaeu's *Theatrum Orbis Terrarum sive Atlas Novus* series (Cams 2020).

While the *Novus Atlas Sinensis* rapidly became a highly coveted, expensive marker of social status amongst wealthy Dutch burghers, *De Bello Tartarico Historia* was far more widely read, initially composed in an accessible style of Latin and later translated into Dutch, French, German, English, Italian, Spanish, Portuguese, Swedish, and Danish (Koeman 1970; Dijkstra 2021). As Trude Dijkstra (2021, 68–70) shows, Martini's various publishers across Europe carefully targeted their editions of the missionary's book to their home markets, paying close attention to emphasising or erasing the fact that the book's author was a Jesuit, depending on their expected readership's religious affiliations. Martini was, above all else, remarkably adept at self-fashioning himself in often contradictory ways to appeal to a heterogeneous variety of audiences. During the Manchu conquest, he reinvented himself as a Manchu-supporting astronomer who could secure the new dynasty the Mandate of Heaven; when he was imprisoned by the VOC in Batavia, he sold himself as a purveyor of valuable commercial information who could help them open up lucrative trade with Qing China; when he returned to Europe,

---

8 *Zeitung Auß der newen Welt* (1654) reveals that Martini had intended to publish his work in Antwerp at the Catholic-friendly Plantin-Moretus press from the outset.

Martini targeted his accounts of China to audiences of different economic and educational status, ensuring as wide a readership as possible.[9]

Martini did not only advertise the Jesuit China mission through his written output; at the start of 1654 the missionary was hosted in Brussels by the governor of Belgium and Burgundy, Archduke Leopold Wilhelm I of Austria, brother of the reigning Habsburg Emperor Leopold I in Vienna. The missionary regaled the Archduke with tales of his adventures in China and the East Indies in the latter's native German tongue, securing a powerful patron for the mission (Golvers 2016). On his same trip to Belgium, Martini astonished novices at the Colegium Sosietatis Iesu Lovaniensis [sic] by giving them a lecture about China using a *laterna magica* operated by the local Jesuit mathematician Andreas Tacquet to display spectacular illustrations of China's wonders (Golvers 2016; Purtle 2010; Weststeijn 2012). As a result of his skilled manipulation of multiple means of communication, Martini's descriptions of the Middle Kingdom came to form *the* defining image of China's history, geography, politics, culture, arts, and sciences over the next century. As late as 1877, the Prussian geographer Baron Ferdinand von Richthofen described Martini's atlas of China as "the most complete geographical description of China that we possess" (Richthofen 1877, 676). Through the pages of the missionary's books, the Chinese Heavens – which had been so important to Martini's survival of the war – reached European readers and created the image of an empire governed by the stars.

De Bello Tartarico Historia quite clearly outlined the political ideology of the Mandate of Heaven, informing its wide readership that the Chinese "in fact, believe that it is Heaven that confers empires, and they believe that they cannot be seized by force or human arts" (Martini 1654, 63). The Dutch translation of the missionary's account, *Historie van de Tartarischen oorloch*, attracted the attention of Holland's "prince of poets", Joost van den Vondel, who adapted the account of the Manchu conquest into his play, *Zungchin of ondergang der Sineesche Heerschappye*, published in Amsterdam in 1667. The play recreates the last night of Zungchin's [Chongzhen's] life, as Lykungzus's [Li Zicheng's] army marches towards Peking. The doomed Zungchin, in a final act of desperation, calls on the Jesuit father Johann Adam Schall to pray for the salvation of his empire. Vondel had converted to Catholicism in 1641 and thus, as Dijkstra (2021, 71) has argued, "turned the fall of the Ming dynasty into a lesson concerning true faith: meaning

---

9 The *Novus Atlas Sinensis* was targeted to wealthy burgher consumers, *De Bello Tartarico Historia* was intended to have a wide, interconfessional and interlinguistic audience, and his account of Chinese chronology, *Sinicae Historiae Decas Prima*, published first by Lukas Straub, a Catholic printer in Munich, and then by Joan Blaeu in Amsterdam, was aimed at an erudite, scholarly audience.

the word of God as given by the Jesuits".[10] However, Vondel's references to "heaven" in the play appear to be (mis)translated, transformed, and transposed allusions to the Chinese Heaven (*tian*) rather than the Christian Heaven.[11] Consider the following lines, near the beginning of Act 2:

> But nature, when it moulds man,
> Creates no window into the bosom, to know the heart
> Where hypocrisy and mistrust lurk in dark corners,
> And to espy whatever misfortune may brew there.
> Even though his majesty bears the name of heaven's son.
> Knowledge of the heart remains a pearl in the crown
> Of the supreme being, who discerns men's thoughts.
> With him, neither disguise, nor deception or dissembling have any value.
> Heaven threatens us with a hurricane of war
> On the city. Now is the decisive moment.
> Here, no eye shall be shut through the entire night.
> The whole court, stirred up, seems to break out of its ranks.
> There the emperor approaches, in magnificent yellow robes,
> Which he alone may wear. How shall my tongue find the right tone
> When I speak so that the sound may tickle his ears.
>
> (Vondel 2011)[12]

The reference to "heaven's son" in line 5 indicates that Vondel, having read Martini's source material, was at least superficially familiar with the concept of the Son of Heaven. Line 9, instead, appears to allude to the concept of "calamities from Heaven", which, somewhat like biblical divine retribution in a Christian context, were considered signs of Heaven's discontent with the actions of an emperor (Elvin 1998).

Moreover, comments by the Adam Schall character at the end of the play suggest that Vondel sought to create an image of China as a superstitious land that, unlike Europe, still looked to the (material) heavens to understand human behaviour. In Act 3, the Empress Jasmyn tells Schall that

> A wise Arab saw how our birthstar
> Wisely advised us to avoid an arrow, which,
> Whizzing from afar out of the sky, shone, like a comet
> [. . .]
> If Toangus were alive, experienced in astronomy,

---

10 On the religious interpretations of the Manchu conquest, cf. Romano (2016).

11 On the conflict between uses of the term "tian" for Christian concepts during the Chinese Rites Controversy, cf. Giovannetti-Singh (2022).

12 For the original, cf. Vondel (1667, 20–21).

And astrology, would reveal
By what means one could avoid the arrow,
Before the unexpected blow violated the sacred right to the crown.
(Vondel 2011)[13]

Father Adam replies:

Most august Empress, Europe, enlightened
From above, places no hope for well-being on the sight of spirits,
Soothsaying from the flight of birds, dreams and palmistry,
Ominous rustling of forests and rushing of streams,
Howling, and barking of dogs, and astrology,
Nor does it heed spirits of the night, or lunacy
Of the augur, nor the drivel of sorcerers.
(Vondel 2011)[14]

Despite its damning verdict about the Chinese propensity to rely on what the character of Schall essentially dismisses as judicial astrology, which had been banned for almost a century by the Catholic Church, *Zungchin* strengthened the association between the imperial Chinese polity and the heavens in the European imaginary, extending the connection made by Martini to even wider audiences. Interestingly, Vondel's play marked a *volte-face* compared to Schall's actual attitude towards Chinese astral sciences. As the Director of the Imperial Astronomical Bureau, the Kölner had become deeply invested in *lifa* and, despite appearing to have engaged exclusively in that branch of Chinese astral sciences, he was accused by his confrères Gabriel de Magalhães and Lodovico Buglio of playing with fire by lending credibility to "superstitious" Chinese *tianwen* (Zürcher 2002).

In tandem with the publication of Vondel's *Zungchin*, the Dutch playwright Joannes Antonides van der Goes dramatised *De Bello Tartarico Historia* in his *Trazil of overrompelt Sina*. While Vondel had remained relatively faithful to Martini's source material despite some artistic license, van der Goes deviated from it substantially, combining Li Zicheng's and the Manchus' invasions into a single event, playing out over twenty-four hours. As Theodor Nicholas Foss and Donald Lach argue, "[i]n both dramas the message is conveyed that Peking, like ancient Troy, fell to predatory conquerors who used treachery, betrayal and deceit to win the city" (1990, 174). However, as Dijkstra (2021, 74) points out, *Trazil* "focussed more on state power and the domestic relevance thereof" than *Zungchin*. The concept of the "Mandate of Heaven" appears early in van der Goes's play. When the

---

13 For the original, cf. Vondel (1667, 20–21).
14 For the original cf. Vondel (1667, 21).

plotting traitors Sinkio and Hungüan first enter the scene, the latter tells his comrade that "Heaven favours us. The bright stars shine / In the Heavenly vault the faint moon is sinking / Rolls steeply from the north axis to the other side of the Earth" (van der Goes 1685, 8). The "Son of Heaven" too, once again makes an appearance in *Trazil*:

> Now leave, O Heavenly Son! show thy army lion
> That thou art patron of so many Kingdoms;
> And that Trazil stands here, or his likeness,
> At mirror, which is the wages of the Imperial slayers.
> (van der Goes 1685, 77)

While *Trazil* was a highly orientalist production, in that it projected European concerns about political treachery and personal tragedy onto an almost entirely fictionalised "China", van der Goes's references to the Chinese "Mandate of Heaven" ideology – even if he understood little of it and used it for entertainment purposes – testifies to the impact of Martini's non-orientalist account of the importance of Heaven in Chinese politics (Said 1978).

European literary accounts of China, its people, and their astro-centric political ideology travelled beyond the world of theatre into several romances, predominantly written in the German-speaking lands and England. In 1670, Christian Wilhelm Hagdorn published the first novel about the Manchu conquest in Europe: *Aeyquan*. This Baroque novel, printed in Amsterdam by Jacob van Meurs, who five years earlier had published Nieuhof's account of the VOC embassy to Qing China, told the story of a heroic Indian knight errant called Aeyquan, who, having helped the Manchus to capture Beijing, was awarded with a courtly position. Lach and Foss suggest that *Aeyquan* drew much inspiration from Gautier de la Calprenedes's *Cassandre* (1642–1645) and the French heroic novel tradition more broadly (Lach and Foss 1990, 172). While to some extent, Hagdorn's novel indulged in early modern orientalism by projecting a European narrative style onto a Chinese landscape, it still followed Martini's account relatively closely and its invocation of the Mandate of Heaven did not seek to ridicule the concept in the way that Vondel's play did. As the character of Lycung [Li Zicheng] told the Manchu princess Saphothisphe, "now / when heaven raised me to the throne of China / I want to show my power and hereby order you / to keep your guardian within two days" (Hagdorn 1670, 264). The political circumstances of China had, by this time, become inseparable from Heaven in the European imaginary.

As Lach and Foss (1990, 172) explain, *Aeyquan* became a standard reference point for most German courtly novels about Asia. In 1673, Eberhard Werner Happel published *Der asiatische Onogambo* in Hamburg, describing the adventures of an

Asian prince who becomes the ruler of the Qing dynasty and a student of Adam Schall, who – despite having died in true life a few years earlier – remained a key intermediary between the unknown, exotic world of Chinese astral sciences and European astronomy and Catholicism for European audiences. These tropes found their way into Daniël Casper von Lohenstein's immense *Arminius*, first published in Leipzig in 1689. Lohenstein's book takes significant artistic license and moves away from the earlier novels in its depiction of the Chinese-Manchu conflict, but actually ends up closer to Martini's account than its fictional predecessors. *Arminius* states that

> Enmity has been ingrained in certain peoples as well as in some animals. From several
>> thousand years ago
> or rather from its origin
> this has made a separation between the Serers [Chinese] and Scythians
> or
> as they call us
> the Tartars. I don't know
> whether this deadly hostility is due to the differences between the countries
> or to the repugnance of our stars.
>
> (Lohenstein 1689, 1:598)

The concept of the "repugnance of our stars" between the Chinese and the Manchus coheres well with the historical fact that different claimants of the throne all had their own astronomical labourers working to establish their respective dynasty's Mandate of Heaven. Lohenstein's book, again, shows how ingrained astral thinking had become in European fictional depictions of political conflict in China.

Returning to England, the playwright and poet Elkanah Settle published his play *The conquest of China, by the Tartars* in London in 1676. Settle's account, unlike Vondel's influential stage adaptation of the Manchu war, did not pit the opposing sides as representing the clash between civilisation and barbarism, but rather depicted them as courtly rivals seeking to undermine each other's control of China (Lach and Foss 1990, 174). Nevertheless, despite the shifted moral of Settle's play, he referred, like all his European contemporaries, to Heaven as being a determinant of political power in China. As the character Zung, the "son of the king Theinmingus of Tartary", announced in the opening act, "When Peace the prop of sluggish Kings secur'd / The *Chinan* Empire from the *Tartars* Sword: / And Heaven did by that Charm this Crown support, / I went a Guest to the *Taymingian* Court" (Settle 1676, 6). Settle, who was a staunch anti-Catholic and took part in anti-popish activism in England, demonstrates quite how international and interconfessional the reach of Martini's account and its promulgation of the Chinese "Mandate of Heaven" political ideology was.

The impact of the emergence and cementing of the stereotype that connected the Chinese to Heaven in the European imaginary is visible in the peculiar case of the fictional East Asian persona who took early modern England by storm: George Psalmanaazaar.[15] In fact a lapsed French Catholic, Psalmanaazaar presented himself in England in 1704 as a native Formosan convert to Christianity who had fled from his brutal Jesuit overlords in Taiwan. Psalmanaazaar published *An Historical and Geographical Description of Formosa* in London in 1704, in which he described – among a great many other things – the religious and political system of the native Formosans. In part due to the anti-papism flourishing in England at the time, Psalmanaazaar's story got him astonishingly far, earning him a professorship at Christ Church, Oxford, and attracting the attention of the Royal Society's then-president, Isaac Newton, who summoned Psalmanaazaar for interview (Schaffer 2009). Simon Schaffer (2009, 248) has convincingly argued that Psalmanaazaar "used the conventions of [the early Royal Society's] information order – of probability, conjecture and assay – to make his story credible."[16] But beyond the impostor's astute deployment of "literary technologies" that would have rendered his story credible to the natural philosophers of the Royal Society, Psalmanaazaar also played on the trope that connected the Chinese cultural sphere – which included Formosa – to the astral realm.[17] As he wrote, describing the Formosan religion,

> there appeared two Philosophers, who had led a Pious and Austere kind of Life in the Deserts, and pretended that God had appear'd to them, and spoke to them, to this purpose; *I am much Troubled for the Blindness of this People because they Worship the Sun, Moon and Stars so devoutly, as the Supreme Deity; go and tell them, I am the Lord of the Sun, Moon and Stars, of the Heaven, the Earth, the Sea and all things that are in them, I Govern the Creatures by the Sun and Moon and the 10 Stars, and without me they cannot exist: Go and tell them, that God has appear'd to you, and said, if they will worship and adore him, he will be their Protector, and will appear to them in the Churches, which they Build to his honour, and promise them in my Name, that if they Worship and obay* [sic] *me, they shall receive great rewards after this Life.* (Psalmanaazaar 1704, 169–170)

In all likelihood, Psalmanaazaar, who was actively seeking to fabricate his credibility, wrote of the astronomical nature of the native Formosan religion to play into a trope that had, since Martini's publication of *De Bello Tartarico Historia*, become ever stronger in the European imaginary.

---

15 On Psalmanaazaar, cf. Needham (1985).
16 On the conventions of the early Royal Society's information order, cf. the classic Shapin and Schaffer (1985).
17 On literary technologies, cf. Shapin and Schaffer (1985) and Shapin (1984).

4 Conclusion

Since the Manchu conquest of the Ming empire in the 1640s, most factual, embellished, and entirely fictional descriptions of China published in Europe emphasised the importance of Heaven in the state's political ideology. Martino Martini's *De Bello Tartarico Historia* of 1654, which was read by a wide, interconfessional, and international European audience throughout the seventeenth century, had an enormous impact on European images of China and Tartary. Martini's book, in part as a consequence of the fact that the missionary himself had engaged closely with Chinese astral sciences to survive the Manchu war, brought the Chinese "Mandate of Heaven" ideology to European audiences. During the war, Martini had presented himself as an astronomical expert to the Manchus, highlighting that his expertise could help the latter acquire and maintain the Mandate of Heaven with his and other Jesuits' help in the Imperial Astronomical Bureau.

The many retellings of the Manchu conquest – in novels, poems, and plays – across Europe served to amplify Martini's (non-orientalist) description of the importance of the Mandate of Heaven in Chinese imperial politics. Some authors, such as the Dutch poet Joost van den Vondel, used the superstitious nature of Chinese astral sciences to discredit them compared to European astronomy, thereby creating orientalist images of a backwards, unscientific China. Others, such as the Silesian Daniël Casper von Lohenstein, simply recounted that the heavens played a crucial role in determining the conflict between the Chinese and the Manchus, superimposing an accurate statement from Martini onto a courtly romance. All the different literary accounts of the importance of heaven in shaping "China's" politics and culture further influenced European stereotypes of the Middle Kingdom, to the extent that the notorious impostor George Psalmanaazaar, in an attempt to gain credibility for his story, told of the centrality of the heavens in shaping the religion and culture of his alleged Formosan compatriots. While scholarship on the role of China (as well as the fictitious construct of "China") in early modern Europe increasingly discusses the importance of the Manchu conquest in shaping European images of the orient, this chapter has drawn attention on the overwhelming importance and centrality of the heavens in shaping European accounts about China.

Bibliography

Begheyn, Paul. "Dutch Publications on the Jesuit Mission in China in the Seventeenth and Eighteenth Centuries." *Quaerendo* 49 (2019): 49–65.

Begheyn, Paul. "The Contacts of Martino Martini S. J. with the Amsterdam Printer Joan Blaeu." *Archivum Historicum Societatis Iesu* 81.1 (2012): 219–232.

Berg, Maxine. "In Pursuit of Luxury: Global History and British Consumer Goods in the Eighteenth Century." *Past and Present* 182.1 (2004): 85–142.

Biagioli, Mario. *Galileo Courtier: The Practice of Science in the Culture of Absolutism*. Chicago: University of Chicago Press, 1993.

Cams, Mario. "Displacing China: The Martini-Blaeu *Novus Atlas Sinensis* and the Late Renaissance Shift in Representations of East Asia." *Renaissance Quarterly* 73.3 (2020): 953–990.

Cassirer, Ernst. *The Philosophy of the Enlightenment*. Princeton: Princeton University Press, 2009 [1932].

Chu Pingyi, "Archiving Knowledge: A Life History of the Calendrical Treatises of the Chongzhen Reign (Chongzhen Lishu)." *Extrême-Orient, Extrême-Occident* 29 (2007): 159–184.

Collani, Claudia von. "Two Astronomers: Martino Martini and Johann Adam Schall von Bell." *Martino Martini: Man of Dialogue*. Eds. Luisa M. Paternicò, Claudia von Collani and Riccardo Scartezzini. Trento: Università di Trento, 2016. 65–92.

Cordier, Henri. "La Chine en France au XVIIIe siècle." *Comptes rendus des séances de l'Académie des Inscriptions et Belles-Lettres* 52.9 (1908): 756–770.

Cullen, Christopher. "Astronomy in China." *The Cambridge Concise History of Astronomy*. Ed. Michael Hoskin. Cambridge: Cambridge University Press, 1999. 48–50.

Cullen, Christopher. *Heavenly Numbers: Astronomy and Authority in Early Imperial China*. Oxford: Oxford University Press, 2017.

Deane, Thatcher. "Instruments and Observation at the Imperial Astronomical Bureau During the Ming Dynasty." *Osiris* 9 (1994): 127–140.

Dijkstra, Trude. *Printing and Publishing Chinese Religion and Philosophy in the Dutch Republic, 1595–1700*. Leiden: Brill, 2021.

Dunne, George H. *Generation of Giants*. Notre Dame: University of Notre Dame Press, 1962.

Elman, Benjamin. *On Their Own Terms: Science in China, 1550–1900*. Cambridge, MA: Harvard University Press, 2005.

Elvin, Mark. "Moral Meteorology in Late Imperial China." *Osiris* 13 (1998): 213–237.

Eun Kyung Min, *China and the Writing of English Literary Modernity, 1690–1770*. Cambridge: Cambridge University Press, 2018.

Giovannetti-Singh, Gianamar. "Astronomical Chronology, the Jesuit China Mission, and Enlightenment History." *Journal of the History of Ideas* 84.3 (2023): 487–510.

Giovannetti-Singh, Gianamar. "Rethinking the Rites Controversy: Kilian Stumpf's *Acta Pekinensia* and the Historical Dimensions of a Religious Quarrel." *Modern Intellectual History* 19.1 (2022): 29–53.

Golvers, Noël. "Jesuit Libraries in Beijing and China in the Perspective of the Communication between Europe and China in the Seventeenth and Eighteenth Centuries." *Foreign Devils and Philosophers: Cultural Encounters between the Chinese, the Dutch, and Other Europeans, 1590–1800*. Ed. Thijs Weststeijn. Leiden: Brill, 2020. 238–253.

Golvers, Noël. "Martino Martini in the Low Countries." *Martino Martini: Man of Dialogue*. Eds. Luisa M. Paternicò, Claudia von Collani and Riccardo Scartezzini. Trento: Università di Trento, 2016. 113–135.

Golvers, Noël. "Martino Martini, S. J., His Stay in the Jesuit College of Brussels (1654) and the Production of his *Novus Atlas Sinensis*." *Quatre siècles de présence jésuite à Bruxelles / Vier eeuwen Jezuïeten te Brussel*. Eds. Alain Deneef and Xavier Rousseaux. Leuven: Kadok, 2012. 125–137.

Hagdorn, Christian Wilhelm. *Aeyquan, oder der Große Mogol. Das ist / Chineische und Indische Stahts-Kriegs- und Liebes-geschichte*. Amsterdam: Jacob von Mörs, 1670.

Happel, Eberhard Werner. *Der asiatische Onogambo*. Hamburg: Joh. Naumanns and Georg Wolffen, 1673.

Hsia, Adrian. *Chinesia: The European Construction of China in the Literature of the 17th and 18th Centuries*. Tübingen: Niemeyer, 1998.

Hsia, Florence. "Chinese Astronomy for the Early Modern European Reader." *Early Science and Medicine* 13.5 (2008): 417–450.

Hsia, Ronnie Po-chia. *A Jesuit in the Forbidden City: Matteo Ricci, 1552–1610*. Oxford: Oxford University Press, 2010.

Jacobsen, Stefan Gaarsmand. "Chinese Influences or Images? Fluctuating Histories of How Enlightenment Europe Read China." *Journal of World History* 24.3 (2013): 623–660.

Jami, Catherine, Peter Engelfriet, and Gregory Blue, eds. *Statecraft and Intellectual Renewal in Late Ming China: The Cross-Cultural Synthesis of Xu Guangqi (1562–1633)*. Leiden: Brill, 2001.

Jami, Catherine. "Revisiting the Calendar Case (1664–1669): Science, Religion, and Politics in Early Qing Beijing." *Korean Journal of History of Science* 27.2 (2015): 459–477.

Janku, Andrea. "'Heaven-Sent Disasters' in Late Imperial China." *Natural Disasters, Cultural Responses: Case Studies toward a Global Environmental History*. Eds. Christof Mauch and Christian Pfister. Plymouth: Lexington, 2009. 233–264.

Kepler, Johannes. *The Harmony of the World*. Eds. E. J. Aiton, A. M. Duncan, and J. V. Field. Philadelphia: American Philosophical Society, 1997.

Koeman, Cornelis. *Joan Blaeu and his Grand Atlas*. London: G. Philip, 1970.

Lach, Donald, and Edwin Van Kley. *Asia in the Making of Europe, Volume III: A Century of Advance. Book 4: East Asia*. Chicago: University of Chicago Press, 1993.

Lach, Donald, and Theodor Nicholas Foss. "Images of Asia and Asians in European fiction, 1500–1800." *China and Europe in Sixteenth to Eighteenth Centuries, Images and Influences*. Ed. Thomas H. C. Lee. Hong Kong: Chinese University of Hong Kong Press, 1990. 165–188.

Lindgren, Uta. "Martini, Neuhof, und die Vereinigte Ostindische Compagnie der Niederländer." *Martino Martini: Man of Dialogue*. Eds. Luisa M. Paternicò, Claudia von Collani and Riccardo Scartezzini. Trento: Università di Trento, 2016. 137–158.

Lohenstein, Daniël Casper. *Großmütiger Feldherr, Arminius oder Herrmann*. Leipzig: Johann Friedrich Bleditschen, 1689.

Martino Martini, *De Bello Tartarico Historia*. Antwerp: Ex Officina Plantiniana Balthasaris Moreti, 1654.

Maverick, Lewis A. *China, a Model for Europe*. San Antonio, TX: Paul Anderson Company, 1946.

Mungello, David. *Curious Land: Jesuit Accommodation and the Origins of Sinology*. Honolulu: University of Hawai'i Press, 1985.

Needham, Rodney. *Exemplars*. Berkeley: University of California Press, 1985.

Pinot, Virgile. *La Chine et la formation de l'esprit philosophique en France*. Paris: Paul Geuthner, 1932.

Psalmanaazaar, George. *A Historical and Geographical Description of Formosa*. London: Dan Brown, 1704.

Purtle, Jennifer. "Scopic Frames: Devices for Seeing China c. 1640." *Art History* 33.1 (2010): 54–73.

Ratio Atq. Institutio Studiorum Societatis Iesu. Dillingen: Johannes Mayer, 1600.

Richthofen, Ferdinand Freiherr von. *China. Ergebnisse Eigener Reisen*. Vol. 1. Berlin: D. Reimer, 1877. 5 vols.

Romano, Antonella. *Impressions de Chine: l'Europe et l'englobement du monde (XVIe-XVIIe siècls)*. Paris: Fayard, 2016.

Rule, Paul. "The Jesuits and the Ming-Qing Transition: How Did Boym and Martini Find Themselves on Opposite Sides?" *Monumenta Serica* 59.1 (2011): 243–258.

Said, Edward. *Orientalism*. New York: Pantheon Books, 1978.

Schäfer, Dagmar, Shih-pei Chen and Qun Che. "What Is Local Knowledge? Digital Humanities and Yuan Dynasty Disasters in Imperial China's Local Gazetteers." *Journal of Chinese History* 4.2 (2020): 391–429.

Schaffer, Simon. "Newton on the Beach: The Information Order of the *Principia Mathematica*." *History of Science* 47.3 (2009): 243–276.

Settle, Elkhanah. *The Conquest of China, by the Tartars*. London: T. M. for W. Cademan, 1676.

Shapin, Steven and Simon Schaffer. *Leviathan and the Air-Pump: Hobbes, Boyle, and the Experimental Life*. Princeton: Princeton University Press, 1985.

Shapin, Steven. "Pump and Circumstance: Robert Boyle's Literary Technology." *Social Studies of Science* 14.4 (1984): 481–520.

Statman, Alexander. "The First Global Turn: Chinese Contributions to Enlightenment World History." *Journal of World History* 30.3 (2019): 363–392.

Statman, Alexander. *A Global Enlightenment: Western Progress and Chinese Science*. Chicago: University of Chicago Press, 2023.

Sun Xiaochun. "Observation of Celestial Phenomena in Ancient China." *Handbook of Archaeoastronomy and Ethnoastronomy*. Ed. Clive Ruggles. New York: Springer, 2015. 2043–2049.

van der Goes, Joannes Antonides. *Trazil, of overrompelt Sina*. Amsterdam: Jan Rieuwertsz., Pieter Arentsz and Albert Magnus, 1685.

Van Kley, Edwin. "An Alternative Muse: The Manchu Conquest of China in the Literature of Seventeenth-century Northern Europe." *European Studies Review* 6 (1976): 21–43.

Van Kley, Edwin. "Europe's 'Discovery' of China and the Writing of World History." *The American Historical Review* 76.2 (1971): 358–385.

Van Kley, Edwin. "News from China; Seventeenth-Century European Notices of the Manchu Conquest." *The Journal of Modern History*, 45.4 (1973): 561–582.

Väth, Alfons. *Johann Adam Schall von Bell S.J. Missionar in China, kaiserlicher Astronom und Ratgeber am Hofe von Peking 1592–1666*. Nettetal: Steyler Verlag, 1991.

Vondel, Joost van den. *Vondels Zungchin (1667) met Engelse vertaling*. Transl. Ton Harmsen, Christopher Joby, Dick van der Mark and Manjusha Kuruppath. https://www.let.leidenuniv.nl/Dutch/Cene ton/VondelZungchin1667English.html. Leiden: Leiden University Department of Dutch Language & Literature 2011 (21 June 2023).

Vondel, Joost van den. *Zungchin of ondergang der Sineesche Heerschappye*. Amsterdam: Abraham de Wees, 1667.

Wakeman Jr., Frederic. "The Shun Interregnum of 1644." *Telling Chinese History*. Ed. Frederic Wakeman Jr. Berkeley: University of California Press, 2009. 59–97.

Wakeman Jr., Frederic. *The Great Enterprise*. Berkeley: University of California Press, 1986. 2 vols.

Weil, Dror. "Chinese-Muslims as Agents of Astral Knowledge in Late Imperial China." *Overlapping Cosmologies in Asia: Transcultural and Interdisciplinary Approaches*. Eds. Bill M. Mak and Eric Huntington. Leiden: Brill, 2022. 116–138.

Weststeijn, Thijs. "The Middle Kingdom in the Low Countries: Sinology in the Seventeenth-Century Netherlands." *The Making of the Humanities, Volume II: From Early Modern to Modern Disciplines*. Eds. Rens Bod, Jaap Maat, and Thijs Weststeijn. Amsterdam: Amsterdam University Press, 2012. 209–241.

Wu Huiyi and Zhang Cheng. "Transmission of Renaissance herbal images to China: the Beitang copy of Mattioli's commentaries on Dioscorides and its annotations." *Archives of Natural History* 47.2 (2020): 236–253.

Wu Huiyi, *Traduire la Chine au XVIII* siècle: Les jésuites traducteurs de textes chinois et le renouvellement des connaissances européennes sur la Chine (1687–ca. 1740).* Paris: Honoré Champion, 2017.

Zeitung Auß der newen Welt oder Chinesischen Königreichen: So P. Martinvs Martini Der Societet JESU Priester ohnlängst auß selbigen Landen in Hollandt anlangendt mit sich gebracht hat . . . Augsburg: Andreas Aperger, 1654.

Zhang Baichun. "The Transmission of the European Clock-Making Technology into China in the 17th–18th Centuries." *Explorations in the History of Machines and Mechanisms. History of Mechanism and Machine Science.* Eds. Teun Koetsier and Marco Ceccarelli. Dordrecht: Springer, 2012. 565–577.

Zürcher, Erik. "In the Yellow Tiger's Den: Buglio and Magalhães at the Court of Zhang Xianzhong, 1644–1647." *Monumenta Serica* 50.1 (2002): 355–374.

Andrea Bréard

"Heavenly Patterns" and Everyday Life in a Nutshell: Astronomy in Pre-Modern Chinese Handy Encyclopaedias

Abstract: *Complete Books of Myriad Treasures* 萬寶全書 were an encyclopaedic genre popular in China from the seventeenth to early twentieth century. They showcased particularly useful classified knowledge claimed to dispense the readers from having to seek help from others in all matters of daily concern. These texts, comparable to a backscratcher[1] so-to-say, provided in their very first chapter illustrations, memorization verses, and short texts on how to understand "heavenly patterns" (*tianwen* 天文) and eventually even exert influence on them through human action. These chapters covered a wide range of themes related to the sky: constellations, cosmogony, weather forecasting, uncommon celestial and meteorological phenomena interpreted as omens and other "heavenly patterns."

The chapter analyses how, through specific narrative and intertextual devices applied in these booklets, classical astronomical knowledge was reorganized in its relation to both enduring and extraordinary patterns of heaven, earth, and man, thus integrating the readers into cosmic theories. Given the high popularity of the *Complete Books* over three and a half centuries, my diachronic analysis of the transformations of the astronomical knowledge contained therein will also shed light on the resistance to "scientific" knowledge, not only transmitted since Antiquity but also brought to China during this time.

The narrative is threefold. First, by giving agency to heaven as a writer ("When heaven is writing"), the term *tianwen* 天文 was a concept understood as both a pattern in the sky and a written trace from heaven, a script that can be read and interpreted just like a text. In a second section ("Writing the heavens"), the chapter adopts the inverse perspective and looks at the *Complete Books'* chapters on "heavenly patterns" and their "astronomical" content. The third section ("Re-writing the heavens") describes the processes and modalities of writing the heavens in the *Complete Books* in terms of processes of recycling, exclusion, and recombination of readily available textual and visual material on "heavenly patterns."

---

1 The modern Chinese term for backscratcher, *bu qiu ren* 不求人, literally "seek no help from others," is often found in the titles of *Complete Books of Myriad Treasures*. See for example *Dingqie* [late Ming].

1 When heaven is writing

The modern Chinese term for astronomy is *tianwen* 天文. As in other cultures, astronomy was for a long time inseparable from astrology. *Tianwen* 天文 accordingly covered a wide range of connotations in China's past, encompassing all knowledge related to the sun, moon, stars, and other celestial bodies observable in the sky, their position in the universe, their movements and their auspicious and inauspicious repercussions with worldly affairs. Such broad definition is reflected in the prime bibliographic taxonomy from medieval China: "天文者, 所以察星辰之變, 而參於政者也" [As for items in the *tianwen* 天文 section, they are concerned with the exploration of the changes of the stars and planets, which are placed in a triadic relation with politics]" (天文者, 所以察星辰之變, 而參於政者也).[2]

While the first character in *tianwen*, *tian* 天, clearly refers to concepts of "heaven," the second character, *wen* 文, was and still is more polysemantic: pattern, to embellish, ornament, to write, written character, literary text,[3] also meaning politics and the arts. In the following, my narrative is based on its connotation of both "pattern" and "writing." By understanding *tianwen* as "heavenly patterns" and the result of "heavenly writings"[4] I will show that there was an explicit ontological link between the two with respect to the heavenly and human realm.

Among others, Wang Chong (27–c.97 AD), a Han dynasty astronomer, meteorologist, and philosopher, establishes a clear analogy between heavenly patterns and written production by scholars. That he is critical of his contemporaries' use of beautiful words for attaining fame is not our concern here. What is interesting is that in the following passage, he constantly parallels the patterns (*wen* 文) of heaven with the textual patterns (*wen* 文) of man. Although I was tempted to translate the word consistently with (heavenly) "patterns" and "writings" (by Man), it was not always possible to do so, in particular in the last sentence in which heaven is even credited for manipulating ink and a brush:

---

2 Cf. the Bibliography Monograph n° 3 of the *Book of Sui* 《隋書·經籍志三》. The dynastic history of the Sui dynasty (581–617) "was compiled in 629–636, and its monographs added in 656." Cf. Twitchett (1979, 42). For changes in the classification of "Heavenly Patterns" from the Han to the Sui, cf. Morgan (2017a, 12). On "Heavenly Patterns" understood as portent astrology, cf. Nakayama (1966).

3 Kern (2001, 54) argues convincingly that *wen* took on the meaning of a written text only around the late first century BCE and may even have referred to a very specific textile pattern before.

4 Morgan (2017, 20) suggests "skywriting" as an alternative translation without further justification.

候氣變者, 於天不於地, 天文明也。衣裳在身, 文著於衣, 不在於裳, 衣法天也。 [. . .] 《易》
曰: [. . .] "觀乎天文, 觀乎人文。"此言天人以文為觀, 大人君子以文為操也。 [. . .] 天文人
文, 文豈徒調墨弄筆, 為美麗之觀哉?

[Meteorologists look up to the sky, but not down onto earth,[5] for only the patterns (*wen* 文)
of heaven are decisive. Outer and inner garments cover the body, but the ornaments (*wen*
文) are on the outer garment, not on the inner ones, therefore only outer garments are modelled after heaven. [. . .] The *Classic of Changes*[6] says: [. . .] "We look at the patterns (*wen*
文) of Heaven, and we look at the writings (*wen* 文) of Man." This means: Heaven and Man
are to be judged by their writings (*wen* 文), and writings (*wen* 文) are the proper measures
and practices of the great man and the superior man. [. . .] As regards the patterns (*wen* 文)
of Heaven and the writings (*wen* 文) of Man, how can their writings (*wen* 文) merely be the
pointless homogenizing of ink and twiddling with a brush, with the aim of producing beautiful and elegant looks?] (Wang 1935, 863–867)[7]

Understanding the patterns of celestial bodies as a celestial script was not unique
to ancient China. Rochberg shows that for ancient Mesopotamian literati of the
middle of the first millennium BCE "heavenly writing" was

a poetic metaphor occasionally used in Babylonian royal inscriptions to refer to temples
made beautiful 'like the stars' [. . .]. In these Babylonian inscriptions, the metaphor is not
used explicitly for astrology or celestial divination, but the notion of the stars as a heavenly
script implies their capacity to be read and interpreted. (Rochberg 2004, 1)

But let me come back to China. Besides Wang Chong's depiction of heaven as an
active writer, the beginning of the above quote suggests yet another analogy. In traditional Chinese thought there is a textile parallel between "patterns" of heaven
and those of woven and embellished cloth. The warp-weft (*jingwei* 經緯) fabric is a
common trope that links literature to weaving, and, in turn, weaving to a spatial
organization, just as in heavenly patterns with their horizontal and vertical meas-

---

5 The common conceptual pendant of *tianwen* is *dili* 地理, literally "terrestrial structures," now
the modern term for the science of geography. Wang Chong here probably hints to the famous
passage in the Great Appendix (*Xici* 繫辭) to the *Classic of Changes* (see fn. 7) which states:

仰以觀於天文, 俯以察於地理 [*yang yi guan yu tianwen, fu yi cha yu dili*; [The sage], in accordance with [the Changes], looking up, contemplates the patterns of heaven, and, looking
down, examines the structures of the earth].

For a slightly different translation of this passage, cf. Legge (1963, app. III, chap. IV.21, 353).
6 The *Classic of Changes* (*Yijing* 易經), one of the Confucian classics, is a prestigious and fundamental collection of oracles with a long textual and exegetic history in China. Cf. Smith (2008).
7 *Balance of Discourse* (*Lun Heng* 論衡), Chapter 'Lost Writings' (*Yi wen pian* 佚文篇). Translated
from the original text in Wang (1990). For an excellent introduction to Wang Chong's intellectual
context and writings on destiny, providence and divination, cf. Wang (2011, I–CLXXX).

ures.[8] As Zürn (2020, 371) points out, "that writings are textual fabrics" was "an image that gained prominence during the Han dynasty [206 BCE–220 CE]." He shows how intertextuality can actually be understood (and visualized) as the outcome of threads of words woven together into a seamless and ordered textual composition.[9] In Han dynasty astronomy, warp and weft take on a more technical yet not unrelated meaning. They refer to the regular spatial and timely patterns of the apparent movements of the "warp stars" (the twenty-eight constellations fixed along the ecliptic) and the "weft stars" (sun, moon, and the five planets). Yet, regularity, just as forward movement, was only an ideal. When heaven wrote a different pattern, deviating from the norms, the phenomenon required expert interpretation by a diviner. The Grand Historian of the Han, Sima Qian 司馬遷 (145–86 BCE) explains with a statistical argument that retrograde movement was no longer to be considered an irregularity:

故甘、石曆五星法, 唯獨熒惑有反逆行; 逆行所守, 及他星逆行, 日月薄蝕, 皆以為占。余觀史記, 考行事, 百年之中, 五星無出而不反逆行, 反逆行, 嘗盛大而變色; 日月薄蝕, 行南北有時: 此其大度也。[. . .] 水、火、金、木、填星, 此五星者, 天之五佐, 為經緯, 見伏有時, 所過行贏縮有度。[. . .] 凡天變, 過度乃占。[10]

[In the old Gan and Shi[11] methods for calendrical computations of the Five Stars only [Mars] had retrogradation. They took the [asterisms] it guards in retrograde, the retrogradation of other [planets], and the veilings and eclipses of the Sun and Moon all as the objects of omen-interpretation. I have looked at the clerks' records and investigated what concerns the movements [of the stars]. In a hundred years the five [planets] have never once emerged without going into retrograde; and when they retrograde they invariably are at their fullest and change colour. The Sun and the Moon have their proper times for when they are veiled or eclipsed, and for when they move North or South. These are their general rules. [. . .] Mercury, Mars, Venus, Jupiter, and Saturn – these five stars are the five assis-

---

8 As for terrestrial structures, the *Huainanzi* 淮南子 (*Master of Huainan*, before 139 BCE) states: "凡地形, 東西為緯, 南北為經" [In the [fabric of] the earth's shape, east and west are the weft north and south are the warp.] (translated in Liu et al. 2010, 159). On the term warp (*jing* 經) in the *Huainanzi* cf. Liu et al. (2010, 262) and Zürn (2020).
9 Warp and weft literature furthermore refers to two literary genres, the canonical Confucian Classics and the deuterocanoncial Apocrypha. Cf. Liu Xiuming 劉修明. "Jing, Wei yu Xi Han wangchao 經緯與西漢王朝 (The Classics and Apocrypha, and the Western Han Dynasty)." *Zhongguo zhexue* 中國哲學 (Chinese Philosophy) 9 (1983): 81–102.
10 Sima Qian 司馬遷, *Historical Memoirs* 史記, Monography of Celestial Offices 天官書, 武英殿二十四史 edition, *juan* 27, f. 40r–41r. Translation adapted from (Morgan 2017b, 22). Cf. also Chavannes (1967, 409–410) for a French translation.
11 Two of the three astronomical schools (*shi* 氏) that developed the Chinese system of constellations during the Han: the Shi 石, Gan 甘 and Wuxian 巫咸 schools. All three were named after real or legendary historical figures. Cf. Sun and Kistemaker (1997, 26).

tants of Heaven, they constitute the warp and the weft.[12] They have their proper times for when they rise and set, for their transit and movements, their advance and retreat; all has its rules. [. . .] Whenever heaven's changes transgress the rules, one divines.] (adapted from Morgan 2017b, 22).

Iregularities of the patterns of the celestial tissue, when warp and weft are not in order, were thus an integral part of the realm of "astronomy" (*tianwen* 天文). Such was even the case until the end of the imperial era. There was no classificatory distinction made in official histories, neither in the astronomical memoirs nor in the pre-Qing bibliographic sections,[13] between the regular warp and weft written into the sky ("astronomy") and disturbances in these orderly patterns.[14] Literature on the interpretation of the latter ("astrology") focused on rarely seen eclipses, extraordinary meteorological phenomena, and comets.

In the next section – while remaining with *tianwen* 天文 – I will inverse my perspective and the subject of "When Heaven is Writing" becomes the object. "Writing the Heavens" looks at omen literature concerned with the "perturbations of heavenly patterns" (*tianwen xiangyi* 天文祥異).[15] Such was the title of a subsection in the *tianwen* chapter, the very first chapter in an encyclopaedic genre popular in China between the late sixteenth to the early twentieth century. In contrast to officially approved or imperially commissioned compilations, these sources incorporated a large repository of non-standardized discourse and reference of the time and provided shortcuts to access knowledge otherwise reserved for the scholarly elite. Structural elements, in particular classification schemes remain closely connected to Confucian literati's literary production.

---

12 Morgan (2017b, 24) interprets the warp and the weft here as celestial coordinates, i.e., declination and right ascension. My reading of *jingwei* 經緯 here is more general, mainly due to the parallel syntax with the previous sentence.

13 The bibliographic treatise of the *Draft History of the Qing*, compiled in the early twentieth century, is the only dynastic history which more clearly separates between more technical "writings on calculations" (算書之屬) in a section entitled "Heavenly Patterns & Calculation Methods" (*tianwen suanfa lei* 天文算法類) from "Artistic Computations" (*shushu* 術數). The latter contains titles from the numerological genre often linked to impostors and to heterodoxy, whereas the former lists mathematical and astronomical texts and compilations of Chinese and foreign origins. Cf. Zhao et al. (1981, chap. 志一百二十二, 藝文三, 子部, 4343–4348).

14 Cf. for example the historiographic introduction to the Astronomical Treatise of the *History of the Qing*, vol. 26 清史 卷二十六 天文志一 and 天文志十四 for records of comets and the respective omen interpretations.

15 Brook (1998, 167) translates briefly as "Heavenly Perturbations."

2 Writing the heavens

Since the Song dynasty the official bibliographic monographs referenced an independent category of works generally translated as "encyclopaedias" (*leishu* 類書, lit. "books topically arranged"). The *leishu* sections included a heterogeneous range of collections of examination literature, biographical dictionaries, primers in how to read classical literature, handbooks on the art of letter writing, pharmacopoeias, geographical surveys, administrative and procedural manuals, and the like. Although the first work officially considered as a *leishu* in this category was compiled during the third century under official patronage, it was only during the Song dynasty that the widespread use of printing facilitated the task of compiling such topically arranged compilations of books or excerpts from a number of sources.

It was then that also more general encyclopaedias destined for daily reference began to appear, but they first remained for use of the literati and their needs to prepare for the imperial examinations. With the economic boom under the Ming (1368–1644) that allowed higher social mobility, the spread of literacy, and the production of paper and books at a lower cost, a new encyclopaedic genre emerged at the end of the sixteenth century and developed all the way into the early twentieth century when the last editions were printed. These daily-life encyclopaedias, in which the kinds of knowledge traditionally reserved for the upper class were presented in a distinctly different way – explicitly by their title and also by their enlarged content beyond orthodox Confucian learning – began to address a wider reading public that did not necessarily have examination degrees. The editors' intended audiences were the "four folks" (*simin* 四民) – scholars (*shi* 士), farmers (*nong* 農), artisans (*gong* 工), and merchants (shang 商) – and showed a female presence with assumed interests in their publications.

Printed in two registers, one on top of the other, the *Complete Books* were carved in a distinct and stable textual layout: blocks of seemingly unrelated texts were printed in two or even three separate horizontal sections per page, one above the other, for the reader to choose.[16] Thus in a chapter entitled "Traveling Merchants" (*shanglü* 商旅) the rubric "Warning Instructions for the Traveling Merchant"

> offers tips on coping with strangers, establishing temporary relationships with fellow travellers, and protecting oneself against robbery and fraud [. . .] The upper register of the page

---

16 Such page layout can be seen for the first time in the *Copper Coins for the Purpose of Unsealings and Submissions* (*Qi zha qing qian* 啓劄青錢), first printed in 1436.

contains "Norms in Brothels" [. . .], which consists of separate discourses on whoremonger-ing cast in the form of conduct books.[17] (Shang 2005, 69–70)

In a scroll on Fate Calculation of one other late Ming *Complete Book,* the *Dragon Head Carved in an Ancient Vessel – An Ocean of Learning at a Single Glance without Consulting the Help of Others* (*Dingqie longtou yilan xuehai bu qiu ren* 鼎鍥龍頭一覽學海不求人), one even finds a treatise on physiognomy on the upper half of the pages, and a manual on the calculation of different kinds of geometric surfaces on the lower half.[18] Some of the surfaces below have striking similarities with the kind of face shapes discussed and illustrated in the upper half of the chapter.[19]

Although the popularity of the daily-life encyclopaedias spread all through the early twentieth century and into Japan, only few of these compilations were pre-served in Chinese libraries, not to speak of the dynastic bibliographies and imperial collections in which they are not included. Printed on cheap paper, abounding in typos, using the popular or simplified form of Chinese characters, several scholars of their time expressed a negative attitude towards these encyclopaedias. They ac-cused them of being distorted[20] and "irregular" (*bu zheng* 不正) in terms of Confu-cian orthodoxy.[21] Another aspect that might have upset the higher social classes that so far had privileged access to certain kinds of knowledge was secrecy. Examples of secret knowledge included in the *Complete Books* are the description of rules and odds for games of chance, and mathematical procedures explicitly presented as "se-cretly transmitted" (*mizhuan* 秘傳) methods. The 1772 imperial edict concerning the criteria of collection, inclusion and exclusion of books and manuscripts in the Qian-long Emperor's (r. 1736–1796) encyclopaedic project to compile a *Complete Library of the Four Treasuries* (*Siku quanshu* 四庫全書)[22] clearly states that books like the daily-life encyclopaedias were considered useless for his purpose:

---

17 Shang concludes that "the inclusion of these texts in daily-life encyclopaedias shows that visit-ing brothels was recognized as part of the experience of the traveling merchant's life" (Shang 2005, 70).

18 "Methods for measuring the surfaces of fields" (*Zhangliang tianmu fa* 丈量田畝法) in scroll 17 of Dingqie [late Ming].

19 Cf. Figure 6.1 in Bréard (2019, 144).

20 Cf. for example Lu Bi's 廬璧 preface (f. 1v) *Kaozheng zhupu daliie*考證諸譜大略 (Survey of all treatises [on chrysanthemum]) to the *Dongli pin huilu* 東籬品匯錄 (2 *juan*), where he complains that he had not seen any complete text of a Song dynasty treatise (*pulu*譜錄) on chrysanthemum, but only versions abbreviated arbitrarily (*renyi*任意) by authors of encyclopaedias (*leishu*) like the *Comprehensive Records of a Forest of Matters* (*Shilin guangji* 事林廣記) and two others. I am much indebted to Martina Siebert for this reference. Cf. Siebert (2006, 268–269; 281).

21 *Baoding fuzhi* (Gazetteer of Baoding prefecture 1607, *juan* 40, f. 26v).

22 On this compilation project which lasted from 1772 to 1780 cf. Guy (1987). The *Complete Li-brary* itself contained a number of encyclopaedias. Kaderas (1998b: chap. 4) describes all the

We thus command the provincial governors, commissioners of education and others to generally order their subordinates to apply their mind particularly to the acquisition and search [of books]. Discarded should be what is sold in the commercial bookstores, like essays written for state examinations and the useless popular (*minjian* 民間) genealogies, letter guides, greeting scrolls, and the like. Furthermore do not merit to be selected all those by authors who basically possess no knowledge of "practical learning" (*shixue* 實學), who merely hunt for fame, and who only to please others assemble trifling and inappropriate excerpts from drinking and singing prose and lyrics.[23]

There certainly was an issue with regard to "scholarly" knowledge in the *Complete Books*. Although the topics and headings of their first three chapters always reflect the classical trilogy of heaven (*tian*天), earth (*di*地) and man (*ren*人): "Heavenly patterns" (*Tianwen* 天文), "Earthly territories" (*Diyu* 地輿) and "Human biographies" (*Renji* 人紀),[24] we will not find what is covered under that same heading in the official dynastic histories. Looking at the content of the chapter on celestial patterns in particular, we find techniques of foretelling the future and interpretation of celestial phenomena as omens, yet nothing about astronomical observations and instruments, nor tables filled with data.[25] Divination was restricted knowledge, for only the emperor was permitted to divine heaven's will. Brook interprets such "illegal" inclusions of omens which "presumably circulated as secret lore among the people" in the light of commercial publishing. The editor being "an entrepreneur willing to make information of any provenance available for a price without concern for legal consequences" responds with his omens to common anxieties, such as droughts, famines, and vulnerability to price fluctuations (1998, 167):[26]

As mentioned earlier, comets (lit. "brush-stars") were traditionally considered inauspicious signs with only a few exceptions. Contrary to the *Complete Books* which only state the worldly response to heaven's perturbations, official historiographic literature abounds in additionally recording actual observations and

---

sixty-five *leishu*-"encyclopaedias" which are included in the imperially commissioned text collection. For early Song dynasty encyclopaedias and anthologies cf. Kurz (2003).

23 A German translation and the Chinese text of the edict is given in Kaderas (1998a, 354 and 358–359).

24 In the earliest model of a daily-life encyclopaedia, the *Comprehensive Records of a Forest of Matters* (*Shilin guangji* 事林廣記, 1266) we can find the same sequence, although each of the three topics is treated in two subsequent chapters.

25 Cf. for example the official *History of Yuan dynasty* (*Yuanshi* 元史), *juan* 48 & 49, Astronomical Treatise (*Tianwen zhi* 天文志).

26 Brook quotes from Yu (1599). Another example for rising prices from the omen section given by Brook is the following: "Wind on Qingming portends that the price of paper will rise; a wind on Guyu portends that the wheat harvest will fail; a wind on Chongyang portends that the price of grain will rise" (1998, fig. 24, 164).

"慧星入箕, 天下人民疫飢, 米谷貴。"

[If a comet enters the Sieve constellation, the people of the realm will suffer hunger and disease and the price of rice will rise.]

Figure 1: (*Miaojin* 1612, *juan* 1, f. 16b).

the harmful effects they had on earth.[27] Sima Qian, for example, recounts – albeit in a slightly exaggerated manner – that:

秦始皇之時, 十五年彗星四見, 久者八十日, 長或竟天。其後秦遂以兵滅六王, 并中國, 外攘四夷, 死人如亂麻, 因以張楚并起, 三十年之閒兵相駘藉, 不可勝數。自蚩尤以來, 未嘗若斯也。[28]

[at the times of the First Emperor of the Qin, in the fifteenth year of his reign [232 BCE], comets made four appearances. The longest one lasted for eighty days and it was of such size that it filled heaven nearly entirely. And then, *Qin* proceeded with the forces of his army to wipe out the six kings, reunite the Middle Kingdom, and beyond the frontiers stood off the surrounding barbarians. The dead soldiers were like scattered stalks of hemp. Then,

---

27 I do not imply that official histories contain veritable historical records only. Cf. for example pure omen texts in the Monography of Heavenly Patterns in the *Book of Sui* 隋書 (636 AD), 卷二十一志第十六 天文下 雜氣:

敵軍上氣如囷倉, 正白, 見日逾明, 或青白如膏, 將勇。大戰氣發, 漸漸如雲, 變作此形, 將有深謀。

[When there are vapours above the enemy's army like a round granary, truly white, and the sun appears to be more and more bright or green and white like an ointment, the generals will be brave. In a major battle when vapours develop gradually into clouds and transform into such a shape, the generals will have a profound plan.]

城營上有雲如人頭, 赤色, 下多死喪流血。

[When there are clouds above the city campment as many as peoples' heads, red in color, then many will die and blood will flow on earth.]

28 Sima Qian 司馬遷, *Historical Memoirs* (*Shiji* 史記), Monography of Celestial Offices (*tianguan shu* 天官書) *Wuyingdian* ed. 武英殿二十四史, *juan* 27, f. 39b–40a.

when the "Greater Chu" provoked a general uprising, the soldiers for thirty years trampled on each other and piled up in insurmountable numbers. Since *Chiyou*, something like this has never been seen![29]

But instead of history, we find visual material in the *Complete Books*. What their *tianwen* chapters do include are images and diagrams, something entirely lacking in official histories. Sometimes preceded by a map of the Northern sky (see Figure. 2), the patterns of all perturbations of heaven are equally illustrated in the *Complete Books*. In Figure 1 above, for example, we saw a comet passing through a certain constellation with the corresponding omen text right below, but no historical record of a precise connection between celestial and terrestrial events is quoted. Only occa-

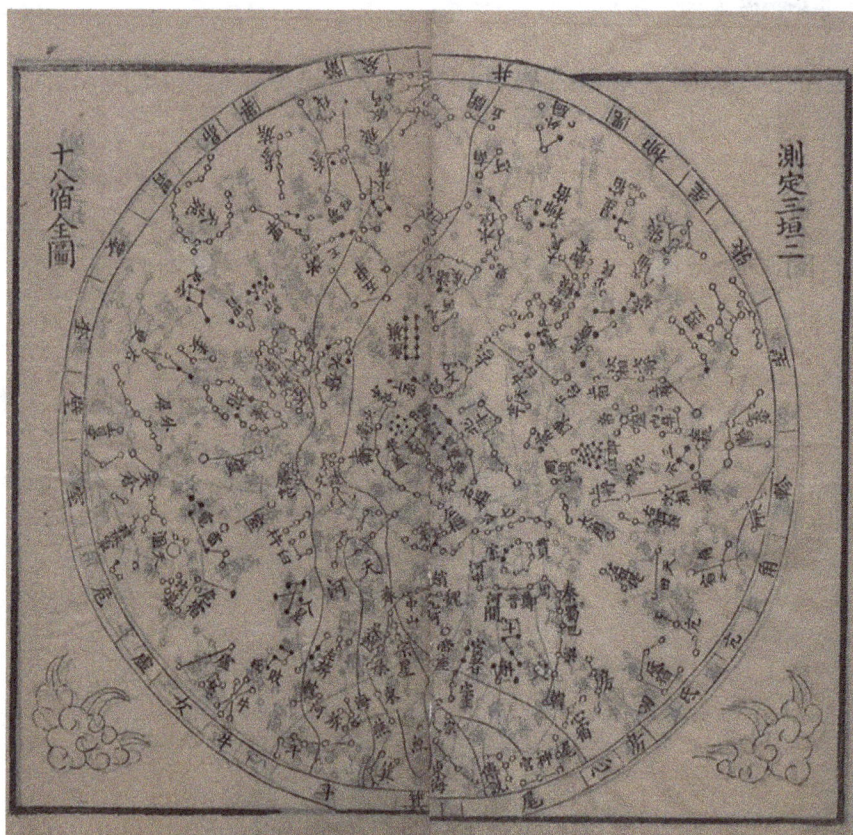

Figure 2: "Complete Map of the twenty-eight lunar lodges" illustrating the Northern sky (Yanshui n.d.).

---

29 Cf. the French translation in Chavannes (1967, T. 3 406–407).

sionally does one find a probabilistic argument which could be an indication of an empirical basis. Such is the case of the "comet in the Dawn constellation, which portends an epidemic among the people which will strike one in ten."[30]

Rare, too, are quotes from classical texts. Unlike the common explicit (and implicit) intertextuality in scholarly works or imperially approved writings,[31] there are only few instances of references to older texts found in the *Complete Books'* omen sections. In the following example, rather than making an argument by authority, the quote seems to justify the kind of illustration seen in Figure 3.[32] It is not thunder that is shown without clouds, but a mythic bird-like creature stomping on drums:

"無雲而雷者五行傳曰雷者天鼓也 無雲而雷當有暴兵"
[Thunder in the absence of clouds: The *Five Phases Tradition* says "thunder is the beating of heaven's drums." Thunder in the absence of clouds means that violent soldiers will come.]

Figure 3: (*Miaojin* 1612, *juan* 1, f. 2r).

In a proper encyclopaedic fashion, the *Complete Books* subdivided the subsection "perturbations of heavenly patterns" (*tianwen xiangyi* 天文祥異) further, generally into five, and in later editions into only the first four of the following categories:

1. Heaven (*tian lei* 天類)
2. Sun (*taiyang lei* 太陽類)
3. Moon (*taiyin lei* 太陰類)

---

30 *Huixing ru yizhu renmin bing shi fen zhi yi* 彗星入翌主人民疫十分之一. Quoted from *Miaojin* (1612, *juan* 1, f. 18v). Reprint Sakade et al. (vol. 12, 48).

31 Cf. the examples of cloud oracles from the Song dynasty *Predictions Based on Images of Heavenly Patterns* (*Tianwen xiang zhan* 天文象占) which heavily quote from the *Book of Sui*. Some pages from a manuscript copy from ca. 1573–1620 reproduced in Feist et al. (2021, 202–205).

32 The omen text in Figure 3 most likely refers implicitly to the *Five Phases Tradition on the Great Plan* (*Hongfan wu xing zhuan* 洪範五行傳), a commentary to the *Book of Documents* (*Shangshu* 尚書) ascribed to Liu Xiang 劉向 from Han dynasty, a famous librarian and historian. Only fragments of this text survived. It is also quoted in the interpretation of "Rain in the absence of clouds" (*wu yun er yu* 無雲而雨) in *Miaojin* (1612, *juan* 1, f. 2v).

4. Stars and Planets (*xing lei* 星類)
5. Twenty-eight Lunar Lodges (*ershiba xiu lei* 二十八宿類)

While the first section concerns meteorological perturbations (e.g. rain or thunder in the absence of clouds), the second and the third relate more specifically to colored shapes (spots, flames, etc.) or animals (such as snakes, birds, toads or rabbits) appearing in or around the sun or the moon: "Should a yellow toad appear at the bottom of a full moon and move upward, it portends cannibalism."[33] Unusual shapes of planets and stars are dealt with in section 4, where comets equally cross their trajectories. In the fifth section, their appearance in the twenty-eight lodges is dealt with systematically and in the usual order of the lodges which divide the ecliptic into (unequal) parts. This allowed readers to memorize more easily the names and sequence of the lodges. Furthermore, in the *Complete Books*, the interpretation of comets appearing in the twenty-eight lodges shows a fairly stable linguistic pattern. It follows the formula: "If a comet enters the x [constellation], in response there will be y" (慧星入 x, y 應之).[34] Also, the number of stars in the constellation of each lodge is shown in the drawing, it is even mentioned explicitly in one of the earliest editions of a *Complete Book* (Yu 1599). At a single glance, readers could thus handily acquire some basic astronomical knowledge.

If the *Complete Books* were supposed to have an educational outlook – I will come back to this point again in the third section –, a clear division into categories could certainly have helped to clarify confusions of the layperson that might arise from terminological issues. For example, *Dou* 斗 referred to both, the Northern Dipper constellation[35] and the Southern Dipper lunar lodge.[36] Differentiating between the two was particularly important for the readers since their interpretation was diametrically opposed to each other:

---

33 "月大作黃蟾現下陵上, 人民相食" *Miaojin* (1612, *juan* 1, f. 9r).

34 In other chapters and textual genres, versification played an important role in transmitting knowledge. The *Rhapsody on the Great Images of Heavenly Patterns* (*Tianwen da xiang fu* 天文大象賦), "a piece of rhyme-prose narrating the names and meanings of asterisms as an aide-mémoire to the student of the night sky" (Morgan 2019, 33), for example, proceeds slowly from one constellation to the next in regular verse (and images). Cf. Li (1856 卷上 f. 33r–34r): "If then you look towards the Northern Palace in the Dark Warrior, you will anchor at the Southern Dipper and the Led Ox" (若乃眺北宮於玄武, 泊南斗於牽牛). The Led Ox is one of the 28 lunar lodges, a constellation of six stars which has Altair (β Cap) as its determinative star. On the history of the *Rhapsody* and its authorship, cf. Chen (2003, 339–340).

35 Big Dipper (*ursa major*) constellation in the Northern sky.

36 One of the "twenty-eight lunar lodges" into which the ecliptic is (irregularly) divided; it corresponds to six stars in the *Sagittarius* constellation: ζ Sgr, τ Sgr, σ Sgr, φ Sgr, λ Sgr, and μ Sgr.

Stars and Planets Category		Twenty-eight Lunar Lodges Category	
	"If a comet enters the [Northern] Dipper [constellation], there will be tranquillity in every direction and peace everywhere." 慧星入斗四方太平天下平安		"If a comet enters the [Southern] Dipper [lodge], in response there will be robbers and great disorder within no more than one year." 慧星入斗盜賊大乱不出一年應之

Figure 4: (*Miaojin* 1612, *juan* 1, f. 2r).

The illustrations certainly also helped to swiftly make the difference between the two Dippers. In the Chinese sky they do not contain the same numbers of stars. Yet, as the longue durée history of the editions of the *Complete Books* shows (see tab. 1), the drawings of the Southern Dipper soon degenerated from a constellation of six stars into one of seven.

3 Rewriting the heavens

In this section, I do not intend to discuss how man could "rewrite" the heavenly patterns. Resonance, after all, was believed to be mutual. Moral behavior and rituals could act upon the signs of heaven and change its patterns. I leave considerations of "rewriting" in this acceptation to intellectual historians and will rather understand "rewriting" here as an authorial act. For the astronomical chapters in the *Complete Books*, rewriting is even the only authorial act they are subjected to. No new content is added from the late sixteenth to the early twentieth century, in spite of the growing importance of foreign astronomical knowledge transmitted to China through the Jesuits, in particular for predicting eclipses and other calculable phenomena. Rewriting, as will be shown below, was an act of appropriation – and I do not mean to imply that it was in its radical form an act of plagiarism – of a very specifically selected part of astronomical knowledge, with minimalist changes in the subsequent editions once the code for the astronomical chapters' style was defined.

Looking at the different editions of the *Complete Books*, one notices the variety of book titles indicated at the beginning of each chapter. It is obvious that for their compilations the printers participated in a process of recycling readily available print material. During the Ming dynasty entire woodblocks went from one commercial printer family to another, even located in different provinces.[37] Nevertheless, for the mathematical chapters, for example, we can observe a wide variety of wording for versified algorithms found therein, and content is arranged in various combinations.[38] The *tianwen* chapters, on the contrary, display a fairly high degree of intersection of written and visual content over three centuries following the same distinct stylistic code.

For example, for the case of a comet crossing the Southern Dipper lodge – a case already encountered in Figure 4 and systematically translated and illustrated for sixteen editions from 1599 to 1913 in tab. 1[39] – the omen text basically has two slightly different versions. By the beginning of the seventeenth century, the earlier phrasing seems to be entirely replaced by a later one.[40] Whether this is due to language change, local variations, or a phenomenon of oral knowledge circulation, I cannot tell. What is certain is that the editors of the *Complete Books* over time provided more and more simplifications of content and abstraction of images.

It would go beyond the scope of this chapter to trace the genealogy of the omen texts for comets crossing each of the twenty-eight lodges.[41] In order to understand the processes of rewriting that were at the basis of the final version which we find in the *Complete Books*, a juxtaposition with other genre writings shall serve our purpose. Here, I will limit myself to point out two kinds of simplifications which the *Complete Books* exhibit by providing the relevant passage for the Southern Dipper from two kinds of sources. First, from an important Tang dynasty compendium. The *Prognosticatory Classic of the Opening Epoch Reign* (*Kaiyuan zhanjing* 開元占經) from 729, and rediscovered in the early seventeenth century, was an anthology of astronomical and astrological knowledge. It quotes heavily from other, often otherwise lost classical and apocryphal texts. The fol-

---

37 Cf. Chia (2002, 31–32, 217–220).

38 Cf. Bréard (2010).

39 Mine is by no means an exhaustive account. Judge (2017, 364) estimates that alone between 1894 and at least 1928 "there were well over 20 editions of *wanbao quanshu* [*Complete Books of Myriad Treasures*] in print."

40 I am not taking into account what is grammatically incorrect and thus probably a misprint due to graphic proximity between two characters in Chen (1871): "一年之四見之" instead of "一年之內見之" (they will be seen within one year). Zhao (n.d.) also has a misprint, inverting the last two characters 應 and 之.

41 Wu (2021) has analyzed the 'Heavenly Perturbations' subsection in eighteen Ming and eight Qing dynasty editions of *Complete Books* in terms of language, misprints, and illustrations and shown possible links between them and to previous literature covering a similar topic.

lowing passage shows the variety of interpretations that circulated for the case of a comet colliding with the Southern Dipper (斗彗孛犯南斗):

甘氏曰彗孛干犯南斗度其國必亂兵大起期一年
　陳卓曰彗茀長干犯南斗王者病疾臣謀其君子謀其父弟謀其兄是謂無理諸侯不通
天下易政大亂兵起期百八十日遂不出一年 按宋書天文志曰 [. . .]
　春秋緯曰彗星出南斗天下有勢皆謀
　孝經雌雄圖曰彗星出南斗中宮中失火燒寶
　甘氏曰彗星出南斗大臣謀反兵水並起天下亂將軍有戰若流血 星若滅斗其國主
亡若星明反臣受殃近三年中五年遠七年
　荆州占曰彗星出南斗天下皆謀上
　石氏曰彗星茀於斗房赤帝之後受令人主凶有亡國

[The Gan 甘 astronomical school says: when the body of a comet collides with the extension of the Southern Dipper [lodge], the country will fall into disorder, wars will rise greatly within one year.

Chen Zhuo 陳卓[42] says: When the long body of a comet offends the Southern Dipper, the king will have a disease. The officers will counsel their dukes, sons will counsel their fathers, younger brothers will counsel their elder brothers. Call it without structure! The feudal princes will not have a complete grasp of how to conduct government with ease over the whole world. There will be great disorder and wars will rise for a period of 180 days within one year. Commentary: The Treatise on Heavenly Patterns of the *Book of Song*[43] says: [. . .]

The *Wefts of the Spring and Autumn Annals* says: when a comet comes forth in the Southern Dipper, the powerful under heaven all will consult [about the case].

The *Diagrams ci and xiong in the Classic of Filial Piety* says: when a comet comes forth in the middle of the Southern Dipper, a fire will accidentally start in the centre of the palace and its treasures will be burnt.

The Gan astronomical school says: when a comet comes forth in the Southern Dipper, high officials will plot a rebellion, armies and waters will rise, and the whole world will be in great disorder. The generals will conduct war with rivers foaming with much blood. If a planet occults [one of the six stars in] the [Southern] Dipper, the country will be ruined. When the star returns to shine, officials will face calamity within the third, the fifth, and the seventh years.

The *Jingzhou Divination* says: when a comet comes forth in the Southern Dipper, all under heaven will plot against those above.

---

42 See Sun and Kistemaker (1997, 26): "an astronomer of the Sanguo 三國 (Three Kingdoms) period (AD 220 – 280)" who studied and conflated the constellations of the three astronomical schools (see footnote 11). Chen's works are lost but "all the information concerning the constellations of the three schools has been passed on to us through the astrological works of the Tang dynasty."

43 Official dynastic history of the Liu-Song 劉宋 dynasty (420–479).

The Shi 石 astronomical school[44] says: when a comet overgrows the [Southern] Dipper or the Chamber [lodge], the Red Emperor's heir will receive orders to rule his men. Inauspicious, he will ruin the state.][45]

Compared to such breadth of sources and readings of one specific heavenly pattern disturbance, the *Complete Books* look pale. What they reproduce corresponds roughly only to the very first excerpt from the Gan astronomical school. I said roughly: no reference to the school is given, linguistic expression is simplified, lexicon is reduced, and the army replaced by robbers. One explanation for such rewriting is, as I have argued for the mathematical chapters in the *Complete Books* (Bréard 2010), that these everyday encyclopaedias were also catering for a less proficient reading public. (Para-)scientific knowledge served the purpose of language acquisition, to be more precise: it served the purpose of basic language acquisition. Short and formulaic sentences were easy to memorize and reproduce. A single straightforward answer to strange phenomena also reassured the book consumer in everyday life. When one further looks at the images, either the woodblock carver was sloppy or acquisition of factual knowledge was not of central concern for the later editors of the *Complete Books*: the number of stars in the constellation of the Southern Dipper evolves from six to seven, and, in the latest editions, the drawings become so schematic that it is even difficult to distinguish the "brush-star" from the stars in the depicted constellation.

My second example that contrasts the heavenly disturbances section in the *Complete Books* is a colorful anonymous manuscript from Ming dynasty. Entitled *A Rhapsody of Perturbations of Heavenly Patterns, commented and illustrated* (*Tianwen tuzhu xiangyi fu* 天文圖註祥異賦), with illustrations in the upper half and only little text below – thus the same format as in the *Complete Books* but scaled to an entire page – the manuscript quotes omen interpretations from older texts. For comets crossing lunar lodges, we have two pages for each lodge, one showing the constellation with a short sentence indicating the number of stars and its length in degrees, the other showing again the same constellation but with a comet passing through it. What can be seen in Figure 5 (to the right) is that, even for only stating that there are six stars in the constellation of the Southern Dipper, an authoritative text is quoted. To the left in the same figure, we see the incident visualized in the upper register. In the lower register, an interpretation is quoted from the famous Neo-Confucian scholar Zhu Xi 朱熹 (1130–1200), and another one from the *Song [History Astronomical] Memoirs* (*Songzhi* 宋志). The

---

44 See footnote 11.
45 Quoted from Gautama (1782, 卷 89, 8a–8b), the edition in the *Complete Library of the Four Treasuries* (*Siku Quanshu* 四庫全書), see footnote 23.

first announces robbers and disorder, the second a swarm of robbers to come, precisely what we find in the *Complete Books* in the earlier and later editions. Comparison also reveals that the sentences are constructed in nearly the same way in both works, with the *Complete Books* being closer to a single normal form of the omen-statement:

	Variant 1 / Zhu Xi quote	Variant 2 / *History of the Song* quote
Complete Books	慧星入斗盜賊大乱不出一年應之 If a comet enters the [Southern] Dipper [lodge], in response there will be robbers and great disorder within no more than one year.	慧星入斗盜賊大起一年之內見之 If a comet enters the [Southern] Dipper [lodge], swarms of robbers will rise, they will be seen within one year
A Rhapsody of . . .	朱文公曰 慧[星]入斗盜賊大乱不出一年應之 Zhu, the Duke of Culture, said: If a comet enters the [Southern] Dipper [lodge], in response there will be robbers and great disorder within no more than one year.	宋志曰 斗宿慧星入之一年之內盜賊蜂起 The *Song* [*History Astronomical*] *Memoirs* say: The [Southern] Dipper lodge, if a comet enters it, within one year swarms of robbers will rise.

Rewriting the Heavens in this case seems like a combinatorial game of merging the pieces of two sentences and eventually permuting grammatical parts of them. That this truly happened, I doubt. Oral circulation of astrological knowledge certainly took part in such combinatorial processes. But what one can conclude here is that rewriting mainly meant condensing the text into one short sentence, cutting out the reference, and compressing two images into one. After all, paper was costly and the *Complete Books* were meant to be handy and pocket-sized so that one could have them always at one's elbow, like a packet of white pebbles to find one's way back home.

4 Time to conclude

In this chapter, I questioned the notion of "astronomical" writings in the late imperial era in China. I showed in particular that the *Complete Books of a Myriad Treasures* genre contributed an alternative body of astronomical knowledge when compared to the content of the transmitted specialized literature. They had a clearly delineated content resulting in a definite shape from processes of rewriting and they remained impermeable to outside influences over their three and a

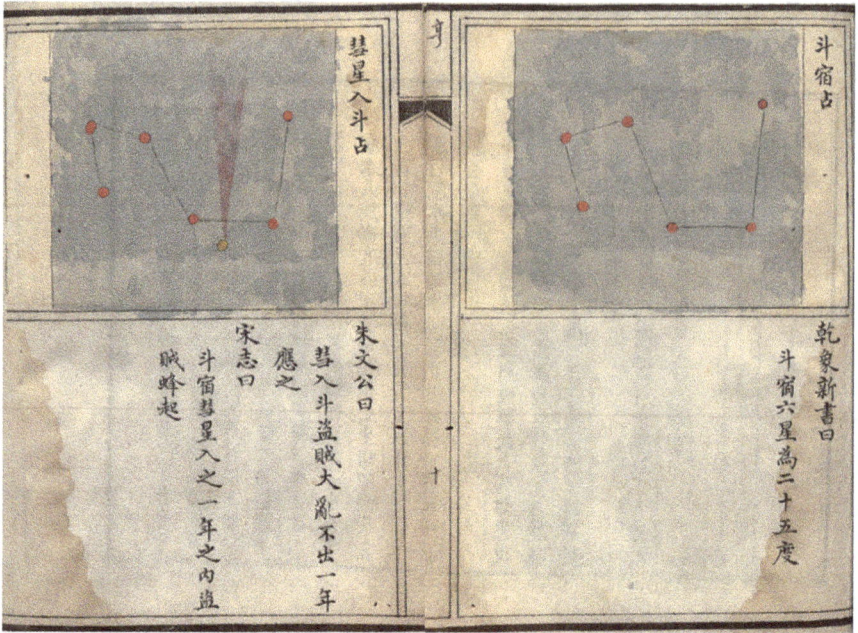

Figure 5: Omen interpretation of a comet entering the [Southern] Dipper. (*Tianwen* [n.d.], *juan* 2, f. 10), manuscript copy from the Chinese Rare Book Collection (Library of Congress)

half centuries of existence. Integrating the *Complete Books* into the study of the history of astronomical literature in China can thus contribute to understanding the place of astronomical knowledge cultures in the Chinese late imperial intellectual historical landscape beyond the picture that we get from the canonical texts of calendrical astronomy and translations of Western astronomical theories and geometric models by the Jesuits. Although the *Complete Books* were an audience-oriented and market-dependant set of discourse practices, they nevertheless shed light on the social differentiation in the knowledge and practice of astronomy between popular encyclopaedias and the literati kinds of writings.

Astronomical literature, I argued further, is not only written by scientists or professional observers of the sky and rewritten by commercial publishers, but astronomy, or rather the metaphoric trope of *tianwen,* both a "heavenly pattern" and "heavenly writing," denotes a literary genre in itself. With heaven as its sole author inscribing perpetual and irregular patterns, the written signs and encoded messages in the observable sky were to be read and interpreted by the astrologically literate. Not unlike my own undertaking in this chapter, turning upside-down the title of the volume and looking at "Heaven's Writings" as the inverse to

Table 1: Texts and illustrations from sixteen editions of Complete Books related to the Southern Dipper.

(Yu 1599)	(Xu 1600)	(Xu 1610)	(Miaojin 1612)
in a section listing omens for the 28 lodges:		in a section listing omens for the 28 lodges:	in a section listing omens for the 28 lodges:
"慧星入斗盜賊大亂 不出一年應之 斗六星" [If a comet enters the [Southern] Dipper [lodge], in response there will be robbers and great disorder within no more than one year. There are six stars in the [Southern] Dipper [lodge]."]		"慧星入斗盜賊大亂 不出一年應之" [If a comet enters the [Southern] Dipper [lodge], in response there will be robbers and great disorder within no more than one year. There are six stars in the [Southern] Dipper [lodge].]	"慧星入斗盜賊大亂 不出一年應之" [If a comet enters the [Southern] Dipper [lodge], in response there will be robbers and great disorder within no more than one year.]
	(7 stars shown)		in a section listing omens for the 28 lodges (but contains only 27):
			"慧星入斗盜賊大亂 不出一年應之" [If a comet enters the [Southern] Dipper [lodge], in response there will be robbers and great disorder within no more than one year.]

(continued)

Table 1 (continued)

(Zhao [n.d.])	(Xu 1614)	(Chen 1628)	(Ai [late Ming])
in a section listing omens for the 28 lodges:	in a section listing omens for the 28 lodges:	in a section on stars (星類):	in a section on stars:
"慧星入斗盜賊大乱 不出一年之應"46 [If a comet enters the [Southern] Dipper [lodge], in response there will be robbers and great disorder within no more than one year.]	"慧星入斗盜賊大乱 不出一年應之 斗六星" [If a comet enters the [Southern] Dipper [lodge], in response there will be robbers and great disorder within no more than one year. There are six stars in the [Southern] Dipper [lodge].]	"慧星入斗盜賊大起 一年之內見之" [If a comet enters the [Southern] Dipper [lodge], swarms of robbers will rise, they will be seen within one year.]	"慧星入斗盜賊大起 一年之內見之" [If a comet enters the [Southern] Dipper [lodge], swarms of robbers will rise, they will be seen within one year.]

46 I suspect that there is a misprint here and the last two characters should be reversed in order in the sentence, to obtain a correct sentence identical to other editions of the *Complete Books*.

(Xinqie 1694)	(Yanshui [n.d.])[47]	(Zhuming 1739)	(Zhuming 1823)
in a section on stars:	in a section on stars:	in a section on stars:	in a section on stars:
"慧星入斗盜賊大起不出一年應之"	"慧星入斗盜賊大起一年之內見之"	"慧星入斗盜賊大起一年之內見之"	"慧星入斗盜賊大起一年之內見之"
[If a comet enters the [Southern] Dipper [lodge], in response swarms of robbers will rise within no more than one year.]	[If a comet enters the [Southern] Dipper [lodge], swarms of robbers will rise, they will be seen within one year.]	[If a comet enters the [Southern] Dipper [lodge], swarms of robbers will rise, they will be seen within one year.]	[If a comet enters the [Southern] Dipper [lodge], swarms of robbers will rise, they will be seen within one year.]
(7 stars shown)	(7 stars shown in spite of the map showing six, see Figure 2)	(8 stars shown)	(7 stars shown)

(continued)

47 Although this edition is undated, I have placed it next to (Xinqie 1694) for textual and visual proximity.

Table 1 (continued)

(Mao 1828)	(Chen 1871)	(Qixin ca. 1912)	(Zhang 1758)
in a section on stars:	in a section on stars:	a section on stars:	Scroll 1 ends without a section on comets
"慧星入斗盜賊大起一年之內見之"	"慧星入斗盜賊大起一年之內[48]見之"	"慧星入斗盜賊大起一年之內見之"	
[If a comet enters the [Southern] Dipper [lodge], swarms of robbers will rise, they will be seen within one year.]	[If a comet enters the [Southern] Dipper [lodge], swarms of robbers will rise, they will be seen within a one year.]	[If a comet enters the [Southern] Dipper [lodge], swarms of robbers will rise, they will be seen within one year.]	
(7 stars shown)			

48 The original has "四" instead of "內." Cf. footnote 40 for the emendation of the character here.

discussing "Writing the Heavens," the *Complete Books of a Myriad Treasures* packed into the category of "Heavenly Patterns" precisely what officially approved astronomical writings had excluded from it.

Bibliography

Ai Nanying 艾南英, ed. *Xinke Ai xiansheng tianluge huibian caijing bianlan wanbao quanshu* 新刻艾先生天祿閣彙編採精便覽萬寶全書 (Complete Book of Myriad Treasures for the convenient perusal), 34 vols. 卷. [Jianyang, Fujian:] Tanyi Shulin Sanhuaitang Wang Taiyuan ed. 潭邑書林三槐堂王泰源刊本, [n.d., late Ming].

Bréard, Andrea. "Knowledge and Practice of Mathematics in Late Ming Daily life Encyclopedias." *Looking at It from Asia: The Processes that Shaped the Sources of History of Science*. Ed. Florence Bretelle-Establet. Dordrecht: Springer, 2010. 305–329.

Bréard, Andrea. "Fate Calculation 算命: The Mathematics of Divination." *Nine Chapters on Mathematical Modernity: Essays on the Global Historical Entanglements of the Science of Numbers in China*. Cham: Springer, 2019. 143–167.

Brook, Timothy. *The Confusions of Pleasure. Commerce and Culture in Ming China*. Berkeley, Los Angeles and London: University of California Press, 1998.

Chavannes, Édouard, trans. "Les gouverneurs du ciel." *Les mémoires historiques de Se-Ma-Ts'ien*. Paris: Librairie d'Amérique et d'Orient Adrien-Maisonneuve, 1967. Vol. 3, part 1, 339–412.

Chen Huaixuan 陳懷軒, ed. *Wanbao quanshu* 萬寶全書 (Complete Book of Myriad Treasures. Complete title: 新刻眉公陳先生編輯諸書備採萬卷搜奇全書), 37 vols. 卷. Cunren tang ed. 存仁堂, 1628 (Housed in Keio University).

Chen Meidong 陳美東. *Zhongguo kexue jishu shi: Tianwenxue juan* 中國科學技術史: 天文學卷 (History of Chinese Science and Technology: Astronomy) Beijing: Kexue chubanshe 科學出版社, 2003.

Chen, Meigong 陳眉公, ed. *Zengbu wanbao quanshu* 增補萬寶全書 (Complemented Complete Book of Myriad Treasures), 8 vols. 卷. Jiqing tang ed. 積慶堂, 1871.

Chia, Lucille. *Printing for Profit. The Commercial Publishers of Jianyang, Fujian (11th-17th Centuries)*. Cambridge, MA, and London: Harvard University Press, 2002.

Chou, Anpang 周安邦. "Jingdian de tongsuhua yu tongsu de jingdianhua: Song Ming shumin daode tongmeng jioacai de tongsuhua qingxiang 經典的通俗化與通俗的經典化: 宋明庶民道德童蒙教材的通俗化傾向" (The Popularization of the Classical Scriptures and the Classicization of Popular Literature: the Trend of Popularization of Moral Educational Primers for Common People during the Song and Ming Dynasties). *Xingda zhongwen xuebao* 興大中文學報 33 (2013). 43–44, 46–76.

Cui, Jinming 崔金明, ed. *Mingdai tongsu riyong leishu jikan* 明代通俗日用類書集刊 (Compendium of Popular Daily Life Encyclopaedias from Ming Dynasty). Chongqing: Xinan shifan daxue chubanshe 西南師範大學出版社, 2011.

Di Giacinto, Licia. *By Chance of History: The Apocrypha under the Han*. PhD thesis, Ruhr-Universität Bochum, Fakultät für Ostasienwissenschaften, 2007.

Dingqie longtou yilan xuehai bu qiu ren 鼎鍥龍頭一覽學海不求人 (A Dragon Head Carved in an Ancient Vessel – An Ocean of Learning at a Single Glance without Consulting the Help of Others). [China: n.p.], [n.d., late Ming dynasty (1368–1644)]. Reprint in Cui 2011, 14:139–318.

Espesset, Grégoire. "Portents in Early Imperial China: Observational Patterns from the "Spring and Autumn" Weft Profoundly Immersed Herptile (*Qiantan ba* 潛潭巴)." *International Journal of Divination & Prognostication* 1 (2019): 251–287.

Feist, Marie-Therese, Michael Lackner, and Ulrike Ludwig, eds. *Zeichen der Zukunft: Wahrsagen in Ostasien und Europa / Signs of the Future. Divination in East Asia and Europe.* Nuremberg: Germanisches Nationalmuseum, 2021.

Gautama Siddhārtha 瞿曇悉達. *Kaiyuan zhanjing* 開元占經 (Prognosticatory Classic of the Opening Epoch Reign [713–741]). Wenyuange Siku quanshu 文淵閣四庫全書 ed., 1782. Reprint Taibei: Taiwan shangwu yinshuguan, 1983–1986.

Guy, Kent R. *The Emperor's Four Treasures: Scholars and the State in the Late Ch'ien-Lung Era.* Boston, MS: Harvard University Press, 1987.

Hsu Kuang-Tai 徐光台. "Yixiang yu changxiang: Ming Wanli nianjian xifang huixing jianjie dui shiren de chongji 異象與常象: 明萬曆年間西方彗星見解對士人的衝激" (= Starry Anomaly and Meteorological Phenomenon: The Impact of the Aristotelian View of Comets on Xu Guangqi and Xiong Mingyu in the Ming Wanli Period). *Qinghua xuebao* 清華學報 39.4 (2009), 529–566.

Judge, Joan. "Science for the Chinese Common Reader? Myriad Treasures and New Knowledge at the Turn of the Twentieth Century." *Science in Context* 30.3 (2017): 359–383.

Kern, Martin. "Ritual, Text, and the Formation of the Canon: Historical Transition of *Wen* in Early China." *T'oung Pao* 87.1 (2001): 43–91.

Kory, Stephan N. *Cracking to Divine: Pyro-Plastromancy as an archetypal and common mantic and religious practice in Han and Medieval China.* PhD thesis, Department of East Asian Languages and Cultures, Indiana University, 2012.

Legge, James, trans. *The I Ching.* New York: Dover, 1963.

Li Bo [隋] 李播 (撰), Miao Wei 苗為 (comm. 注), Sun Zhilu 孫之騄 [style name Jing Chuan 晴川] (compl. 補), Liuyan 六嚴 (ed. 校). *Tianwen da xiang fu* 天文大象賦 (Rhapsody on the Great Images of Heavenly Patterns), 2 vols. 卷. Jiangyin 江陰: n.p., preface dated 1856 [咸豐丙辰].

Li Yilu 鹿憶鹿. "Wan Ming *Shanhaijing* tuxiang zai riben de liuchuan – yu *Guaiji niaoshou tujuan* yu *Yiguo wuyu* wei li 晚明《山海經》圖像在日本的流傳-以《怪奇鳥獸圖卷》與《異國物語》為例" (The Spread of Illustrations from the *Classic of Seas and Mountains* in Japan in the Late Ming Dynasty: The *Illustrations of Strange Birds and Beasts* and the *Exotic Tales* as Examples). *Zhongguo xueshu jikan* 中國學術年刊 41.2 (2019): 1–34.

Liu, An (King of Huanian); Major, John S. and Queen, Sarah A. and Meyer, Andrew Seth and Roth, Harold D., trans. and eds. *The Huainanzi. A Guide to the Theory and Practice of Government in Early Han China.* New York: Columbia University Press, 2010.

Magnani, Arianna. *Gewu bu qiu ren, un'enciclopedia popolare cinese nella biblioteca dei Gesuiti a Genova: un caso studio nella dinamica dei rapporti tra Europa e Cina in età barocca.* PhD thesis, Università Ca'Foscari Venezia, 2019.

Mao Huanwen 毛煥文, ed. *Zengbu wanbao quanshu* 增補萬寶全書 (Expanded Complete Books of Myriad Treasures), 4 vols. Guiwen tang ed. 貴文堂, 1828.

Miaojin wanbao quanshu 妙錦萬寶全書 (Complete Book of Myriad Treasures, Magnificently Embroidered). Complete title: *Xinban quanbu tianxia bianyong wenlin shajin wanbao quanshu* 新板全補天下便用文林紗錦萬寶全書). Jianyang, China: n. p., 1612. Housed in the Institute of Oriental Culture, University of Tokyo, Niida Collection (Tōkyō Daigaku Tōyō Bunka Kenkyūjo Niida Bunko 東京大學東洋文化研究所仁井田文庫). Reproduced in Sakade, Ogawa, and Sakai 2004, vols. 12–14.

Morgan, Daniel Patrick. *Astral Sciences in Early Imperial China. Observation, Sagehood and the Individual.* Cambridge: Cambridge University Press, 2017[a].

Morgan, Daniel Patrick. *Sciences Astrales en Chine des Han aux Tang. Master. Histoire des sciences en Asie*, France. 2017[b]. https://shs.hal.science/cel-01684870/document (15 September 2023).

Pankenier, David W. *Astrology and Cosmology in Early China: Conforming Earth to Heaven*. Cambridge: Cambridge University Press, 2013.

Nakayama, Shigeru. "Characteristics of Chinese Astrology," *Isis* 57 (1966): 442–454.

Qixin shuju 啓新書局. *Zui xin huitu zengbu zhengxu wanbao quanshu* 最新繪圖增補正續萬寶全書 (Newest, illustrated, expanded, and truly continued *Complete Book of Myriad Treasures*). 8 vols. Shanghai: Qixin shuju 啓新書局, ca. 1912.

Rochberg, Francesca. *The Heavenly Writings. Divination, Horoscopy, and Astronomy in Mesopotamian Culture*. Cambridge: Cambridge University Press, 2004.

Sakade, Yoshinobu 坂出祥伸, Yōichi Ogawa, 小川陽一, and Tadao Sakai, 酒井忠夫(監修) (eds.). *Chūgoku nichiyō ruisho shūsei* 中國日用類書集成 (Compendium of Chinese Daily Life Encyclopaedias). Tokyo: Kyūko shoin 汲古書院, 1999–2004.

Shang, Wei. "The Making of the Everyday World. *Jin Ping Mei cihua* and Encyclopedias for Daily Use." *Dynastic Crisis and Cultural Innovation. From the late Ming to the Late Qing and Beyond*. Eds. David Der-wei Wang and Wei Shang. Vol. 249. Harvard: Harvard University Asia Center (2005). 63–92.

Smith, Richard J. *Fathoming the Cosmos and Ordering the World. The* Yijing (I-Ching, *or* Classic of Changes) *and its Evolution in China*. Charlottesville and London: University of Virginia Press, 2008.

Sun, Xiaochun and Kistemaker, Jacob. *The Chinese Sky During the Han: Constellating Stars and Society*. Leiden, New York and Cologne: Brill, 1997.

Tianwen tuzhu xiangyi fu 天文圖註祥異賦 (Heavenly Patterns, Illustrated and Commented with Odes for Perturbations). [n.p.], [n.d., late Ming dynasty (1368–1644)].

Twitchett, Denis, ed. *The Cambridge History of China*. Vol. 3: *Sui and T'ang China, 589–906, Part I*. Cambridge: Cambridge University Press, 1979.

Wang Chong 王充 and Huang Hui 黃暉, eds. *Lunheng jiaoshi* 論衡校釋 (*The Balance of Discourse*, critically edited and explained). Taipei: Taiwan Shangwu yinshuguan 台灣商務印書館, 1935.

Wang Chong 王充 and Marc Kalinowski, trans. *Balance des Discours. Destin, Providence et Divination*. Paris: Les Belles Lettres, 2011.

Wu Changzheng 吳昌政. *Ming Qing* Wanbao quanshu *"tianwen xiangyi" yiwen yanjiu* 明清《萬寶全書》"天文祥異"異文研究 (= On the Variants of *Tianwen Xiangyi* of *Wanbao Quanshu* in Ming and Qing Dynasties). Master thesis, East China Normal University 华东师范大学, School of International Chinese Studies 国际汉语文化学院, 2021.

Wu Huifang 吳蕙芳. "Shanghai tushuguan suo cang *Wanbao quanshu* zhuben – jian lun minjian riyong leishu zhong de pincou wenti 上海圖書館所藏《萬寶全書》諸本–兼論民間日用類書中的拼湊問題" (On the editions of *Complete Books of Myriad Treasures* in the Shanghai Library: A discussion on the piecing together of daily life encyclopaedias). *Shumu jikan* 書目季刊 36.4 (2003): 53–58.

Xu Huiying 徐會瀛, ed. *Xinqie yantai jiaozheng tianxia tongxing wenlin jubao wanjuan xingluo* 新鍥燕臺校正天下通行文林聚寶萬卷星羅 (*Treasures of Literary Collections* in ten thousand scrolls scattered like stars in use all over the world, newly engraved and revised by Yantai), 39 vols. 卷. Shulin jingguanshi ed. 書林靜觀室, printed by Yu Xianke 余獻可, 1600 (date of preface). Reprint: Beijing: Shumu wenxian chubanshe 書目文獻出版社, 1988 (北京圖書館古籍珍本叢刊, 76), 113–405.

Xinqie zengbu wanbao quanshu 新鍥增補萬寶全書 (Newly collated and augmented *Complete Book of Myriad Treasures*), 13 vols. 卷. Wumen 吳門: Sizhi tang re-ed. 四知堂重訂, (carved 梓 by) Yang Ruiqing 楊瑞卿, 1694.

Xu Qilong 徐企龍, ed. *Wanshu yuanhai* 萬書淵海 (Abyss of an Ocean of a Myriad Books, complete title: Xinke quanbu shimin beilan bianyong wenlin huijin wanshu yuanhai 新刻全補士民備覽便用文林彙錦萬書淵海), 1610. Housed in Maeda Ikutokukai-Sonkeikaku Bunko 前田育德會尊經閣文庫 (Sonkeikaku library of the Maeda Ikutokukai Foundation in Meguro City, Tokyo). Reproduced in Ogawa 2000, vols. 6–7.

Xu Qilong 徐企龍, ed. *Wuche wanbao quanshu* 五車萬寶全書 (Complete Compendium of Five Chariots of a Myriad Treasures). Cunren tang ed. 存仁堂, 1614. Housed in Kunaichō shoryōbu 宮内庁書陵部 (Imperial Household Agency, Tokyo). Reproduced in Ogawa 2000, vols. 8–9.

Yanshui Shanren 煙水山人, ed. *Jingtang dingbu wanbao quanshu* 敬堂訂補萬寶全書 (Complete Book of Myriad Treasures, revised and augmented, from the Hall of Dedication. Title on cover: 文會堂增訂不求人眞本), 24 vols. 卷. Zuihuaju ed. 醉花居藏板, [n.d.].

Yu Xiangdou, ed. 余象斗.*Santai wanyong zhengzong* 三台萬用正宗. Complete title: *Xinke tianxia simin bianlan Santai wanyong zhengzong* 新刻天下四民便覽三台萬用正宗 (Santai's orthodox instructions for myriad uses for the convenient perusal of all the people in the world, newly engraved), 1599. Housed in Tōkyō Daigaku Tōyō Bunka Kenkyūjo Niida Bunko 東京大學東洋文化研究所仁井田文庫 (Institute of Oriental Culture, University of Tokyo, Niida Collection). Reproduced in Sakade, Ogawa, and Sakai 2000, vols. 3–5.

Zhang Taishi 張太史 (1538–1588). [*Ju jia bao yao*] *Bai bei quanshu* [居家寶要]百備全書 (Complete Book Prepared for the Hundred [Families]. [A Must Have for Everyone]).

Zhang Tianru 張天如, ed. *Xinke Tianru Zhang xiansheng jingxuan Shiliang huiyao wanbao quanshu* 新刻天如張先生精選石渠彙要萬寶全書 (A new edition of the collected essentials of the Complete Book of Myriad Treasures carefully selected by Mr. Zhang Pu [as if they were] from the Stone Channel Pavilion), 32 vols. 卷. [Guangzhou 廣州]: Zhengzu huixian zangban ed. 正祖會賢藏板, 1758.

Zhao Erxun 趙爾巽 et al. *Qingshi gao* 清史稿 (Draft History of the Qing Dynasty). Taibei: Guofang yanjiuyuan 國防研究院, 1981.

Zhao Zhiwu 趙植吾, ed. *Xinke simin bianlan wanshu cuijin* 新刻四民便覽万書萃錦 (Newly engraved Brocade Fan of a Myriad Books for the Convenient Perusal for the Four People), 36 vols. Printed by Zhan Linwo 詹林我, Qixi tang wenku ed. 棲息堂文庫, [n.d.].

Zhuming jia hexuan 諸名家 合選 ("coedited by several families"). *Zengbu wanbao quanshu* 增補萬寶全書 (Augmented Complete Book of Myriad Treasures), 6 vols. 卷. [China]: Shide tang ed. 世德堂刊本, 1739.

Zhuming jia hexuan 著名家 huiji ("coedited by several families"). *Zengbu wanbao quanshu* 增補萬寶全書 (Augmented Complete Book of Myriad Treasures), 20 vols. 卷. [China]: Jinchang jingyitang ed. 金閶經義堂, 1823.

Zürn, Tobias Benedikt. "The Han Imaginaire of Writing as Weaving: Intertextuality and the Huainanzi's Self-Fashioning as an Embodiment of the Way." *The Journal of Asian Studies* 79.2 (2020): 367–402.

List of Contributors

Andrea Bréard was Professor for History of Science at the Université Paris-Saclay, Faculté des Sciences d'Orsay before being awarded an Alexander von Humboldt-Professorship at the Friedrich-Alexander-Universität Erlangen-Nürnberg in 2021, where she holds the *Chair for Sinology with a Focus on the Intellectual and Cultural History of China*. Trained as a sinologist, computer scientist and mathematician, she works at the interface between mathematical sciences and sinology, with research topics ranging from ancient times to the twenty-first century. She has published extensively on topics from early conceptual history to the transcultural history of twentieth-century science. She has recently published, with Les Belles Lettres (*Bibliothèque Chinoise* vol. 40, 2023), an annotated translation and bilingual edition of Li Shanlan's 李善蘭 number theoretical book written entirely in natural language, the *Comparable Categories of Discrete Accumulation* 垛積比類 (1876). With her research group sin-aps, she currently pursues research on *Quantification from Europe to East Asia* and *The language of algorithmic mathematics*.

Gianamar Giovannetti-Singh is the Lumley Research Fellow in History at Magdalene College and a Leverhulme Trust Early Career Fellow in the Faculty of History at the University of Cambridge. His forthcoming first monograph, *The Tartar Moment*, explores how the Manchu conquest of China transformed cultures of knowledge in early modern Europe. Gianamar's current research explores how European settlers and scientific travellers drew on epistemic resources from China and the East Indies to make sense of unfamiliar natures and peoples in colonial southern Africa. His work has been published in venues including *Isis*, *History Workshop Journal*, the *Journal of the History of Ideas*, the *Journal for the History of Knowledge*, and the *Los Angeles Review of Books*. Gianamar has been a Freer Prize Fellow of the Royal Institution, and a visiting fellow at the Max Planck Institute for the History of Science in Berlin and the Descartes Centre in Utrecht. In both 2023 and 2024, Gianamar was shortlisted for the BBC's New Generation Thinker Award.

Aura Heydenreich is an Associate Professor for German Literature at the Department of German and Comparative Studies, University Erlangen-Nürnberg in Germany. Her research interests include literature and science, mainly focussing on astronomy, optics, relativity theory, quantum field theory, postmigrational postmemory studies, and subversive memory cultural practices. In 2014, together with Klaus Mecke, she founded ELINAS (Erlangen Center for Literature and Natural Science). Since 2014, she is editor of the series, "Literature and Natural Science" with De Gruyter. Since 2019, she is the acting president of the European Society for Literature, Science and Arts (SLSAeu). In 2022, she was a Distinguished Max Kade Visiting Professor at the German Department of the University of Illinois, Urbana-Champaign; and in 2023, Visiting Scholar at the Department for Germanic Languages at the Columbia University in New York. Her publications include *Wachstafel und Weltformel. Erinnerungspoetik und Wissenschaftskritik im Spätwerk Günter Eichs* (Göttingen: Vandenhoeck & Ruprecht, 2007) and *Literatur und Naturwissenschaft: Interformation und epistemische Transformation. Literatur in Wechselwirkung mit Astronomie, Mikrobiologie, Relativitätstheorie und Quantenphysik* (Berlin and Boston: De Gruyter, 2024; Open Access publication available from https://www.degruyter.com/document/doi/10.1515/9783110729887/), as well as the volume, co-edited with Klaus Mecke, *Physics & Literature. Concepts, Transfer, Aestheticization and Popularization* (Berlin and Boston: De Gruyter, 2021).

Alexander Honold is Professor of German Literature at the University of Basel, Switzerland, and director of the *Forum Basiliense*. His main research interests include narratology, Swiss and Austrian literature, nature writing and landscape, cultural concepts of time and calendar. He has published numerous articles on classicism and romanticism, on modernist and contemporary literature. His latest book publications are: *Grenzenlose Verwandlung. Hugo von Hofmannsthal. Biographie* (with Elsbeth Dangel-Pelloquin, Frankfurt am Main: S. Fischer, 2024); *Poetik der Infektion. Zur Stilistik der Ansteckung bei Thomas Mann* (Berlin: Vorwerk 8, 2021); *Thomas Hürlimann* (co-ed.) (Munich: *Text + Kritik*, 2021).

Walker Horsfall is Assistant Professor at the Department of Germanic Languages & Literatures at the University of Illinois Urbana-Champaign, where he teaches courses on Norse mythology, the Icelandic saga tradition, and the history of sexuality and literature in the premodern world. His research is on the intersections of science, religion, and literature in the medieval period, with a focus on Middle High German religious and love poetry from the twelfth to the fourteenth centuries, with additional interest in Arthurian courtly romances and mystical texts. He received his PhD in 2022 from the Centre for Medieval Studies at the University of Toronto, with a dissertation on the medieval German poet Frauenlob, and his rhetorical encodings of contemporary natural philosophy into his enigmatic religious poetry.

Florian Klaeger is Professor of English Literature at the University of Bayreuth, Germany. He is the author of *Forgone Nations* (Trier: WVT, 2006) and *Reading into the Stars* (Heidelberg: Winter, 2018) and the (co-)editor of several volumes on representations of Europe in literature, diasporic writing, and literary form. His research focuses on the intersections of literary form, knowledge, and collective identities, with a special focus on the interface of astronomy and literature in early modern England. He headed the DFG research project, *Cosmopoetic Form-Knowledge: Astronomy, Poetics, and Ideology in England, 1500–1800* (2019–2023) and is currently a co-PI in the joint AHRC/DFG research consortium, *Scientific Poetry and Poetics in Germany and Britain, from the Renaissance to the Enlightenment* (2024–2027).

Sophie Knapp studied medieval and modern German literature and history of art at Ludwig-Maximilians-University Munich and Zurich University/ETH. She works at the German Department of Heidelberg University as a lecturer and research assistant. Here, she also completed her doctorate with a thesis on the forms and functions of intertextual references in Middle High German *Sangspruchdichtung*. Her research focuses on the vernacular literature of the thirteenth and early fourteenth centuries, especially secular epic and lyrical texts and their intertextual references to literary texts as well as learned discourses. Before her studies, she completed an apprenticeship as a goldsmith.

Daniel Könitz works as a research assistant at the Institute for Medieval German Philology at Philipps University of Marburg. He received his doctorate in 2012 in the field of Early German Literature at Philipps University of Marburg and has been head of the project *Handschriftencensus – An inventory of the manuscript tradition of German-language texts from the Middle Ages* funded by the Akademie der Wissenschaften und der Literatur | Mainz since 2017. His research and teaching focus on the field of German-language literature of the Middle Ages, its manuscript tradition and editorial studies.

Gábor Kutrovátz is an associate professor at the Budapest University of Technology and Economics (BME), Department of Philosophy and History of Science. He formerly worked at the Eötvös Loránd University (ELTE), Department of History and Philosophy of Science, and later at the Department of Astronomy. He graduated from ELTE in physics, astronomy and philosophy, and received his PhD from BME in history and philosophy of science. He has been teaching history of astronomy for two decades, and doing research in this field more recently. He authored a Hungarian book on the history of parallax measurements and related topics, and he co-authored two books containing new Hungarian translations of classical astronomical texts, accompanied by extensive annotations and introductions. Presently, his main research area is stellar identification and identity in the Western astronomical tradition, with a recent focus on journal article references to stars in the early modern era.

Klaus Mecke, since 2004 full professor for Theoretical Physics at the Universität Erlangen-Nürnberg, studied philosophy and physics and received his PhD at LMU Munich in 1993 with a thesis on integral geometry in physics. After research stays in Austin and Boston, he worked in Wuppertal and at the MPI Stuttgart on liquids on the molecular scale and the geometric characterization of spatially complex materials. Recently, he developed a theory of quantum spacetime based on finite projective geometry of event processes. An important aspect of his research at the Erlangen Center for Literature and the Sciences (ELINAS) are the manifold exchange modes between physics and literature. Here, his research goal is a narratology of physics as well as its process-ontological foundation.

Helge Perplies studied medieval and early modern German literature and history at Bremen University and completed his doctorate at Greifswald University with a thesis on a collection of early modern travel accounts of the Americas. He worked at the German Department of Heidelberg University on a project on the edition and commentary of Johann Fischart's translation of Jean Bodin's *Démonomanie des sorciers*. His research has been focused on travelogues and cosmographies as well as the postcolonial Middle Ages. Most recently, he published a book on the romantic legacy of medieval German studies. In between his academic jobs, he has worked in planetariums in Bremen, Potsdam and Berlin, presenting mainly children's and astro-historical shows. He now works at Leipzig University Library.

Reto Rössler studied German Studies and Philosophy at the University of Trier and received his PhD in 2018 in the bi-national PhD-program, *The Knowledge of Literature* at Humboldt University of Berlin, with a thesis on the interrelationships between cosmological knowledge and literature in the seventeenth and eighteenth centuries. He was a research assistant in German Literature at the Universities of Trier, Innsbruck and Flensburg, in the Collaborative Research Centre 600, *Strangers and Poor People. Changing Forms of Inclusion and Exclusion from Antiquity to the Present*, in the DFG project, *Essay and Experiment. Concepts of experimentation between natural science and literature (1700–1960)*, and Junior Fellow in the DFG Collegiate Research Group *Imaginaries of Force (Imaginarien der Kraft)* at the University of Hamburg. His research interests include the history of literature and knowledge, intercultural literary studies (with a focus on representations of Europe in literature), literary theory, 'metaphorology,' as well as 'similarity' in literature and cultural theory. His publications include the monograph, *Weltgebäude. Poetologien kosmologischen Wissens der Aufklärung* (Göttingen: Wallstein, 2020) and the volume, co-edited with Tim Sparenberg und Philipp Weber, *Kosmos & Kontingenz. Eine Gegengeschichte* (Paderborn: Wilhelm Fink, 2016).

Hania Siebenpfeiffer is Full Professor of Literature and Culture of the Early Modern Period and European Enlightenment at Philipps-University of Marburg. She has worked extensively on literature and law in the modern and pre-modern period as well as on literature and astronomy in the early modern period. Her further research interests focus on gender studies, discourse analysis, rhetoric, genre poetics and literary materiality and mediality. Her publications include *Die literarische Eroberung des Alls. Literatur und Astronomie 1593–1771* (Göttingen: Wallstein, 2025); *Handbuch Literatur und Recht*, ed. with Peter Schneck and Claudia Lieb (Berlin and Boston: De Gruyter, forthcoming in 2025); *Überschreitungen/Überschreibungen: Zum Werk von Sibylla Schwarz (1621–1638)* in *Daphnis. Zeitschrift für Mittlere Deutsche Literatur und Kultur der Frühen Neuzeit (1400–1750)*, vol. 44, issue 1/2 (2016); *Diversity Trouble? Vielfalt – Gender – Gegenwartskultur*, ed. with Peter Christian Pohl (Berlin: Kadmos, 2016); *Materie. Grundlagentexte zur Theoriegeschichte*, ed. with Sigrid G. Köhler and Martina Wagner-Egelhaaf (Berlin: Suhrkamp, 2013) *and Böse Lust. Violent Crime in Discourses of the Weimar Republic* (Cologne [et al.]: Böhlau, 2005). In addition, she is the editor of the series *Literarische Weltraumreisen*, published by Wehrhahn.

Agata Starownik graduated in Polish Philology (2016) and Art History (2017) and earned her PhD in literary studies from the University of Warsaw (2023). Since 2019 she has been a lecturer at the Artes Liberales Faculty (University of Warsaw). She conducts her research there, leading a project by National Science Centre "The Warsaw pageant of planets (1580) described by Martin Gruneweg – organization, genre and iconography of an Early Modern spectacle" since 2022. As of 2024, she works on a commentary on Jan Kochanowski's "Psałterz Dawidów" at The Institute of Literary Research of the Polish Academy of Sciences, as part of a project "Completion of the parliamentary edition of Jan Kochanowski's "Dzieła wszystkie" ["Collected Works"]", funded by National Humanities Development Program. Her research interests include astronomical and cosmic motifs in literature and art, as well as reception of the Bible in early modern culture. She is the author of a monograph on astronomical motifs in the works of Edward Stachura (2021).

Dirk Vanderbeke is professor of English studies at the Friedrich Schiller University in Jena. His doctoral thesis, *Worüber man nicht sprechen kann* (*Whereof One Cannot Speak*), explored the unrepresentable in philosophy, science and literature. His habilitation study, *Theoretische Welten und literarische Transformationen* (*Theoretical Worlds and Literary Transformations*) examined the 'science wars' and the debate on science's role(s) in contemporary literature. He has published on a variety of topics, e.g., science and literature, evolutionary criticism, James Joyce, Thomas Pynchon, John Milton, science fiction, fantasy, crime fiction, self-similarity, vampires and graphic novels. In addition, he has co-edited an annotated edition of the German translation of James Joyce's *Ulysses*, published in celebration of the Bloomsday centenary 2004.

Maximilian Wick studied comparative and German literature at Goethe-Universität (Frankfurt am Main), where he also completed his doctorate in 2020. From 2016 to 2019, he was as a research fellow at LMU München with the DFG-funded FOR 1986 project, *Natur in politischen Ordnungsentwürfen: Antike – Mittelalter – Frühe Neuzeit*. From 2019 to 2021, he was a research assistant at the Institut für deutsche Literatur und ihre Didaktik at Goethe-Universität, and from 2021 to 2023 at the Germanistisches Institut at Ruhr-Universität Bochum. In the course of the latter two employments, he was an (associate) member of the research group *Dimensionen der techne in den Künsten. Erscheinungsweisen – Ordnungen – Narrative*. He is currently a post-doctoral researcher at the Institut für deutsche Literatur und ihre Didaktik at Goethe-Universität.

Jörn Wilms studied physics and astrophysics at Universität Tübingen and at the University of Colorado, Boulder. After his diploma in physics (1996), his PhD (1998) and his habilitation (2002), he was a lecturer in the Department of Physics, University of Warwick, Coventry (UK; 2004–2006). Since 2006, he has been professor and chair (since 2021) of astronomy and astrophysics at Dr. Karl Remeis-Observatory and Erlangen Centre for Astroparticle Physics, Friedrich-Alexander-Universität Erlangen-Nürnberg. His research interests are high energy astrophysics of black holes and neutron stars, space instrumentation, and the history of astronomy.

Index of Names